ALAN PORTER
404 384 6295
alan.porter@isye.gatech.edu

# Innovation, Technology, and Knowledge Management

**Series Editor**

Elias G. Carayannis
School of Business
George Washington University
Washington, DC, USA

More information about this series at http://www.springer.com/series/8124

Tugrul U. Daim · Denise Chiavetta
Alan L. Porter · Ozcan Saritas
Editors

# Anticipating Future Innovation Pathways Through Large Data Analysis

Springer

*Editors*
Tugrul U. Daim
Department of Engineering and Technology Management
Portland State University
Portland, OR
USA

Denise Chiavetta
Search Technology, Inc.
Norcross, GA
USA

Alan L. Porter
Georgia Institute of Technology
Roswell, GA
USA

Ozcan Saritas
National Research University Higher School of Economics
Moscow
Russia

ISSN 2197-5698 ISSN 2197-5701 (electronic)
Innovation, Technology, and Knowledge Management
ISBN 978-3-319-39054-3 ISBN 978-3-319-39056-7 (eBook)
DOI 10.1007/978-3-319-39056-7

Library of Congress Control Number: 2016940793



Printed on acid-free paper

This Springer imprint is published by Springer Nature
The registered company is Springer International Publishing AG Switzerland

# Series Foreword

The Springer book series *Innovation, Technology, and Knowledge Management* was launched in March 2008 as a forum and intellectual, scholarly "podium" for global/local, transdisciplinary, trans-sectoral, public–private, and leading/"bleeding"-edge ideas, theories, and perspectives on these topics.

The book series is accompanied by the Springer *Journal of the Knowledge Economy*, which was launched in 2009 with the same editorial leadership.

The series showcases provocative views that diverge from the current "conventional wisdom" that are properly grounded in theory and practice and that consider the concepts of ***robust competitiveness***,[1] ***sustainable entrepreneurship***,[2] and ***democratic capitalism***,[3] central to its philosophy and objectives. More specifically, the aim of this series is to highlight emerging research and practice at the dynamic intersection of these fields, where individuals, organizations,

---

[1]We define *sustainable entrepreneurship* as the creation of viable, profitable, and scalable firms. Such firms engender the formation of self-replicating and mutually enhancing innovation networks and knowledge clusters (innovation ecosystems), leading toward robust competitiveness (E.G. Carayannis, *International Journal of Innovation and Regional Development* 1(3), 235–254, 2009).

[2]We understand *robust competitiveness* to be a state of economic being and becoming that avails systematic and defensible "unfair advantages" to the entities that are part of the economy. Such competitiveness is built on mutually complementary and reinforcing low, medium, and high technology and public and private sector entities (government agencies, private firms, universities, and nongovernmental organizations) (E.G. Carayannis, *International Journal of Innovation and Regional Development* 1(3), 235–254, 2009).

[3]The concepts of *robust competitiveness* and *sustainable entrepreneurship* are pillars of a regime that we call "*democratic capitalism*" (as opposed to "popular or casino capitalism"), in which real opportunities for education and economic prosperity are available to all, especially—but not only—younger people. These are the direct derivative of a collection of top-down policies as well as bottom-up initiatives (including strong research and development policies and funding, but going beyond these to include the development of innovation networks and knowledge clusters across regions and sectors) (E.G. Carayannis and A. Kaloudis, *Japan Economic Currents*, p. 6–10 January 2009).

industries, regions, and nations are harnessing creativity and invention to achieve and sustain growth.

Books that are part of the series explore the impact of innovation at the "macro" (economies, markets), "meso" (industries, firms), and "micro" levels (teams, individuals), drawing from such related disciplines as finance, organizational psychology, research and development, science policy, information systems, and strategy, with the underlying theme that for innovation to be useful it must involve the sharing and application of knowledge.

Some of the key anchoring concepts of the series are outlined in the figure below and the definitions that follow (all definitions are from E.G. Carayannis and D.F. J. Campbell, *International Journal of Technology Management*, 46, 3–4, 2009).

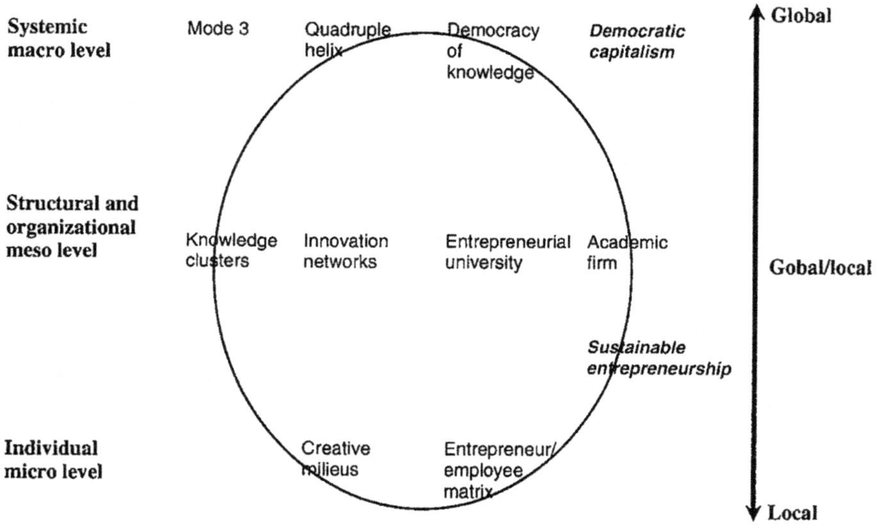


Conceptual profile of the series *Innovation, Technology, and Knowledge Management*

- The "Mode 3" Systems Approach for Knowledge Creation, Diffusion, and Use: "Mode 3" is a multilateral, multinodal, multimodal, and multilevel systems approach to the conceptualization, design, and management of real and virtual, "knowledge stock" and "knowledge flow," modalities that catalyze, accelerate, and support the creation, diffusion, sharing, absorption, and use of cospecialized knowledge assets. "Mode 3" is based on a system-theoretic perspective of socioeconomic, political, technological, and cultural trends and conditions that shape the coevolution of knowledge with the "knowledge-based and knowledge-driven, global/local economy and society."
- Quadruple Helix: Quadruple helix, in this context, means to add to the triple helix of government, university, and industry a "fourth helix" that we identify as

the "media-based and culture-based public." This fourth helix associates with "media," "creative industries," "culture," "values," "lifestyles," "art," and perhaps also the notion of the "creative class."

- Innovation Networks: Innovation networks are real and virtual infrastructures and infratechnologies that serve to nurture creativity, trigger invention, and catalyze innovation in a public and/or private domain context (for instance, government–university–industry public–private research and technology development coopetitive partnerships).
- Knowledge Clusters: Knowledge clusters are agglomerations of cospecialized, mutually complementary, and reinforcing knowledge assets in the form of "knowledge stocks" and "knowledge flows" that exhibit self-organizing, learning-driven, dynamically adaptive competences and trends in the context of an open systems perspective.
- Twenty-First Century Innovation Ecosystem: A twenty-first century innovation ecosystem is a multilevel, multimodal, multinodal, and multiagent system of systems. The constituent systems consist of innovation metanetworks (networks of innovation networks and knowledge clusters) and knowledge metaclusters (clusters of innovation networks and knowledge clusters) as building blocks and organized in a self-referential or chaotic fractal knowledge and innovation architecture,[4] which in turn constitute agglomerations of human, social, intellectual, and financial capital stocks and flows as well as cultural and technological artifacts and modalities, continually coevolving, cospecializing, and cooperating. These innovation networks and knowledge clusters also form, reform, and dissolve within diverse institutional, political, technological, and socioeconomic domains, including government, university, industry, and nongovernmental organizations, involving information and communication technologies, biotechnologies, advanced materials, nanotechnologies, and next-generation energy technologies.

*Who is this book series published for?* The book series addresses a diversity of audiences in different settings:

1. *Academic communities*. Academic communities worldwide represent a core group of readers. This follows from the theoretical/conceptual interest of the book series to influence academic discourses in the fields of knowledge, also carried by the claim of a certain saturation of academia with the current concepts and the postulate of a window of opportunity for new or at least additional concepts. Thus, it represents a key challenge for the series to exercise a certain impact on discourses in academia. In principle, all academic communities that are interested in knowledge (knowledge and innovation) could be tackled by the book series. The interdisciplinary (transdisciplinary) nature of the book series underscores that the scope of the book series is not limited a priori to a specific

[4] E.G. Carayannis, *Strategic Management of Technological Learning*, CRC Press, 2000.

basket of disciplines. From a radical viewpoint, one could create the hypothesis that there is no discipline where knowledge is of no importance.

2. *Decision makers—private/academic entrepreneurs and public (governmental, subgovernmental) actors*. Two different groups of decision makers are being addressed simultaneously: (1) private entrepreneurs (firms, commercial firms, academic firms) and academic entrepreneurs (universities), interested in optimizing knowledge management and in developing heterogeneously composed knowledge-based research networks, and (2) public (governmental, subgovernmental) actors that are interested in optimizing and further developing their policies and policy strategies that target knowledge and innovation. One purpose of public *knowledge and innovation policy* is to enhance the performance and competitiveness of advanced economies.
3. *Decision makers in general*. Decision makers are systematically being supplied with crucial information, for how to optimize knowledge-referring and knowledge-enhancing decision making. The nature of this "crucial information" is conceptual as well as empirical (case-study-based). Empirical information highlights practical examples and points toward practical solutions (perhaps remedies), and conceptual information offers the advantage of further-driving and further-carrying tools of understanding. Different groups of addressed decision makers could be decision makers in private firms and multinational corporations, responsible for the knowledge portfolio of companies; knowledge and knowledge management consultants; globalization experts, focusing on the internationalization of research and development, science and technology, and innovation; experts in university/business research networks; and political scientists, economists, and business professionals.
4. *Interested global readership*. Finally, the Springer book series addresses a whole global readership, composed of members who are generally interested in knowledge and innovation. The global readership could partially coincide with the communities as described above ("academic communities," "decision makers"), but could also refer to other constituencies and groups.

Washington, DC, USA

Elias G. Carayannis
Series Editor

# Preface

Tech Mining (TM) is a special form of large data analytics (LDA). It concentrates on mining global ST&I publication/patent databases normatively by searching on a target emerging technology (or key organization) of interest in global databases, or investigates in trends and developments in STI domains in an explorative way. One then downloads, typically, thousands of field-structured text records (usually abstracts) and analyzes those for useful competitive technical intelligence (CTI) on existing and emerging trends and developments in the areas under investigation. Tech Mining has now been widely recognized by public and private research institutions, policy and strategy makers, universities, as well as corporations identifying future opportunities and threats leading to future innovations. Publication and patent databases are among the most frequently used sources for Tech Mining. Besides these structured data sources, more recently Tech Mining have been extended to analyze semi-structured and unstructured data including social network data, Web sites, blogs, reports, and even speech. New concepts and approaches with the combination of quantitative and qualitative methods are continuously introduced into the field. There are also advancements in computational tools which, besides generating bibliometric/scientometric data in more efficient and visually powerful ways, help to analyze text semantically to obtain useful insights from large volumes. To this end of anticipating future innovation pathways through LDA, the present book draws from authors who presented their cutting-edge approaches in the recent leading conferences including:

- Portland International Conference on Management of Engineering and Technology (PICMET; Aug., 2015).[5] PICMET is the leading conference in engineering and technology management bringing scholars, industry, and government representatives together.

[5] http://www.picmet.org/new/conferences/16/.

- Global Tech Mining Conference (GTM; Sep., 2015). The goal of this conference is to engage cross-disciplinary networks of analysts, software specialists, researchers, policymakers, and managers to advance the use of textual information in multiple science, technology, and business development fields.[6]
- International Conference on Future-Oriented Technology Analysis (FTA; Nov., 2014). FTA brings together those studying foresight and various forecasting and assessment methods, in a policy-oriented context.[7]

In the following 18 chapters, we are pleased to present advances in:

1. frameworks to consider Tech Mining and related analytical interests,
2. multiple methods able to treat science, technology, and innovation (ST&I) data effectively—plus means to incorporate human expertise and interests,
3. translation of analyses to useful intelligence on likely future developments of particular emerging ST&I targets,
4. informative indicators and compelling visualizations.

Applying the practice of generating indicators and visualizations to our own content, we imported the keywords accompanying chapter abstracts into the VantagePoint text mining software to generate an Aduna map of chapter–keyword relationships. Each of the 18 yellow dots represents a chapter, and the "arms" of each yellow dot extend out to their chapter's primary keywords (number in parentheses indicates the number of chapters with keyword). We can see from this visualization that topical attention varying from patent/publication databases to scenarios indeed provides an integrated discussion on large data and innovation in terms of uses, methods, processes, and outputs.

[6]http://www.gtmconference.org/index.html.

[7]https://ec.europa.eu/jrc/en/event/site/fta2014/about.

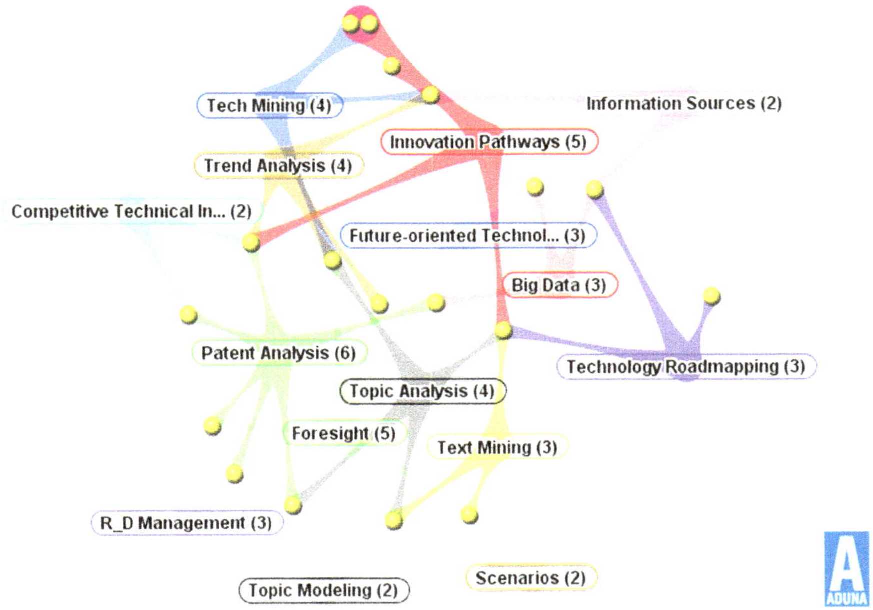


This book has three parts. The first part reviews the fundamental concepts. The second part provides a set of emerging methods. The final part demonstrates the concepts through applied cases.

## Part I Data Science/Technology Review

Chapters in the review part are mainly concerned with the use of data and Tech Mining for the purpose of future-oriented technology analysis (FTA), as well as the evolution of the FTA and Tech Mining concepts over time. In the first paper, entitled "How can Future-oriented Technology Analysis (FTA) be improved to take advantage of emerging information resources?", Loveridge and Cagnin suggest lessons to be garnered from business "due diligence" processes. Foresight and FTA strive to identify future possibilities. Those reflect complex balances of emerging capabilities and needs, informed by a mix of human knowledge and perceptions—plus data. Big data and analytics could serve FTA aims. The results suggest FTA-transforming potentials of diverse data resources and interconnections, such as autonomous systems, energy management, logistics, social networking, and forecasting institutions. The combination of accessible data and potent algorithms can foster FTA "due diligence" spanning a wide spectrum of potential factors (e.g., social, economic, ecological, political). It also poses staunch challenges re: balance between automated decision systems and roles for human judgments.

Following the opening paper, Li presents a more specific conceptual framework to consider text analyses in support of FTA—expressly, forecasting innovation pathways (FIP). "Tech Mining"—text analyses in support of FIP and FTA—is empirical in nature and has not been informed by a clear knowledge structure concerning objectives and methods. Li addresses objectives in terms of engineering processes, set in a broader management/policy context of attendant issues concerning development. He turns an engineering perspective to reflect on the elements of a systematic Tech Mining process. He divides the Tech Mining process into three connected components: models and tools; processes and technical standards; and process management factors. Drawing on analogy with software engineering, Li offers a top-down model to manage Tech Mining engineering systematically via needs assessment, strategic planning, and implementation planning. He provides a "big picture" perspective on using Tech Mining in support of innovation processes, with quality controls.

The next three papers use Tech Mining to investigate the evolution of Foresight and Tech Mining itself. Andersen and Alkaersig analyze changes in FTA and Foresight empirically through a bibliometric analysis of journal papers. They analyze journal special issues with publications emanating from FTA conferences in 2004, 2006, 2008, and 2011, along with abstracts submitted for the 2014 conference. Consideration of topical emphases in the FTA conferences over time finds increasing attention to "systems thinking." Hypothesized increases in emphasis on participatory methods and regional (vs. national) issues were not supported (rather, these seemed stable over time). The data indicate that the special issue publications exert comparable influence to other papers in the key FTA and Foresight journals.

Similarly, Mikova uses "Tech Mining" to examine trends in the content of the Global Tech Mining (GTM) conferences, 2011–2015. She benchmarks that activity against general trends of literature concerning Tech Mining (TM). Main streams of TM research are led by methodological development (e.g., bibliometrics, citation analyses, clustering, and network analyses), followed by consideration of uses in FTA. Special interests include focus on research and technology profiling, devising technology life cycles, and development of indicators. Visualization is noted as a keen interest. Mikova catalogues emerging methods, data types, and software tools used in TM; these can stimulate ideas to enrich established users' approaches. Semi-automated TM algorithms may combine with alternative Web-based data to open new possibilities.

In the final chapter of this part, Efimenko, Khoroshevsky, and Noyons offer a novel approach to anticipate future science, technology, and innovation pathways called Map of Science (Sect. 5.2). They perform extensive searches in Web of Science and Scopus on Scientometrics and related terms (e.g., science indicators) and on Tech Mining and related topics (e.g., technical intelligence). They analyze which R&D fields are addressed in those papers to map them, with co-occurrence providing a basis for establishing associations (i.e., fields being analyzed together in scientometric or TM studies). The name Map of Science (Sect. 5.2) reflects that these are visualizations based on the analyses of research fields (in Scientometrics and TM). They examine such VOSviewer maps over time to show evolving cross-disciplinary connection shifts. They also report topical evolutions in Scientometrics and in TM over time. This is an interesting hybrid approach using text and statistical analyses to extract information on R&D concentration, connection, and movement.

## Part II Text Analytic Methods

The second part of the book focuses on the methods of Tech Mining and text analysis. There are a variety of approaches proposed. These range from the innovative approaches based on the use of new automated processes, algorithms, and ontologies, as well as the combination of Tech Mining approaches with other frequently used FTA methods to connect intelligence into decision-making processes.

The chapter by De Spiegeleire, van Duijne, and Chivot presents the metafore approach developed by The Hague Centre for Strategic Studies. Foresight 3.0, as the authors call it, is an attempt to distill more insights into future by combining all available data systematically. The chapter presents the main steps in the metafore research protocol that was used in a recent project for the Dutch government's "Strategic Monitor," which tries to anticipate the future in the area of foreign, security, and defense policy. The project had partners from multiple countries.

Benson and Magee introduce a new method called the classification overlap method (COM) which provides a reliable and an automated way to divide the patent database into understandable technological domains where progress can be measured. The authors conclude that there is now an overall objective method named Patent Technology Rate Indicator (PTRI) for using just patent data to reliably estimate the rate of technological progress in a technological domain. Thus, the first link between the patent database information and the dynamics of technological change is now firmly established; robustness and back-casting tests have shown that the assertion of reliability is meaningful and that the estimate has predictive value.

Courseault Trumbach, McKesson, Ghandehari, DeCan, and Eslinger introduce ontologies which are used in text mining processes to better understand text from a specific domain. Authors present a broad ontology for the innovation and design process. Through an example within the shipbuilding domain, the authors take steps toward building an innovation and design process ontology which can be applied to the forecasting innovation pathways (FIP) framework as a means of capturing and understanding the influences on the technology delivery system.

Huang, Zhang, Ma, Porter, Wang, and Guo combine topical analysis, patent citation analysis, and term clumping analysis to gather technology intelligence. The method identifies key subdomain patents, associated with particular component technology trajectories and then extracts pivotal patents via citation analysis. The authors compose evolutionary pathways by combining citation and topical intelligence obtained through term clumping. The case of dye-sensitized solar cells (DSSCs) is used to demonstrate the approach.

Huang, Shang, Wang, Porter, and Zhang use patent information and semantic analysis to identify targets for technology mergers and acquisitions. A case of China's cloud computing industry is analyzed to demonstrate the approach.

Chen, Zhang, and Zhu propose a topic change identification approach based on latent Dirichlet allocation, to model and analyze topic changes and topic-based trend with minimal human intervention. After textual data are cleaned, underlying semantic topics hidden in large archives of patent claims are revealed automatically. The results of a case study show that the proposed approach can be used as an automatic tool to provide machine-identified topic changes for more efficient and effective R&D management assistance.

The two remaining papers in the chapter describe the integration of Tech Mining with well-known FTA methods. Zhang, Chen, and Zhu attempt to develop technology roadmaps (TRM) semi-automatically through multiple data sources. The authors apply the fuzzy set to transfer vague expert knowledge to defined numeric values for automatic TRM generation. They present a case study on computer science-related R&D and show that the approach can assist in the description of computer science macro-trends for R&D decision makers.

In the final chapter of this part, Kayser and Shala propose developing scenarios using text mining. In their approach, text mining automatically processes texts and summarizes the topic. Two different approaches: Concept mapping and speech tagging are applied on two different scenarios which were developed through a Europe-wide project.

## Part III Anticipating the Future—Cases and Frameworks

Following the discussion on the concepts and methods in the first two parts, the third and final part of the book demonstrates the ways of putting ideas into practice through case studies in selected high-technology domains. The first paper, a case study of LDA employed in CTI for national technology strategy purposes, is provided by Salvador, Menendez, and Novillo on the field of additive manufacturing. Their assessment of the potential for additive manufacturing begins with a global technological and market landscape, which is then compared with actual development in Latin America. Market research and expert interviews are rounded out with patent and scientific literature analyses to determine the degree to which Latin America is behind other regions of the world in technological and market development. In such a position, such analyses are useful to pinpoint leading research organizations, research focus, key research networks, key patent holders, and private players investing in process commercialization in order to streamline decisions on where best to enter technology and market development.

Next, Daim, Khammuang, and Garces apply social network analysis (SNA) to the technology area of smart roofing to identify the dynamics of expert networks, so researchers gain a better understanding of the current state of smart roofing research and development programs. Using bibliographic data from Web of Science, the authors generate the SNA attributes of degree of centrality, betweeness, closeness, and number of citations, to pinpoint the top 11 experts in the field as well as their collaborative influence on the R&D network. This in-depth knowledge of experts and networks is particularly useful in such management of technology processes as technology roadmapping (TRM), R&D portfolio selection, R&D project initiation, and strategic technology planning.

New drug development can take a decade or longer, with most compounds not proving fruitful. Analyzing primary patent applications in a therapeutic class can give a picture of which drugs may reach the market in the future, though this picture is severely muddled as at early patenting stages, therapeutic uses are unclear, and multiple patents filed at different stages may include the same compound with different claims. Mendes and Antunes address this matter with a method for gathering and analyzing the primary patent applications for new antibiotics using Derwent Innovations Index Database and VantagePoint text mining software. Employing IPC codes, Derwent Manual Codes, the World Health Organization's Anatomical Therapeutic Chemical (ATC) classification system, and MEDLINE's Medical Subject Headings (MeSH) as resources in the text-mining software, 32,068 antibiotic patents where reduced to a set of 1333 primary patents. These patent applications can be analyzed to show whether antibiotics are from old or new chemical classes, which bacteria they act against, mechanisms of action, and what strategy was used to discover the compounds. Further study can point out antibiotics expected in future markets and if these antibiotics will meet projected critical bacterial disease priorities.

Liu, Sun, Xu, Jia, Wang, Dong, and Chen describe successful institutionalization of the use of text analytic-generated CTI at the National Science Library (NSA), Chinese Academy of Sciences (CAS), for Chinese government and enterprise MOT decision making. By tailoring specific combinations of text analytic-generated indicators from the science literature and patents (bibliometrics, patent metrics, text mining) and expert review, the NSA-CSA is able to address three levels of information needs—micro level (specific technology), meso level (technology field), and macro level (industry). Examples of the CTI products include technology novelty reviews, innovation pathway selections, product development evaluations, competitor monitoring, R&D partner identification, and industry–technology field analyses to support industrial technology and development strategies. The authors provide case studies of a micro level analysis of hydrodynamic cavitation technology for wastewater processing, meso-level analysis of swine vaccine technologies, and a macro-level strategic intelligence analysis of the ionic rare earth industry. The authors also share insights into the specific feedback clients have provided as well as quality control measures adopted over the years.

Finally, the chapter by Bakhtin and Saritas introduces a methodology for the identification of trends through a combination of "thematic clustering" based on the co-occurrence of terms and "dynamic term clustering" based on the correlation of their dynamics across time. In this way, it is possible to identify and distinguish four patterns in the evolution of terms, which eventually lead to (i) emerging, (ii) maturing, and (iii) declining trends, as well as (iv) weak signals of future trends.

Key trends identified are then further analyzed by looking at the semantic connections between terms identified through Tech Mining. This helps to understand the context and further features of the trend. The proposed approach is demonstrated in the field photonics as an emerging technology with a number of potential application areas.

Tugrul U. Daim
Department of Engineering and Technology Management
Portland State University, Portland, OR, USA

Denise Chiavetta
Search Technology, Inc., Norcross, GA, USA

Alan L. Porter
Georgia Institute of Technology, Roswell, GA, USA

Ozcan Saritas
National Research University Higher School of Economics
Moscow, Russia

# Contents

**Part I Data Science/Technology Review**

1 **FTA as Due Diligence for an Era of Accelerated Interdiction by an Algorithm-Big Data Duo** . . . . . 3
Denis Loveridge and Cristiano Cagnin

2 **A Conceptual Framework of Tech Mining Engineering to Enhance the Planning of Future Innovation Pathway (FIP)** . . . . . 25
Munan Li

3 **Profile and Trends of FTA and Foresight** . . . . . 45
Per Dannemand Andersen and Lars Alkærsig

4 **Recent Trends in Technology Mining Approaches: Quantitative Analysis of GTM Conference Proceedings** . . . . . 59
Nadezhda Mikova

5 **Anticipating Future Pathways of Science, Technologies, and Innovations: (Map of Science)$^2$ Approach** . . . . . 71
Irina V. Efimenko, Vladimir F. Khoroshevsky and Ed. C.M. Noyons

**Part II Text Analytic Methods**

6 **Towards Foresight 3.0: The HCSS Metafore Approach—A Multilingual Approach for Exploring Global Foresights** . . . . . 99
Stephan De Spiegeleire, Freija van Duijne and Eline Chivot

7 **Using Enhanced Patent Data for Future-Oriented Technology Analysis** . . . . . 119
Christopher L. Benson and Christopher L. Magee

8 **Innovation and Design Process Ontology** . . . . . 133
Cherie Courseault Trumbach, Christopher McKesson, Parisa Ghandehari, Lawrence DeCan and Owen Eslinger

9 **Generating Competitive Technical Intelligence Using Topical Analysis, Patent Citation Analysis, and Term Clumping Analysis** . . . . . 153
Ying Huang, Yi Zhang, Jing Ma, Alan L. Porter, Xuefeng Wang and Ying Guo

10 **Identifying Targets for Technology Mergers and Acquisitions Using Patent Information and Semantic Analysis** . . . . . 173
Lu Huang, Lining Shang, Kangrui Wang, Alan L. Porter and Yi Zhang

11 **Identifying Technological Topic Changes in Patent Claims Using Topic Modeling** . . . . . 187
Hongshu Chen, Yi Zhang and Donghua Zhu

12 **Semi-automatic Technology Roadmapping Composing Method for Multiple Science, Technology, and Innovation Data Incorporation** . . . . . 211
Yi Zhang, Hongshu Chen and Donghua Zhu

13 **Generating Futures from Text—Scenario Development Using Text Mining** . . . . . 229
Victoria Kayser and Erduana Shala

**Part III Anticipating the Future—Cases and Frameworks**

14 **Additive Manufacturing: Importance and Challenges for Latin America** . . . . . 249
Marisela Rodríguez Salvador, Ana Marcela Hernández de Menéndez and David Alfredo Arcos Novillo

15 **The Application of Social Network Analysis: Case of Smart Roofing** . . . . . 273
Tugrul U. Daim, Monticha Khammuang and Edwin Garces

16 **Building a View of the Future of Antibiotics Through the Analysis of Primary Patents** . . . . . 303
Cristina d'Urso de Souza Mendes and Adelaide Maria de Souza Antunes

17 **Combining Scientometrics with Patent-Metrics for CTI Service in R&D Decision-Making: Practices of National Science Library of CAS** . . . . . 321
X. Liu, Y. Sun, H. Xu, P. Jia, S. Wang, L. Dong and X. Chen

18 **Tech Mining for Emerging STI Trends Through Dynamic Term Clustering and Semantic Analysis: The Case of Photonics** . . . . . 341
Pavel Bakhtin and Ozcan Saritas

# Part I
# Data Science/Technology Review

# Chapter 1
# FTA as Due Diligence for an Era of Accelerated Interdiction by an Algorithm-Big Data Duo

**Denis Loveridge and Cristiano Cagnin**

**Abstract** In the face of the 'digital revolution' and its wide penetration of all aspects of life, FTA needs to consider new approaches and skills to enable it to cope with a 'new' world. An approach based on 'due diligence,' adapted from the business world, is suggested. The paper links the digital world to an algorithm-big data duo, where computation is preferred to human judgment, with its behavioural and intuitive 'baggage', in policy formulation. Turing's 1936 paper enabled the evolution of digital computers capable of using complex algorithms to work with large and uncertain data sets. The current favouring of computation highlights the need for FTA to be based on an appreciation of dynamic situations that face all life on Earth replacing silo-based problem-solving. To cope with these situations, new skills are needed based on excellence in breadth and depth using due diligence concepts that can build a bridge between FTA and policy makers to ensure both the quality and the ability to embrace ignorance are coped with.

**Keywords** Algorithms · Big data · Ignorance · Existence · Extinction · Emergence · FTA skills

## 1.1 Introduction

In its infancy TA was the embodiment of a protest movement before it became institutionalized in the US Office of Technology Assessment and the UK's Programmes Analysis Unit, both of which are now defunct. Huddle (1972) defined TA initially. By 1996, the situation had changed and a revised definition empha-

D. Loveridge (✉)
MIoIR, MBS, University of Manchester, Oxford Road, Manchester M13 9PL, UK
e-mail: denis.loveridge@manchester.ac.uk

C. Cagnin
Center for Strategic Studies and Management (CGEE), SCS Qd 9, Lote C, Torre C, 4° andar, Salas 401 A 405, Ed. Parque Cidade Corporate, Brasília, DF 70308-200, Brazil
e-mail: ccagnin@cgee.org.br



T.U. Daim et al. (eds.), *Anticipating Future Innovation Pathways Through Large Data Analysis*, Innovation, Technology, and Knowledge Management,
DOI 10.1007/978-3-319-39056-7_1

sized uncertainty (Dale and Loveridge 1996). Since 1996, the characteristics of FTA do not seem to have changed much. The fascination with methods seems unending but Wittgenstein's dictum that 'methods pass the problem [situation] by' calls for human judgement to be promoted. In the evolving digital world, engagement with what will later be called 'the algorithm-big data' duo enters in increasing force. The paper explores how this duo has grown in importance. The argument develops through Sect. 1.2, where some primary notions of FTA are set out; Sect. 1.3 describes the relation between FTA, living systems and complexity; Sect. 1.4 highlights some notions about ignorance; Sect. 1.5 describes the relation between human decision-making and computation; Sect. 1.6 illustrates the algorithmic world; Sect. 1.7 does the same for the big data world; Sect. 1.8 explains the duo of algorithms and big data; Sect. 1.9 proposes a different approach to FTA aligned with some new skills; these are described in Sect. 1.10. The paper ends with a brief discussion (Sect. 1.11) and conclusion (Sect. 1.12).

FTAs' conventional concern was the linkage between new technologies and social development which was perceived during the enlightenment well before the notion of 'economics' was born as a cultural invention. In the digital world, advances in communication technologies have quickened the pace of science and technology and has created globalization of world markets. However, a long, slow running unease with the assumption that all S&T were 'good things' and that human mental plasticity would always adapt to them began to split society. Rejection of these assumptions grew from 1970 onwards and has been accompanied by the rejection of technological determinism, through exposure of its hidden social consequences. Soddy (1922) provided a scientific criticism of the conventional economic mantra. However, it was probably the use of nuclear weapons to end World War II and tensions during the Cold War that gave added impetus to the questioning of the role of S&T in human development. The conceptual and methodological basis of FTA was developed in this period and first systematic methods (e.g. Delphi) were developed at that time.

A clamour grew simultaneously for the governance of S&T. New fora for involving the public in the governance of S&T came in many forms: these highlighted the breadth of the situations involved as a cascade of them evolved over the last 40 years. 'Unpredictable' events increased recognition that global systems are uncertain and complex, causing the notions of 'grand challenges' and sustainability to emerge. All of the above occurred during a persistent rearrangement of the world's chessboard of power that has now (2014) moved towards the Pacific Basin in which invention and innovation, and their risks, are an important part of the emerging landscape. Now there is much force to Whitehead's perception that 'Science is concerned with generalities. The generalities apply, but they do not determine the course of history apart from some anchorage in fact'. FTA affects ***all*** life and has a pivotal role to play in assuring the continuation of basic services and infrastructures, human rights, freedom, democracy and privacy, all of which is threatened through risk, regulation and governance. All the above points to the necessity for new skills for FTA practice towards one which encompasses ignorance, complexity and creativity.

## 1.2 Primary Notions of FTA

Cagnin et al. (2012) described the role FTA plays' in informing decision-making, structuring and mobilizing actor networks and capacity-building among innovation actors. FTA is part of foresight which, for the sake of clarity here, will be assumed to endeavour to identify future possibilities from what is known or can be speculated about from current knowledge; this involves *subjective opinion*. The kernel of subjective opinion is the ability for people to project their substantive knowledge into the future to assess and represent uncertainties concerning the future, in a non-empty way. Non-emptiness implies speculation or opinion, based on an incomplete understanding of events (ignorance) since the future is by definition ***unknown***.

Briefly, Dalkey (1969) describes the nature of the knowledge experts have at their disposal: the gradations from knowledge to opinion and from opinion (or speculation) to conjecture are hazardous as opinion or speculation implies the presence of ***incomplete*** evidence: reasoned opinion or speculation can then only be made on a probabilistic or fuzzy reasoning basis, though the expert will usually decline to attach any kind of measure to his opinion. The further their reasoning goes into the future, the further their opinions move towards the fuzzy transition into conjecture, where evidence to support their opinions becomes fragmentary. Amara and Lipinski (1983) demonstrated that most experts are far too confident, when extending their knowledge into the future (so what of non-experts?), frequently leading to 'lock-in' even though what is sought are patterns from all the streams of their experience that ***seem relevant to the situation***.

In most studies opinions, expert or non-expert tend to be regarded as of equal weight; this is not a valid assumption. However, this vexed behavioural question has never been resolved though there is fragmentary empirical evidence that some expert opinions are many times more effective than others (Amara and Lipinski ibid.). Experts and non-experts have to consider two very broad sets of entities and their intersections, neither of which can be clearly identified (Fig. 1.1).

The fuzziness of foresight (and FTA) is evident from Fig. 1.1, so that notions of certainty are misplaced. Rather, it is as well to acknowledge the phenomenon of *ignorance* (discussed in Sect. 1.4). Lastly, the important issue is understanding how ideas do emerge, in a random manner and sometimes fleetingly, through the fuzzy boundary between the unknown and the barely appreciated.

## 1.3 Living Systems and Complexity

The situations that FTA addresses are living systems that evolve, regenerate and self-organize themselves to adapt to changing circumstances. Maturana and Varela (1980) described these as autopoietic complex adaptive systems where change is self-organized creating an emergent structure and pattern without external

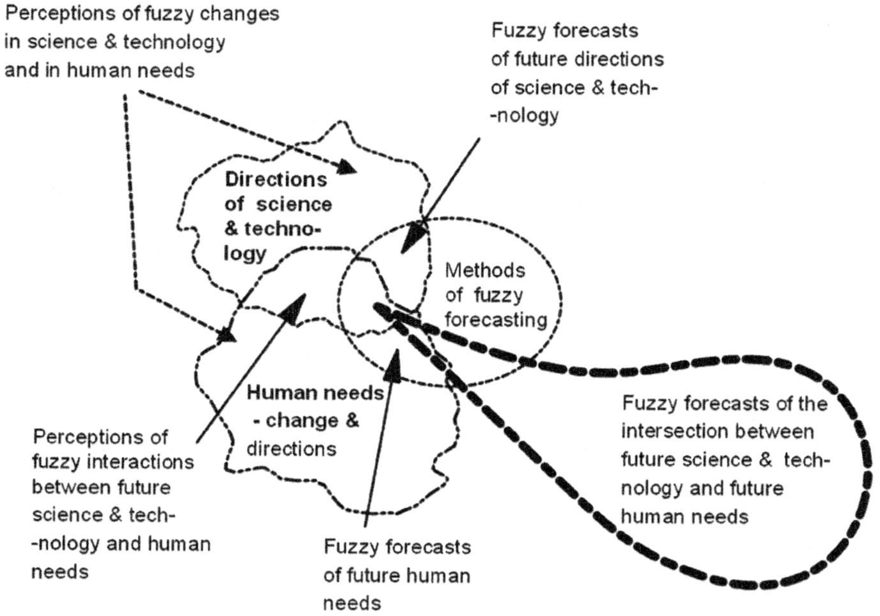


**Fig. 1.1** Intersection of human needs, science and technology and methods of forecasting

intervention. Every organism has the ability to self-generate implying continuous auto-production and reproduction (Maturana and Varela 1997): autopoietic systems are a product of themselves (Rocha 2003), have self-defined boundaries and are organizationally closed. Living systems learn and use new information to alter present and future behaviour to maintain internal homeostasis.

Complex adaptive systems are unpredictable: their emergent behaviour is more than the sum of the properties of parts and this relationship is ill-understood.

Dempster (1998) described as sympoietic, a complex ecosystem which does not have self-defined borders and that are collectively produced as well as organizationally ajar. Dempster (2000) concluded that autopoietic systems are homeostatic, development oriented, centrally controlled, predictable and efficient, whereas sympoietic systems are homeorhetic, evolutionary, distributively controlled, unpredictable and adaptive. Hence, one of the most important differences between autopoietic and sympoietic systems relates to the balance between their ability to maintain their identity despite changes in the environment and to adapt their identity to fit changes.

For FTA, the above descriptions present a useful heuristic to complex living systems that share matter, information and energy with their external environments: there is simultaneous autonomy and interdependence with a requirement for interactivity Rocha (2003). These matters are relevant to the understanding of social systems and their internal connectivity.

The information, knowledge and ignorance that are shared within social systems can lead to individual and collective adaptation and evolution. What one part does to another is indefinitely interpreted and informed to form more complex chains. The system will then be able to make legitimate evolutionary leaps characterized by the appearance of emerging properties. In this context, mutual trust (Maturana 1998; Losada 1999, 2001; Fredrickson and Losada 2005) is crucial for choosing a common path for life or for moving the whole system towards higher levels of sustainability. Dialogue and information sharing, founded on trust (forms of 'handshaking'), are prerequirements for both.

Existence imposes real limits that are controlled practically through extinction events. The combination of existence and extinction has the nature of a feedback loop that produces forms of stability: these only get out of control when that balance fails producing an inequality in the form of 'persistent' feed-forward, which can be either positive or negative, until stability returns but in a different way, a phenomenon called homeorrhesis. Ultimately, an inequality between existence and extinction leads to major crises for living systems. Each crisis stems from the summation of a myriad of individual events. In current parlance, these crises are called 'grand challenges' though history is littered with such traumas for life on Earth, humanity in particular, that have been referred to under different names.

In this context, social change implies that people within a society must change: this happens either through encounters outside the specific social system or via reflections through language (Maturana and Varela 1997), requiring handshaking between policy makers and FTA practitioners, as well as between social actors in general. Basic emotions are the basis of the operationalization of living organisms, and these change as the environment changes, requiring an individual to adapt to his/her environment to avoid disintegration.

FTA becomes key to enable a creative dialogue and the interactions required to allow social systems to behave as sympoietic complex systems (Cagnin and Loveridge 2012). Features of universal ethics or universal principles and those of respect (Zohar 1990) can be linked with notions of high-performance teams and organizations (Losada 1999, 2001).

## 1.4 Notions of Ignorance

Appreciation of complex and dynamic situations lies at the heart of FTA in which the future is logically redundant because if a science, technology or engineering is known or imaginable it is no longer in the future, and only their applications lie there together with their ethical, legal and social influences (ELSI). Due diligence embraces ELSI studies that necessarily encounter various aspects of ignorance: in this frame, *ignorance is not the antithesis of knowledge*. Ignorance penetrates ELSI very deeply both technically and behaviourally. In engineering and invention, ignorance is a well-appreciated matter: it lies at the root of the dilemma of a system being 'fail-safe' (a long-established engineering and risk principle) rather than the

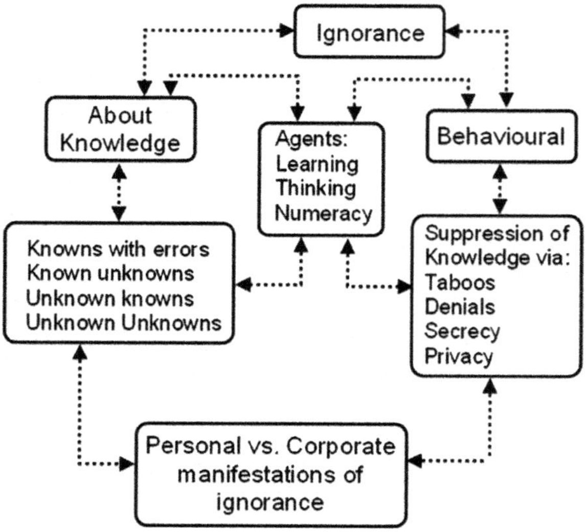


**Fig. 1.2** Summary of Roberts taxonomy of ignorance

ecological principle of a system being 'safe when it fails' (Holling 1977). Roberts (2012) sets out a taxonomy of ignorance, summarized in Fig. 1.2, that indicates the duality of ignorance being 'about knowledge' and 'about the behavioural influences'.

In engineering, ignorance breeds an appreciation of the need for caution in design procedures (fail-safe principle) which, through SEEP and V pressures, over recent decades has become formalized through the 'precautionary principle'. Stirling (2008) introduced important matters concerning science, precaution and politics relating to technological risk in particular. Stirling's key factors were uncertainty (characterized by probability), ambiguity (presumably of information) and ignorance: these can be fitted into Robert's taxonomy beneficially.

For policy, the clash between ignorance and knowledge, and its many grey areas, creates serious dilemmas for policy makers that can be illustrated as in Fig. 1.3. Policy makers tend to resolve these situations by imposing agreed boundaries on them to enable the appreciation of risk; boundaries to these perceptions and their fitness for purpose, valuation and risk. How these 'boundaries' are conceived and drawn then becomes an important matter. If the boundary regards the situation as autopoieotic rather than sympoietic, then the outcome will be markedly different. An autopoieotic situation can be regarded as organizationally closed, effectively becoming a silo, whereas a sympoietic situation will be characteristically ajar and open to outside influences, perhaps acknowledging the nature of the 'real' world devoid of silo features. These notions generate conflicts in the realities of policy making creating a sense of appreciation of 'existence'.

For policy makers, it is essential that FTA creates of a sense of 'handshaking' (Boettinger 1969) and common ground for appreciation of the situation within the taxonomy of ignorance (Fig. 1.2). These steps begin to create a common language for appreciation. In the real world, ignorance can become mired in behavioural traits

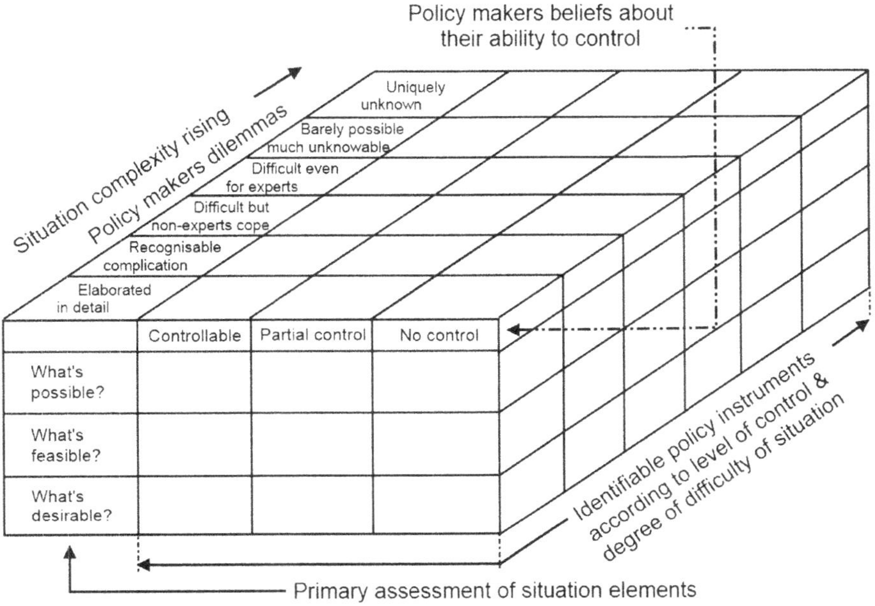


**Fig. 1.3** Policy-makers dilemmas (© Denis Loveridge reproduced with the kind permission of Routledge)

that make themselves apparent in corporate ignorance characterized in the way shown (Fig. 1.2). However, it is the merging of the two streams of ignorance that poses hazards for policy makers and corporate executives. For example, when 'known unknowns' are suppressed in order to ensure that policy outcomes are achieved, this amounts to limiting the policy makers dilemma to 'recognizable complication' that is 'controllable' to achieve 'what is desirable' (Figs. 1.2 and 1.3). Behaviourally, this implies the adoption of a highly constrained appreciation of the dynamic situation and the absence of a common language for policy and leads to a biased and partial model.

## 1.5 Human Decision-Making Versus Computation

In many ways, polities have been lured into accepting a preference for numbers in place of thought without necessarily appreciating either how the numbers were produced or what they mean. Computation and the computational models that create them have grown an aura of their own creating a conflict between important decisions made through human processes rather than those that rely on computer-based models. These conflicts have been characterized journalistically as 'Computers and you or computers or you' (Loveridge 1983) a view expressed similarly by Michael (1962) and more trenchantly by Eric Schmidt chairman of Google (Schmidt 2014) and are referred to again in Sect. 1.9.

Belief in numbers is convenient being seen as a way to remove or at least limit the effects of ignorance. Funtocwiz and Ravetz (1990) devised the NUSAP system for understanding numbers in policy making including their more exotic role of how, why and who created them. Whitehead's notion of the fallacy of misplaced concreteness similarly reveals why overemphasis on numbers, however produced, is unwise (Whitehead 1925) creating conflict from differing beliefs without settling the questions posed through ignorance and different personal 'models' of a situation.

FTA models of situations are bounded and may be qualitative, quantitative or a mixture of both. Qualitative models are unavoidable as they are the precursors to any form of later quantitative model. Qualitative models set out a linguistic appreciation and description of a situation of concern: they need to be investigative (later called 'due diligence'), imaginative and based on the evolution of common ground as referred to earlier. Inevitably, common ground needs to cope with the influences of ignorance rather than to focus exclusively on what is believed to be known. Because of their real-world complexity, models of situations are bounded thus limiting appreciation of their wider world influences, imposing strong demands on how these boundaries are created and the 'handshaking' required in doing so.

Quantitative models are the computable embodiment of the qualitative appreciative models of a dynamic situation: they are necessarily incomplete because the 'entire' situation is beyond understanding of the constraints of artificial boundaries. How computable models are constructed, their pitfalls and what can be learned from them is a matter of both interest and concern discussed next.

## 1.6 Algorithmic World

An algorithm, or a set of them, lies at the heart of any method used in FTA that involves a computable model. What then is an 'algorithm'? An algorithm is a precise step-by-step calculation procedure: it is an effective method expressed as a finite list[1] of well-defined instructions for calculating a function. Algorithms have become ubiquitous as the so-called app revolution has made so many consumer products depend on computation for their operation. Each 'app' is a representation of the product's designer's model of how that product ought to work. To reach this state of affairs, the procedure outlined in Fig. 1.4 has been gone through either knowingly or unknowingly.

The transition from the linguistic appreciation of the situation is complex and, as already pointed out, requires intense handshaking throughout and, if an 'app' is involved, that may involve both software versus firmware or both. Sadly, the rise of

[1]http://en.wikipedia.org/wiki/Algorithm-cite_note-1.

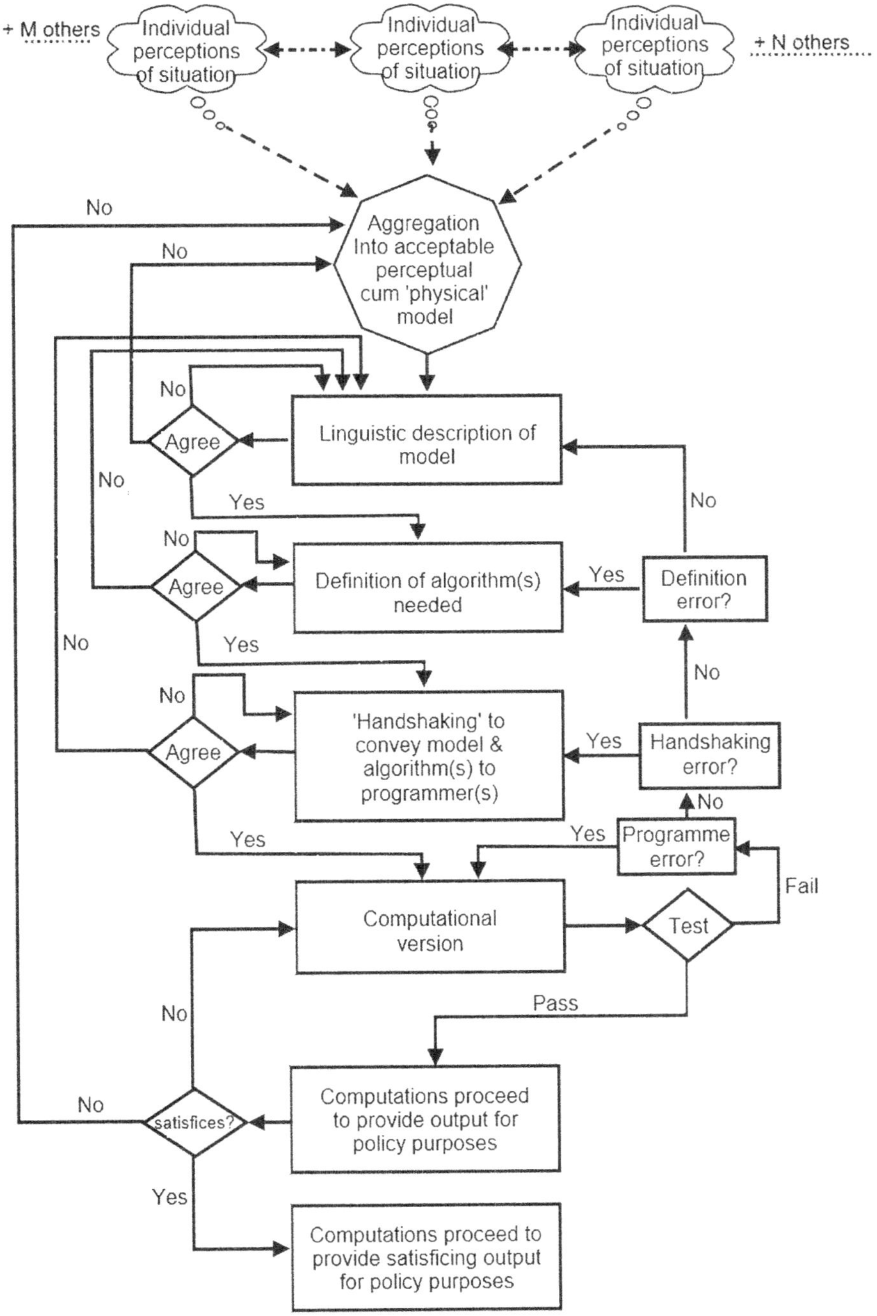


**Fig. 1.4** From individual perceptions to computable policy model—a flow diagram

the world of 'apps' removes the necessity to understand what is going on behind the screen. Approximations and shortcuts are often used by programmers and the nature of the underlying model remains hidden, so that questioning the output lies in the land of intuition else the output is subjected to blind acceptance. The proliferation of 'apps' lies in the human value now placed on immediacy often at the expense of security and privacy, and quality of the information created.

## 1.7 Big Data World

What is 'Big data'? Big data may be described as 'an all-encompassing term for any collection of data sets so large and complex that it becomes difficult to process using traditional data processing applications' (Wikipedia): this amounts to the accumulation of very large amounts of raw textual, numerical and graphical data collected by any means whatsoever that is capable of being digitized. Big data is, in that sense, ubiquitous with the prospect of transforming FTA. The revelations of the 'Snowdon papers' has done a great deal to, somewhat savagely, change the polity's perceptions of itself, of government security agencies throughout the world, and the probability that previously treasured privacy has been given up freely by large segments of people of all ages through the arrival of cybernetic social networks: a much simplified picture of how deeply the appetite of 'big data' is being fed can be gauged from Fig. 1.5.

Perhaps the most astonishing development has been the rise in the number and depth of data-collecting devices. Any individual may now clothe himself/herself so as to become a mobile and unique data source. In addition with the powers now being enacted, governance organizations are changing from local through national to global. Some of these powers are known others are not; falling into the category of unknown knowns: the only surety is that privacy is no longer a feature of human societies. At one time, the huge volume of data in 'big data' would not have mattered greatly as it could not be processed for any real purpose: that too has now become a fiction as the combination of computing power and effective processing procedures (algorithms) of great power has been created. Can FTA now deal with software sciences? Or is it routed elsewhere? It is only a matter of time before at least some if not all of these algorithms are embedded in 'apps' for common use changing the appreciation of security of information, of social interaction and of social control.

It is for the foregoing reasons that 'big data' has become important while the internationalization of all of them has introduced a new dimension through, for example, Webinars that enable management of international affairs in ways quite different to the past (e.g. Issl's management of its corporate intentions and image through the social networks). 'Big data' arrived during World War II when signal interception and deciphering became such an important weapon. Who uses and controls big data is a major preoccupation in all polities now that it is an open 'secret' that companies, governments and NGO's make use of 'big data' for

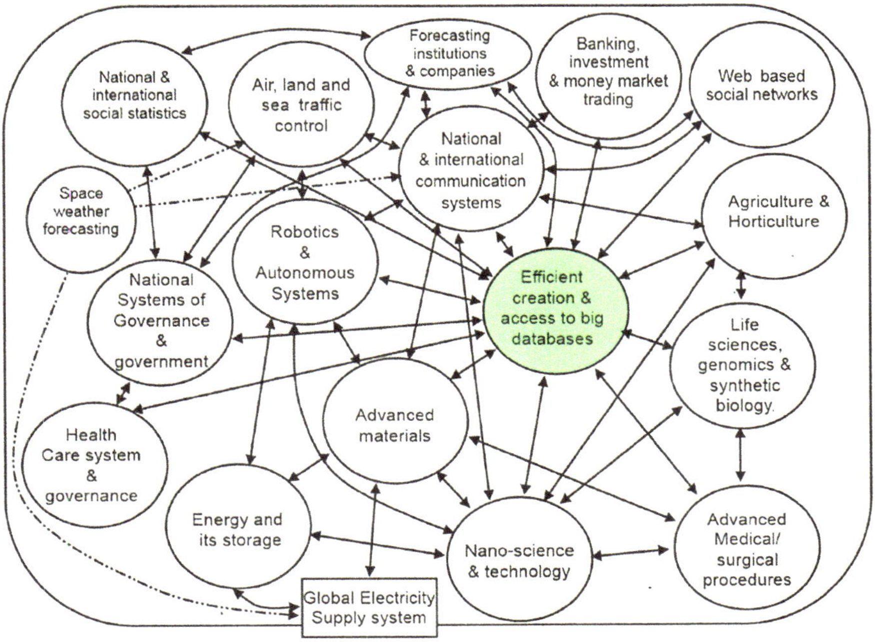


**Fig. 1.5** Some elements 'feeding' to 'big data'

purposes that range from legitimate to questionable. More contentious is that storage and control of access to 'big data' is often in the hands of a relatively select set of organizations, some public and some private, where the latter has considerable financial muscle. 'Big data' runs throughout the STEEPV acronym, so that its values are ubiquitous even if access to it is limited through either commerce or processing and interpretation capabilities. The latter relates to software capability and its quality, neither of which can be taken for granted.

## 1.8 World Duo of Algorithms and Big Data

Earlier comments lead to the inevitable conclusion that the combination of immense computing power, united with similarly huge advances in algorithm design and programming, all married to 'big data', create a duo that may come to dominate human decision-making with influences throughout the human and natural worlds. There is plenty of evidence to that effect. For example, the UK revenue authorities HM Revenue & Customs (HMRC) have long collected an array of information on individual financial affairs. HMRC's IT capabilities, through its Connect system, developed by BAE Systems, can create taxpayer profiles either singly or in groups from 'big data' collected globally (if necessary) to enable identification where tax evasion may be occurring and to do so quickly. Similarly, algorithms are being used to define city areas where criminal activity is likely to be rife to facilitate anticipatory police operations. There is also a plethora of modelling of national economies, climate and weather to mention only three of the better known spheres where the ‘duo’ may be putting human judgement into diminuendo.

The duo has now gained traction through governments and major corporations through conventional modelling: it is clear that the influence of the duo on all life is not a mere assertion. Whitehead’s emphasis on misplaced concreteness points directly to the fallacy embedded in all modelling that it does not and probably cannot represent the real world. The duo cannot create the real world out of the imperfections of modelling nor can it identify more than the model’s framework will allow. The danger is that modelling through its boundaries and the duo close off real-world options while use of the computed options implies attempting to force the real world to conform to the unreal world of computation, a world governed by the nature of algorithm design, programming (with its human fallibilities’) and the frameworks imposed by ‘big data’ structures. In this way, the duo is already setting boundaries, real and unimagined, un-noticed and unconventional, around all life introducing the certainty that the duo’s influence on policy making will endure for decades into the future. The notions of ignorance and precaution are ever present.

Privacy may be only the first casualty, intended or unintended, while opinion and belief formation may not be far behind. Freedom and control of the polity are already deeply involved through matters relating to security, freedom of movement and to many individual privileges and rights, to disease control and to the control of organized crime on all scales.

What all the foregoing means for FTA is explored next.

## 1.9 FTA for the Future

The present section presents two propositions. The first deals with the nature of FTA, and the second concerns the world that FTA is now embedded in and will remain for decades.

The claim to a future orientation for TA is an oxymoron. Once an idea has been expressed in public space, it is no longer in the future but inhabits whatever time space one chooses. It may seem pedantic to make this distinction but the act of recognizing something not previously known has some deep implications for the conduct of FTA. First, the recognized information will lie somewhere in the taxonomy of ignorance described earlier: this will have implications for the directions of learning: the boundaries, which by inference must be sympoietic to allow the entry of new information, will need revision to enable bridging to policy making while embracing uncertainty, complexity and creativity. Second, the boundary between ignorance and knowledge will change implying that the shift is real and not pseudo. Real shifts in the frontiers of ignorance are complex and difficult to recognize.

The second proposition is derived from the struggle between human judgement and the algorithm-'big data' duo. The struggle began with the advent of computing machinery embodied in Babbage's mechanical computing machine but really became obvious following Turing's 1936 paper that set the scene for all modern digital computers. **Will the 'algorithm-big data' age be a frightening jobless dystopia?** Three quotations set the scene:

> Between now …… 1984, business and government …. use extraordinary advances in computer technology to file and collate 'personal' facts about private citizens …. (Michael (about 1962))
>
> …… little questioning is apparent anywhere in relation to computers ……… a major goal for computer technologists is to put new skills back into the hands of individuals rather than to continue to remove them from employment:…. From 'Computers and You' (Loveridge (1983))
>
> 'finding employment for these [displaced] workers will be the "defining issue for the global economy in the decades to come. It's a race between computers and people – and people need to win"' Eric Schmidt, Chairman, Google (2014 Sunday Times Business, 2 February)

The proposition points to some fundamental questions about the future of human life and of ***all*** life as human judgement and/or the duo engage ever more closely. It is this evolving engagement that is reshaping decision-making in many spheres and will shape the nature and content of FTA.

FTA is application oriented, and it is a real-world activity involving threads in all of the STEEPV themes. It is plagued with all the aspects of ignorance along with ambiguity, paradox, complexity that inevitably are simplified through setting boundaries, real or imaginary, to enable appreciation of the situation. Normal accidents (Perrow 1984) of every conceivable kind are only to be expected when the situation is far from equilibrium which those subjected to FTA studies usually are. FTA needs to step away from the time-honoured addiction to methods of analysis that have not changed much in several decades. The need is to move towards the investigative ideas embodied in 'due diligence' with their flexibility; emphasis on the entire STEEPV set and with probing questions that evolve as appreciation of the situation and its dynamics reveal themselves. Appreciation begins and ends in the top level in Fig. 1.4 where probing dialogue occurs before a linguistic model, based on common ground and agreed boundaries, begins to emerge and to be formulated as a model locking out elements of the appreciation that might prove to be the keystone of the situation. Fixed checklists are not suitable for appreciation of a situation.

For FTA, the borders between the themes of the STEEPV set have largely, if not entirely disappeared. FTA therefore needs to become investigative rather than analytical. In the business and investment world, this requirement often is a legal one achieved through ***due diligence***. The proposition is for FTA to embrace the principles of due diligence in a modified form to encompass the virtual disappearance of disciplinary boundaries. If this sounds like a return to the principles set

**Table 1.1** Requirements of due diligence for (F)TA

| STEEPV element | Due diligence representation | Dominant feature |
|---|---|---|
| Social | Expectations and realities of individuals and of groups of individuals | Social cohesion |
| Technical (includes science, technology and engineering) | Influence of engineering, science and their thought processes and outcomes on life | Fail-safe versus safe fail |
| Economics | Business, industry including challenges to conventional economic 'theories' | Ecological economics and purposes of industry |
| Ecology | Principles of sustainability underpinned by laws of thermodynamics | Entropy |
| Politics | Governance, law and compliance | Rules regarding freedom and responsibilities |
| Values | Basis of societies beliefs and its unspoken 'social contract' | Argumentation and legitimization of 'mores' |

out by Huddle (1972) and Dale and Loveridge (1996), then so be it. What then is due diligence? And how does it differ from analytical processes that are essentially defined algorithmically even if that is not declared?

The UK Business Angels Association has advanced notions of due diligence for start-up small businesses: with reasoned modification, these can create an appreciative mind for FTA practitioners. The guidance offered by the UBAA has its necessary limitations which, it is suggested, can be made relevant to FTA as shown in Table 1.1.

Due diligence is a language-based way of systemically and intuitively researching, verifying and appreciating a situation in the context of the taxonomy of ignorance. Sometimes, it is based on legal requirements, often it is not, but may shape legislation later. While the term originated in the business world, where due diligence is required to validate statements about the business, the goal in FTA was to ensure that every endeavour is made to assess the influence of technology within the agreed boundaries with the definition of TA offered by Dale and Loveridge. Will new skills be needed for this form of FTA? That is a 'big maybe' to look at next.

## 1.10 New Skills Required for the Future of FTA

'We think only through the medium of words. Languages are true analytical methods.... The art of reasoning is ... language well arranged' Lavoisier (1790): that is the point of the topmost level in Fig. 1.4. The algorithmic computation that follows limits reasoning unless there are very strong feedback loops: without these algorithms and computers kill reasoning. Until very considerable strides are made in computer-based reasoning, there are subtle but inevitable dangers in slavishly following the output from the duo. Until then, computer-based models will say a great deal about the interaction between the model and how the computer copes with its algorithms, and the programmer's interpretation of them: these subtle influences will remain. ***Reasoning about ignorance is an unfamiliar skill.***

Due diligence is an intensely practical investigation but will remain a prisoner of ignorance in its many guises. It will always depend on a mixture of subjective opinion and quantitative data that ought to be measured against requirements of the NUSAP system of assessing the quality of data. ***Understanding numbers and the quality of data is another required skill.***

Dalkey's description of subjective capability may be reinterpreted as a relation between expertise and creativity; these can be placed at two corners of a triangle. The third apex of the triangle is concerned with interpreting the outcome of the tension between expertise and creativity, into the policy-making processes; this has been termed alignment (see Fig. 1.6), with the implication of interaction (Cameron et al. 1996).

The triangular representation is preferred as none of the three vertices is in opposition, but all work through a creative tension. Effectively Interaction/ Alignment is a process of crossing a bridge between ***two worlds***.

FTA practitioners need the ability ***to engage in speculation, as defined by Dalkey: this is crucial; just as crucial is the ability to articulate that speculation in the form of substantive, but subjective opinion. However, radical insights into the future require the willingness to engage in conjecture, which will involve creativity***. The natural tension between expertise and creativity can bring important shifts in opinion that needs to be introduced into the policy process, effecting some tentative alignment, or bridge-building between radical opinions and the existing legitimated opinions held in the polity: this is the purpose of due diligence.

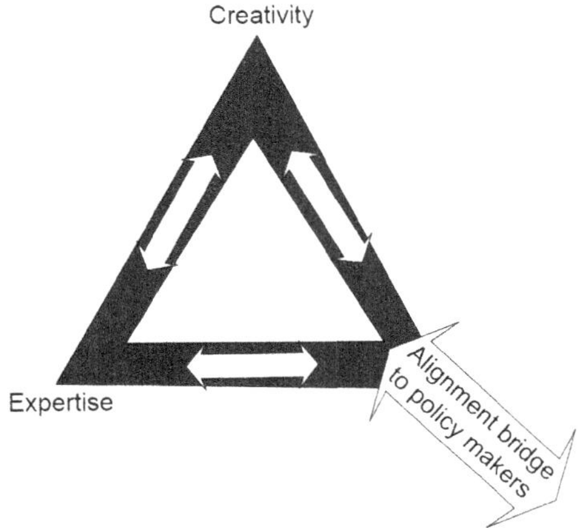


**Fig. 1.6** A notional FTA—policy bridge

FTA as due diligence is concerned with applications as situations. Due diligence in this context is the 'intellectual task of articulating our problems [situations] of living' (Maxwell 1984) with the intention of proposing and criticizing possible solutions and human actions. It is not concerned with reductionist problem-solving but with the dynamism of life itself. It is this dynamism that converts the notions of the problems of living into life as the series of 'situations' that it really is and in which problem-solving is but a small and double-edged procedure. ***Situations are systemic and need to be thought of in the appropriate way***, involving the uncertainties of fuzzy boundaries; interdependencies that convert complications into complexity; the consequent creation of emergent situations that (it is claimed) 'cannot be anticipated.' ***Thinking in terms of dynamic situations is a required skill***.

FTA conducted as due diligence needs an integrator(s) who is/are the key person (s) capable of 'gluing' the many aspects of due diligence together. ***People always come top of the list of required skills in the venturing world which is where FTA really fits***. Often the notion of interdisciplinary or transdisciplinary teams are seen as the required way of working: sadly it is not. Due diligence ***requires excellence in breadth and depth in the core people: this is a scarce resource requiring knowing how to learn, how to think and numeracy in depth***. Figure 1.7 illustrates this using set theory based on nanotechnology.

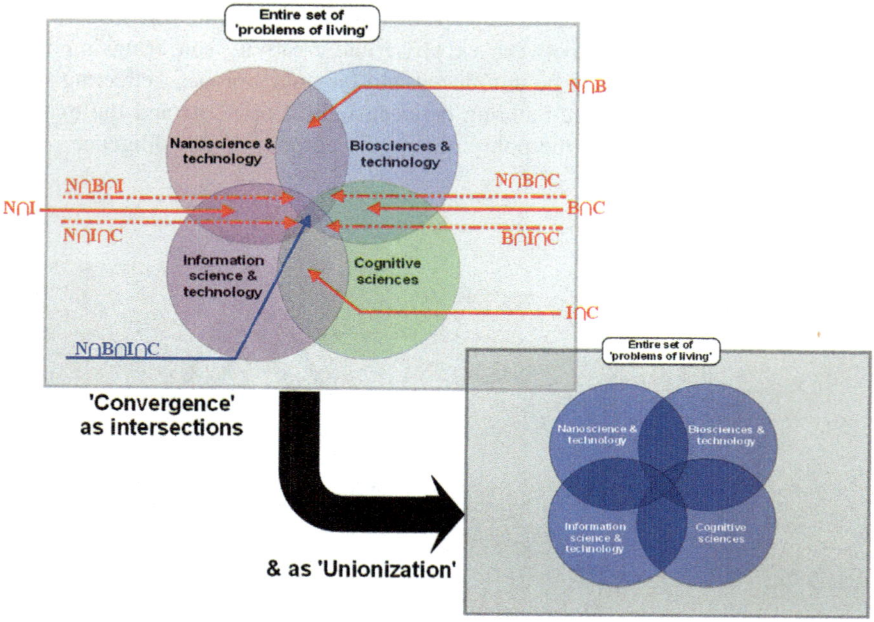


**Fig. 1.7** Convergence through interdisciplinarity versus unification via excellence in breadth and depth

## 1.11 Discussion

Throughout FTA, four dicta need to be in mind. The first and most powerful is that 'the world [situation] is never what it seems'. Conan Doyle, through the 'Sherlock Holmes' stories, revealed himself to be one of the earliest systems thinkers deeply imbued with the principles behind due diligence. Holmes contended that when all other lines of investigation have failed the final, most outlandish lead must be the correct one to follow. So it is with due diligence applied to FTA. The second is Wittgenstein's 'methods pass the problem [situation] by': it is a powerful deterrent to relying on computation alone backed up by Whiteheads' 'fallacy of misplaced concreteness.' In the third, the policy maker's dilemmas (Fig. 1.3) include the appreciative dimensions of what is possible? What is feasible? And what is desirable? How these three relate to the STEEPV themes is illustrated in Fig. 1.8.

The dictum points particularly towards the intersections of the indicated themes of the STEEPV set as these have increasingly important properties in relation to appreciation.

The fourth dictum relates to reasoning from thinking, learning and numeracy, and its outcome as appreciation of a situation (Fig. 1.9).

In FTA using due diligence, reasoning will be an essential step to coping with ignorance as set out in Fig. 1.2 and policy dilemmas (Fig. 1.3) and the three previous dicta when facing the 'algorithm-big data' duo. These are steps that fixed process methods have difficulty with. Changes in how FTA is practiced are needed for the implementation of due diligence in FTA. These will need to enable practitioners to:

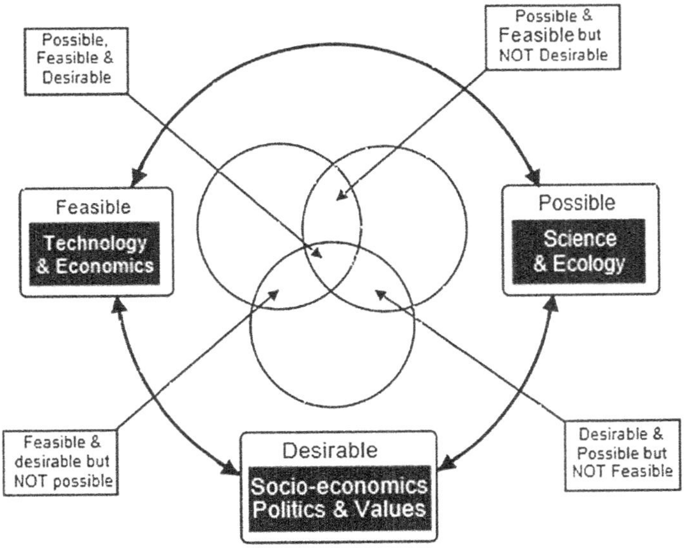


**Fig. 1.8** Interaction between what is possible, feasible and desirable and the STEEPV set

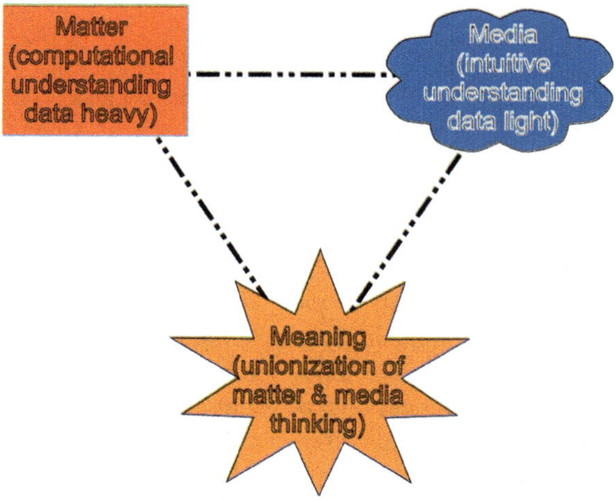


**Fig. 1.9** Unification of modes of reasoning

- Reason about ignorance
- Understand numbers and the quality of data
- Leverage creativity
- Think in terms of systemic situations
- Build excellence in breadth and depth
- Build the necessary bridge to policy and decision makers.

Due diligence principles will guide an evolving understanding of the dynamism of the situations perceived through a process of questioning and reasoning freeing FTA from fixed methodologies, thus allowing new information and learning to take place via a sympoietic understanding of complex systems. Two essential shifts will then take place:

| From… | … To |
|---|---|
| Unrevealed biases and describing extrapolations of the present in the future | Exposing anticipatory assumptions and describing discontinuities and 'unknowns' |
| Addiction to methods and the use of checklists and analytical processes, defined algorithmically | Investigative ideas with emphasis on the entire STEEPV set with probing questions evolving as appreciation of the situation and its dynamics grows |

The relationship between FTA and the 'algorithms and big data' duo needs examination. Due diligence allows questioning and reasoning, according to Fig. 1.1, whereas algorithmic methods may not. Building such ability may support looking outside familiar systems because of the emphasis on specialism in breadth and depth in preference to interdisciplinarity.

In 1996, technology assessment (TA) in its original form faced a crisis (in the sense of a turning point) as indicated by many authors in a special publication

(Loveridge 1996). Even though many of the individual authors claimed cultural diversity influenced how TA was conducted the underlying themes tended to be in problem based and methodological silos. In the past 20 years, it has more than ever become clear that subject silos no longer exist and that their 'invisible walls' have melted away in the face of the almost infinite variety (Ashby 1956) of information and data flowing publicly from the World Wide Web (WWW); the social (and other) networks it enables, and many other sources of data as illustrated earlier (Fig. 1.5). Soddy (1922), Bertalanffy (1960), Ackoff (1974), Checkland (1981), Maxwell (1984) and Loveridge (2009), among others, have all pointed, in different ways, to the need to accept that the STEEPV set is dynamic and that 'solutions to problems' become non-viable quickly and often before a study of them is complete. FTA then needs to:

- Adopt the notion of situations and their dynamism
- Adopt the principle of amelioration
- Reject the notion of problem-solving that yields solutions to well-specified problems that are not typical of the real world
- Encourage people to become, on the basis of learning how to learn, how to think and to appreciate numeracy in appropriate ways
- Create appreciative capabilities in breadth and depth to cope with the crisis in which FTA is already embroiled brought about by the algorithm-big data duo
- Adopt due diligence as a learning-based investigation that eschews structured checklists and similar questionnaires that might constrain what approximates a forensic investigation that requires learning.

## 1.12 Conclusions

Digital technologies have penetrated deeper and ever more quickly all forms of life on Earth: after 50 years of warnings, this has taken human societies by surprise and without much questioning of its implications. For FTA to supply that questioning, its mindset needs to change as indicated in the preceding discussion. The urgency for this is exemplified by Schmidt's express view that it is a race between computers and people and it is one that living systems need to win.

## References

Ackoff, R. L. (1974). *Redesigning the future: A systems approach to social problems*. John Wiley Interscience.
Amara, R., & Lipinski, A. J. (1983). *Business planning for an uncertain future*. Pergamon Press.
Ashby, R. (1956). *An introduction to cybernetics*. Chapman & Hall.
Bertalanffy von, L. (1960). *Problems of life*. Harper 'Torchbooks'.

Boettinger, H. M. (1969). *Moving mountains or the art and craft of letting others see things your way*. Macmillan.

Cagnin, C. H., Amanatidou, E., & Keenan, M. (2012). Orienting EU innovation systems towards grand challenges and the roles that FTA can play. *Science and Public Policy, 39*(2), 140–152.

Cagnin, C. H., & Loveridge, D. (2012). A framework, with embedded FTA, to enable business networks to evolve towards sustainable development. *Technology Analysis & Strategic Management, 24*(8), 797–820.

Cameron, H., Loveridge, D. et al. (1996). Technology foresight: Perspectives for European and International Co-operation. Final Report to CEC DGXII, April.

Checkland, P. (1981). *Systems thinking, systems practice*. John Wiley.

Dale, A., & Loveridge, D. (1996). Technology assessment—Where is it going? *International Journal of Technology Management, 11*(5/6), 715–723.

Dalkey, N. C. (1969). *The Delphi method: An experimental study of group opinion*. RAND Corporation.

Dempster, B. (1998). *A self-organizing systems perspective on planning for sustainability*. B.Sc. Thesis, University of British Columbia, Vancouver, Canada.

Dempster, B. (2000). Sympoietic and autopoietic systems: A new distinction for self-organizing systems. In *Proceedings of the World Congress of the Systems Sciences and ISSS 2000*. Toronto, Canada.

Fredrickson, B. L., & Losada, M. (2005). Positive affect and the complex dynamics of human flourishing. *American Psychologist, 60*(7), 678–686.

Funtocwiz, S. O., & Ravetz, J. R. (1990). *Uncertainty and quality in science for policy*. Theory and decision library, series A. Kluwer Academic Publishers.

Holling, C. H. (1977). The curious behaviour of complex systems: Lessons from ecology. In H. A. Linstone & W. H. C. Simmonds (Eds.), *Futures research: New directions* (pp. 114–129). Addison-Wesley.

Huddle, F. P. (1972). *A short glossary of science policy terms*. Washington: Science policy Research Division, The Library of Congress, US Government Printing Office.

Lavoisier, A. M. (1790). In W.H. Brock, (Ed.), 1992, *The Fontana History of Chemistry*, p 115

Losada, M. (1999). The complex dynamics of high performance teams. *Mathematical and Computer Modelling, 30*(9), 179–192.

Losada, M. (2001). The art of business coaching. In *Second General Conference of the Specialization Course*, Brasilia.

Loveridge, D. (1983). Computers and you: An essay on the future. *Futures, 15*(6), 498–503.

Loveridge, D. (Ed.). (1996). Special issue on technology assessment. *International Journal of Technology Management*, 210 p.

Loveridge, D. (2009). *Foresight: The Art and Science of Anticipating the Future*. Routledge

Maturana, H. R., & Varela F. (1980). *Autopoiesis and cognition: The realization of the living*. Dordrecht, Holland: D. Reidel.

Maturana, H. R., & Varela F. J. G. (1997). *De máquinas e Seres Vivos - Autopoiese: a Organização do Vivo* (3a ed.). Porto Alegre: Editora Artes Médicas.

Maturana, H. R. (1998). *Da Biologia à Psicologia* (3a ed.). Porto Alegre: Editora Artes Médicas.

Maxwell, N. (1984). *From knowledge to wisdom: A revolution in the aims and methods of science*. Basil Blackwell.

Michael, D. (1962). *Cybernation: The silent conquest*. Santa Barbara: Center for the Study of Democratic Institutions, February [Reprinted in Computers and Automation, March 1962, 11, 3, pp. 26–42].

Perrow, C. (1984). *Normal accidents: Living with high risk technologies*. Basic Books.

Roberts, J. (2012). Organizational ignorance: Towards a managerial perspective on the unknown. *Management Learning* Advance Online Publication. doi:10.1177/1350507612443208

Rocha, I. (2003). 'Gestão de Organizações: Pensamento Científico, Inovação, Ciência e Tecnologia, Auto-Organização, Complexidade e Caos, Ética e Dimensão Humana,' São Paulo: Editora Atlas S.A.

Schmidt. E. (2014). Sunday Times Business, 2 February.

Soddy, F. (1922). *Cartesian economics: The bearing of physical science on state stewardship*. Hendersons.

Stirling, A. (2008). Science, precaution, and the politics of technological risk: Converging implications in evolutionary and social scientific perspectives. *Annals of the New York Academy of Sciences, 1128*, 95–110.

Turing, A. M. (1936). On computable numbers, with an application to the Entscheidungs problem. *Proceedings of the London Mathematical Society, 42*, 230–265 (2, 1937).

Whitehead, A. N. (1925). The fallacy of misplaced concreteness, pp. 64, 72 (The error of mistaking the abstract for the concrete).

Zohar, D. (1990). *O Ser Quântico: Uma Visão Revolucionária da Naureza Humana e da Consciência Baseada na Nova Física*. São Paulo: Editora Best Sellers Zohar.

# Chapter 2
# A Conceptual Framework of Tech Mining Engineering to Enhance the Planning of Future Innovation Pathway (FIP)

**Munan Li**

**Abstract** Given the importance of innovation pathway and to meet the rapid growth of tech mining requirements, a novel conceptual framework for tech mining engineering (TME) is proposed to enhance the planning of future innovation pathway. Especially for those small and medium-sized enterprises (SMEs). The framework is intended to improve or guarantee the quality and efficiency of tech mining using engineering methodologies and technical standards. Certain basic elements of TME are defined and illustrated and the enormous potential and promising market for TME are discussed as subjects of future research and applications.

**Keywords** Tech mining · Innovation strategy · TME (Tech mining engineering) · Strategy-oriented methodology · Future innovation pathway · Top-down model of process

## 2.1 Introduction

With the convergence trend of science and technology (S&T), and rapid emergence of the new technologies and materials, future innovation pathway (FIP) has become a critical issue for the enterprises (Harold 2011). Obviously, FIP-oriented decision-making and planning is a definitely complicated engineering. Based on the basic philosophy of tech mining, Guo et al. (2012) ever proposed a systematization of the 'Forecasting Innovation Pathways' analytical approach to facilitate the relevant decisions.

"Tech mining" is defined as the text mining of technological information resources, and its functionality depends on a deep understanding of innovation processes (Porter and Cunningham 2005; Porter 2007). In the traditional, naïve framework of tech mining, the key elements include a TIPM (Technology

M. Li (✉)
School of Business Administration, South China University of Technology, Guangzhou, People's Republic of China
e-mail: limn@scut.edu.cn



T.U. Daim et al. (eds.), *Anticipating Future Innovation Pathways Through Large Data Analysis*, Innovation, Technology, and Knowledge Management,
DOI 10.1007/978-3-319-39056-7_2

Innovation Process Model), FOT (Future-Oriented Technology), R&D data selection, IR (Information Representation), Data Treatment, Innovation Indicators, and so on (Porter 2007). Although tech mining appears to be the application of text mining in technology management and innovation management, it is significantly different from data mining and text mining in both its perspective and methodology. Data mining, text mining and KDD (Knowledge Discovery in Database) focus on analytical models and algorithms for structural, semi-structural and non-structural data based on mathematical modeling. Therefore, text and data mining provide a framework of methods and tools, and their key elements or concerns are the efficiency, accuracy, robustness and flexibility of algorithms and mathematical models. Generally, data and text mining is method-oriented or tool-oriented engineering; however, tech mining is often utilized to support strategic decision-making in technology innovation and R&D management and can therefore be considered to be strategy-oriented engineering. TME (Tech Mining Engineering) itself is a very new concept in the literature, and only a few large-scale organizations, e.g., the strategic departments of governments, MNEs (Multi-National Enterprises) and research institutes, have utilized TME techniques and tools to support management activities related to innovation strategy for any length of time.

Considering the promising value of tech mining for FIP, it should not remain the privilege of MNEs; SMEs (Small and Medium Sized Enterprise) should also be able to harness this capability to enhance their innovation management, planning of FIP and approach competitive advantage by learning or outsourcing the service. Hence, an engineering framework for tech mining appears to be a meaningful and necessary method by which SMEs and even larger scale organizations can gain important guidance on aspects such as team management, work flow or process optimization, evaluation rules for quality and control policies for cost and quality under a uniform engineering framework or model.

## 2.2 Literature Review

According to the strategic pathway and innovation capabilities, Branzei and Vertinsky (2006) argued that the significant connections between innovation pathway and capabilities. Therefore, the planning of FIP should be considered in the level of organizational strategy. However, for those SMEs, the related issues to FIP could not be the easy tasks at all, and tech mining could become an important tool for SMEs to facilitate the planning of FIP (Porter and Newman 2011; Huang et al. 2012; Guo et al. 2012; Mittra et al. 2015).

Using the key word "tech mining" to search for articles published in journals collected in the core database of Web of Science (WoS™) returns almost 75 articles, some of which are noise. Using the key word "text mining" for the same period (2004–2014), however, yields over 3000 records. Considering the critical relationship between "tech mining" and "text mining" several search experiments were performed with different combinations of topics (key words) in such categories as *Management, Operation research management science, Business, Planning development, Industrial engineering, Engineering manufacturing, Economics,* Multidisciplinary engineering, and *Information science and Library science*; the experimental results are shown in Table 2.1.

**Table 2.1** Relevant literature in WoS™ under different combinations of topics (2004–2014)

| Topic (key words) | Search results | Representative authors (Count of publications) | Representative Journals (Count of records) |
|---|---|---|---|
| Tech mining | 75 | Porter A.L. (13), Miyazaki K. (5), Cunningham S.W. (5), Newman N. C. (5) | Technological forecasting and social change (9)<br>Technology analysis strategic management (6)<br>Expert systems with applications (4)<br>Technovation (4) |
| **Tech mining engineering** | **0** | **None** | **None** |
| Tech mining and text mining | 14 | Porter A.L. (7), Guo Y. (3), Kostoff R.N. (2), Newman N.C. (2) | Technological forecasting and social change (3)<br>Technology analysis strategic management (2)<br>Advanced materials research (2) |
| Text mining and patent analysis | 80 | Anderson T.R. (6), Daim T.U. (6), Kocaoglu D.F. (6) | Expert systems with applications (15)<br>Scientometrics (9)<br>Technological forecasting and social change (8) |
| Text mining and bibliometrics analysis | 40 | Kostoff R.N. (12), Porter A.L. (5), Bhattacharya S. (4), Glanzel W. (4) | Technological forecasting and social change (8)<br>Scientometrics (5)<br>Current Science (2) |
| Text mining and technology Roadmapping | 14 | Yoon B. (4), Gomila J.M.V. (3), Phaal R. (3), Porter A.L. (3), Zhang Y. (3) | R&D Management (3)<br>Scientometrics (2)<br>Technological forecasting and social change (2)<br>Technology analysis strategic management (2) |
| Text mining and technology opportunity analysis | 25 | Yoon B. (8), Porter A.L. (4) Yoon J. (3) | Expert systems with applications (4)<br>Scientometrics (3)<br>Technological forecasting and social change (3)<br>Industrial management data systems (2) |
| Text mining and competitive intelligence | 24 | Porter A.L. (4), Gomila J.M.V. (3), Zhang Y. (3), Zhou X. (3) | Scientometrics (3)<br>Decision support systems (2)<br>Data mining VII data text and web mining and their business applications (2)<br>Industrial management data systems (2) |

Beyond the information contained in Table 2.1, we note the interesting phenomenon that citations rarely cross between "text mining" and "tech mining." A count of the records in Table 2.1 shows the number of relevant studies to be fewer than 300, with apparently zero studies on tech mining engineering (TME).

Here, when using the narrow definition of tech mining—the "text mining of technical information resources" (Porter and Cunningham 2005; Porter 2007)—tech mining is an application based on text mining technology that is used in technology and innovation management. Therefore, in most related studies, tech mining is often taken as a tool, process or integrated framework that supports R&D management and innovation strategy planning. For example, Trumbach et al. (2006) described a method of tech mining used to keep small businesses knowledgeable about innovation ideas. Combing tech mining with bibliometrics analysis, Miyazaki and Islam (2007) explored differences between the U.S., Japan and the European Union in terms of the innovation pattern of nanotechnology. Nazrul and Kumiko (2010) analyzed the strengths and weaknesses of different countries in nanotechnology research based on tech mining techniques.

Porter and Newman (2011) proposed a five-stage framework of tech mining to answer typical questions in technology management. Park et al. (2013a, b) adopted TRIZ evolution trends as criteria for evaluating technologies in patents. Zhang et al. (2014) provided six "term clumping" steps that clean and consolidate topical content in such text sources. Becker and Sanders (2006) illustrated how tech mining could profit from innovations in meta-analysis and social impact assessment. Newman et al. (2013) compared alternative ways of consolidating messy sets of key terms. Some researchers have argued that tech mining may present an alternative or potentially complementary way to determine support for emerging technologies using proxy measures such as patents and scientific publications (Hopkins and Siepel 2013). Jose and Fernando (2013) provided a solution for tech mining by combing the semantic–TRIZ for a better technology analysis technique. Based on the patents, other researchers advanced a Subject-Action-Object (SAO) technique for text mining and utilized it to improve the process of technology road mapping (Yoon and Kim 2011; Choi et al. 2013).

Supporting decision-making in innovation pathway, future-oriented technology forecasting is one of the most important tasks in tech mining (Porter 2007). Based on traditional text mining, Ghazinoory et al. (2013) provided a method for locating technology centers of excellence. Aiming at the issue of selecting technology forecasting methods, a multi-criteria fuzzy group decision-making approach was proposed to improve accuracy (Gizem et al. 2013). Guo et al. (2012) discussed the issues surrounding technology forecasting and innovation pathway selection based on text mining information resources.

Actually, tech mining can be understood as an integrated framework or process that can combine many traditional and emerging analytical techniques to enhance planning or decision-making of FIP including technology forecasting and technological opportunity analysis (Newman et al. 2013; Halme and Korpela 2014; Li 2015). Porter ever used the term "supply chain" to describe the process that brings high-quality intelligence to support R&D management (Porter 2007).

Theoretical research on the framework and process of tech mining is still scarce, however; most studies prefer to use tech mining as a tool or method to improve empirical research, e.g., to enhance patent analysis using text mining techniques (Tseng et al. 2007), to identify promising patents or to forecast emerging technology evolutions by combining tech mining with TRIZ (Park et al. 2013a, b; Li 2015), or to identify promising opportunities for products or markets by combining text mining with quality function deployment (Jin et al. 2015).

In summary, aiming to FIP, the basic framework of tech mining brings an integrated solution covering many aspects in technology management and strategic analysis. However, the engineering architecture for the real implementation of tech mining seems insufficient for those different types of organizations, particularly for the SMEs.

## 2.3 Research Questions and Methodology

### *2.3.1 Why Does Engineering Need Tech Mining?*

With the rapid development of emerging technology and the growth of information resources, finding a way to refine innovation strategy, planning of FIP and improve the capability of innovation management has become a significant challenge for all types of organizations. Under the scrutiny of tech mining, there are several incentives for designing an engineering framework for tech mining applications.

First, S&T development is a double-edged sword that can bring both positive effects and negative influences. Halme and Korpela (2014) argued that a responsible innovation pathway should naturally connect to sustainable development. Future-oriented technology forecasting is an important issue in tech miming, although no organization can guarantee that the output of tech mining will be accurate if engineering methodologies and standards are not applied. Clearly, engineering tech mining may moderately reduce the risk inherent in technology innovation.

Second, although each organization develops its innovation pathway independently, the progress of economic globalization ensures that the innovation strategies of nations, territories, industries and enterprises cannot be separated from the world. Competition and cooperation coexist in the issues of development; therefore, the concrete programs of tech mining must confront an environment growing in complexity and competitive issues at both macro and micro levels. In addition to traditional analytical techniques, e.g., patent analysis, technology foresight and forecasting, competitive intelligence collection and so on, the tools of strategy management, such as PEST (Political, Economic, Social and Technological Analytical Model) and SWOT (Strength, Weakness, Opportunity, Threats Analytical Model), should be integrated into the tech mining process. Further, some data mining and text mining techniques, in addition to engineering management tools, are necessary to the actual delivery of tech mining.

Third, innovation pathway planning is not an independent activity; strategic decision-makers should be aware of the harmony and matching issue between innovation strategy and the innovation ecosystem (Adner 2006). Considering the context-dependent preferences in strategic decision-making for disruptive innovations, followers and pioneers can choose different pathways for technological improvement in a dynamic situation (Chen and Turut 2013). Bowonder et al. (2010) made 12 strategic suggestions for a company to obtain a competitive advantage. Even with tech mining, determining an organizational innovation strategy remains a complicated mission, and this process requires an engineering framework to reduce the risk.

In addition, with the rapid growth of text mining technology, an increasing number of analytical methods and techniques can be integrated into the tech mining framework (Tseng et al. 2007; Wang et al. 2012; Wong et al. 2014; Wood and Williams 2014; Yoon et al. 2014). Hence, a real tech mining project should take into account complicated systems engineering, which involves many different technologies and professional experts, e.g., innovation management, information and library science, computer science, mathematical modeling, and so on. In managing a team and coordinating cooperation among experts, engineers need the standard engineering framework to guarantee the schedule and the quality of related activities.

Finally, an engineering framework for tech mining can bridge the theoretical research and the potential market for tech mining services. Although most nations and MNEs (Multi National Enterprises) may have established their tech mining teams, SMEs still lack the related services or products due to costs and their more limited capabilities. Thus, the standardization of tech mining engineering may foster a promising market for tech mining services in the future.

### 2.3.2 Research Questions

The role of engineering in tech mining and the architecture of tech mining engineering (TME) appear to be prominent research gaps based on the above literature review. According to the basic components and activities defined in tech mining (Porter and Cunningham 2005), it can be inferred that strategic decision-makers should be the end-users of tech mining. The previous literature does not detail, however, whether the process of tech mining should be adjusted to meet the different scales of organization (nation/territory, industry and enterprise). After all, innovation strategy or technology development pathways could be very different at the different levels (macro, industrial and micro). Because of this variation in level, i.e., the macro (national), industrial and micro (enterprise) levels, strategy-oriented tech mining engineering could encounter challenges in adaptability and flexibility. Meanwhile, the purpose and main task of TME is to enhance organizations' innovation management capabilities and competitive advantage. The main content and topics of naïve tech mining are dynamic, and many new analytical methods and tools may be integrated into the framework, including social network analysis, cloud computing, big data, and so on.

Based on the analysis of related literature, most researchers see tech mining as a method for exploiting new technology to enhance traditional patent analysis, technology opportunity analysis, CIC (Competitive Intelligence Collecting), TRM (Technology Road Mapping) and so on (Phasl et al. 2004; Salles 2006; Shi et al. 2010). For strategic decision makers, however, several questions must be answered:

- When is tech mining necessary?
- What are the targets and final outputs of tech mining for different types of organizations?
- What types of experts should be pulled into tech mining projects?
- Who are the end-users or real customers of tech mining?
- Who can provide tech mining sourcing services?
- Where is the market for tech mining?
- How do you begin a tech mining project for organizational FIP?
- How do you schedule tech mining activities?
- How do you accurately evaluate and control costs with the right polices and regulations?
- How do you objectively assess the quality of different phases of tech mining in addition to the final product?

Based on the above questions regarding the practices of tech mining, an engineering framework is necessary. The research questions are as follows:

Question 1: What is tech mining engineering? (Definition, goals, implementing team, roles, responsibilities, inputs/outputs, and so on)
Question 2: What is the process model of TME, or how do you regulate and guide the tech mining activities?
Question 3: What is the mechanism of quality assurance?

### 2.3.3 Methodology

According to the basic definition of tech mining, the critical outputs appear to be intelligence, future-oriented forecasting and technology road mapping so forth, which can be integrated into enhancing the planning of FIP, all of which are important to organizations' strategic decision making, especially the innovation strategies of a technology or industry. Recently, some researchers have begun to integrate tech mining into innovation and strategy management; further, the international journal "*Technology analysis and strategic management*" published a special issue on "tech mining and innovation management" in 2013, indicating that it is an attractive and promising methodology for building interactions with innovation strategy planning and management. In turn, organizations' strategic behavior and intentions could influence the targets and processes of tech mining in unseen and profound ways. For example, different perspectives, competitive strategies and marketing campaigns could engender entirely different requirements for tech mining engineering. Facing the rapid development of emerging technology, the choice between exploration and exploitation renders the need to consider many variables quite complicated (Fauchart and Keilbach 2009).

As a key technique in tech mining, technology road mapping is not only an opportunity for technology analysis but also requires the integrated analysis of opportunities in the market (Groenveld 1997; Kostoff and Schaller 2001; Phasl et al. 2004; Lee et al. 2009). Further, technology road mapping provides important decision support for innovation strategies and FIP. In addition, as another important support tool of innovation strategy, technology foresight is facing a similar challenge, i.e., how can we accurately evaluate and improve the quality of the technology foresight process under a certain technical standard (Linstone 2010; Miles 2010; Oliveira and Rozenfeld 2010). The research on the relationship between tech mining, strategic management and innovation performance improvement, however, seems to be just beginning.

When examining the basic definition, processes and framework of tech mining, it becomes clear that the innovation pathway planning or supporting documents for innovation management must be one of critical outputs. Therefore, a strategy-orientation, particularly innovation strategy, is the main methodology of TME framework design. The strategy-oriented methodology for tech mining engineering contains several aspects:

- TME is a complex engineering system that provides an integrated solution for different organizations to improve strategy planning and management.
- The core outputs of TME are organizations' innovation strategies.
- The main goal of TME is to enhance the planning of FIP and improve innovation capability and performance.
- The quality control mechanism in TME comprises the measurements, metrics, and rules in the phases of strategy planning, strategy implementation and strategy adjustment.
- The framework of TME is designed based on the basic engineering methodology in which processes, steps, techniques and tools are integrated to carry out the task of strategy planning and management.

Although we have defined the content and processes of tech mining, we have yet to explore how to embed these processes into the strategic decision-making of organizations. In fact, it is somewhat unclear whether and how tech mining processes will require adjustment to function within different organizations' strategic planning and who would lead the adjustment processes to meet different requirements.

A basic preparatory step before implementing a tech mining project is role configuration, the definition of which is an important element in defining cooperation and efficiency. The following questions concern quality control policies, which contain the definitions of measurements and metrics in addition to engineering evaluation and assurance techniques.

## 2.4 Tech Mining Engineering: Definitions, Targets, Organizations and Roles

### *2.4.1 The Definition of Tech Mining Engineering*

Based on the background analysis and challenges surrounding tech mining described above, tech mining needs an engineering framework or architecture model to support the related management activities. It is clear that tech mining can enhance the FIP planning and operational management of different organizations, especially innovation or technology development strategies. Therefore, according to similar philosophies from other types of engineering, e.g., industrial engineering, software engineering and data mining engineering, tech mining engineering should define and improve the efficiency of tech mining, measure and evaluate the quality of activities related to tech mining, develop a mechanism for quality assurance via qualitative methods and tools, provide technical standards and references to improve the delivery of tech mining, promote the performance of innovation strategies, enhance organizations' sustainable capabilities, and retain the competitive advantages of an organization.

TME (Tech Mining Engineering) is thus defined as an interdisciplinary faculty that integrates multidisciplinary theories, methods, techniques and tools into its architecture, pulling from fields that include library and information science, computer science, management science, and so on. The mission of TME concerns factors such as the environment and resource analysis of innovation strategy, planning of FIP, R&D management, technology management, product innovation and coordination, team management and technical standards regulating and guiding practice, among others. The basic definition of TME leads us to divide it into three connected components:

- Models, algorithms and tools of tech mining based on library and information science, computer science and mathematics;
- Processes, work flows, regulations, rules and technical standards of tech mining based on engineering science and methodologies;
- Scheduling, team motivation, quality assurance mechanisms, cost control and performance evaluations based on management science.

These three components are united into the skeleton of TME, and the main target or perspective of TME is to enhance the strategic management of different organizations and then improve the capability and performance of innovation.

## 2.4.2 The Implementation of TME

According to the basic definition of TME, when implementing a tech mining project, the organizational targets, roles and responsibilities must be illustrated in the engineering framework. The consumers of TME could be nations (territories), industries or enterprises, and the providers could be government departments, universities, other research institutes, third-party companies, and so on. The outputs of TME can be divided into three types of innovation or technology development strategies based on the customer: macro (national), industrial or micro (enterprise).

In terms of the basic concept of tech mining, Porter (2007) did not recognize the potential differences among nations, industries and enterprises when planning an innovation strategy. For example, compared with an enterprise strategy, national strategies are oriented towards long-term development goals, the improvement of public governance, and enhancing national competitive advantage in the process of globalization. In contrast, at the industry level of innovation strategy, the core targets would be key technology innovation and the sustainable development and evolution pathway of industry. At the enterprise level, large enterprises and MNEs in particular differ from SMEs.

Industrial leaders should be willing to, and indeed must, undertake basic research and technology innovation to retain their leadership position. SMEs, however, must focus first on market survival and then on enterprise development. Therefore, the attitude towards investment in basic R&D and the targets of innovation strategy could be very different between MNEs and SMEs. In addition, the internal resources of tech mining could be very different. Many famous high-technology companies, e.g., *IBM, Microsoft, Huawei, SAP* and *Samsung*, have established professional tech mining teams. Most SMEs, in contrast, must seek external resources to meet their tech mining requirements. An illustration of TME consumers, targets, roles and providers is shown in Table 2.2.

Based on the information in Table 2.2, although TME also provides an intellectual product and service compared with software engineering, the output of TME is more difficult to measure and evaluate. Because TME outputs related to innovation-pathway planning or technology development are less tangible than the fruits of software or industrial engineering, they require a much longer evaluation period, if they can be evaluated at all. In contrast, the output of software engineering is much easier to test and evaluate. Therefore, objective evaluation and verification of TME may be a critical challenge.

In addition, compared with traditional software or industrial engineering, there are three levels of TME end-users: national (macro), industrial and enterprise (Micro). The final target could be the acquisition of competitive advantage for any of those levels, but the detailed prospectus, purpose and final outputs of tech mining at each level would remain significantly different. Meanwhile, the macro, industrial and micro strategies of innovation are also interrelated. For example, a particular national strategy for technology innovation would directly or indirectly influence an industry's development policies. The changes in development strategies and the

**Table 2.2** Basic descriptions of consumers, targets and roles in TME

| Consumers of TME | Level of innovation strategy | Targets | Roles (end-user, provider) |
|---|---|---|---|
| Nation (territory) | Macro level | • Planning national innovation strategies<br>• Acquisition of national competition advantage | User: government decision-makers<br>Provider: Research institutions (e.g., universities, S&T development research institutes, etc.), third-party consultants |
| Industry | Industrial level | • Planning industry innovation pathway and strategies<br>• Sustainable development of industry<br>• R&D in key common technologies<br>• Harmonious governance between industry and environment<br>• Development of industry ecosystem | User: industry policy-makers<br>Provider: research institutes or professional third-party consultants (or leading industry enterprises) |
| Enterprise | Micro level | • Planning, implementing and monitoring technology evolution pathways<br>• Improvement of innovation capability and performance<br>• Acquisition of competitive advantage in the market | User: enterprise decision-makers, R&D managers<br>Provider: enterprises' internal organizations, third-party consultants, research institutes |

related policies of nations and industries could affect enterprises, especially SMEs, in a profound and significant way. In turn, the innovation strategies and activities of enterprises may indirectly influence the macro and industrial policies. The interactions among the three types of innovation strategies are presented in Fig. 2.1.

In Fig. 2.1, indirect interaction between macro and micro innovation strategies is presented, which may be the cause of debate. National innovation strategies clearly

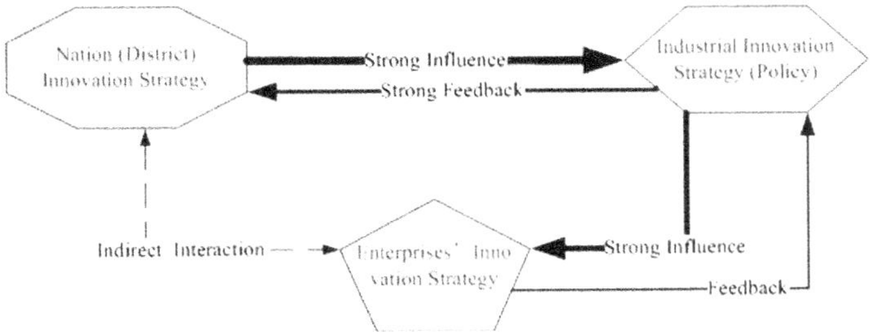


**Fig. 2.1** The relationships among macro, industrial and micro innovation strategies

influence enterprise behaviors through industrial and financial policies. In turn, the significant innovation activities of enterprises provide important feedback for macro and industrial strategy management. In fact, an analysis of the content of various national innovation strategies shows that many words and topics overlap across different nations' innovation strategies. This phenomenon is illustrated in Table 2.3.

In addition to the sampling of national innovation strategies shown in Table 2.3, a majority of countries around the world have established innovation strategies, from which it can be inferred that competitive advantages and sustainable development are among the most prevailing global concerns. Based on national-level innovation strategies, industry and S&T development strategies and policies should

**Table 2.3** Recent national innovation strategies

| Innovation strategy | Nation | Open date | Linked webpage | Prospective |
|---|---|---|---|---|
| Chinese Manufacturing 2025 | China | 2015.5 | http://www.ce.cn/xwzx/gnsz/gdxw/201505/19/t20150519_5402874.shtml | To promote the manufacturing industry via innovation |
| Innovation Driven Development Strategy | China | 2015.3 | http://www.sipo.gov.cn/dtxx/gn/2015/201506/t20150608_1128472.html | To enhance economic development via innovation driven force |
| Strategy for American Innovation | U.S. | 2010.11 | https://www.whitehouse.gov/innovation/strategy/introduction | To motivate innovation for sustainable growth and quality jobs |
| Innovation Nation (White paper) | U.K. | 2008.3 | https://www.gov.uk/government/publications/innovation-nation | To build an innovation nation in which innovation thrives at all levels |
| Japan's Innovation Strategy toward Asia | Japan | 2014.3 | http://www.mof.go.jp/english/pri/publication/pp_review/ppr024/ppr024d.pdf | To enhance innovation cooperation and keep competitive advantages |
| The 6th Plan of industrial innovation (2014–2018) | Korea | 2013.12 | http://1048.edu.pinggu.com/forum/201406/04/41f9e4b5b414/(3)6_(2014-2018)().pdf | To drive the development of key technology innovation in several critical industries in Korea |
| Three-year plan for economic innovation | Korea | 2014.2 | http://www.korea.net/NewsFocus/Policies/view?articleId=117839 | To motivate sustainable and innovative economic development |
| Poles of Competitiveness | France | 2004.9 | http://competitivite.gouv.fr/home-903.html | To develop a competitiveness cluster in France |
| High-Tech Strategy 2020 for Germany | Germany | 2010.7 | http://www.bmbf.de/en/publications/index.php | To promote several high-tech German industries via innovation |

be adjusted to meet the requirements of the relevant macro strategies; furthermore, enterprise and research institutes should consider aligning their strategies with their respective national priorities.

Two interesting and puzzling questions remain: (1) how were these national innovation strategies composed? And (2) how can we evaluate these macro strategies, particularly in terms of their suitability? For example, the latest US innovation strategy, "Reviving the Manufacturing Sectors of the United States," emphasized the development of traditional manufacturing industries; this appears to be a micro verification aimed at resurrecting previous U.S. government strategies. In addition, when compared to the innovation strategy of the U.S. and the German "Industry 4.0" strategy, China's "2025 Chinese Manufacturing" appears to be a deliberate and positive response.

## 2.5 The Process Model of TME

Here, the TME process model is completely different from the tech mining process (Porter and Cunningham 2005; Porter 2007). Based on the philosophy of software engineering and considering the environmental analysis requirements and challenge of strategic decision-making in addition to the characteristics of tech mining activities, a "top-down" process model of TME is proposed, which comprises the following steps or phases.

First, to explore the optional solutions for organizational strategy, the TME team should take planning of FIP and sustainable development as the goal and create a detailed analysis of the organization's strategic environment and current status.

Second, according to the strategic plan and options, the team should formulate a detailed implementation schedule for tech mining, choose the correct methods, techniques and tools, and establish an evaluation mechanism for milestones and stage outputs.

Third, tech mining activities should be implemented, including technical monitoring, competitive intelligence collection, technology forecasting, technical opportunity analysis, technology road mapping and intellectual property strategy analysis.

Finally, aiming at the outputs of tech mining activities, a comprehensive evaluation of the strategic planning proposal should be developed based on multiple objective decision making, multiple attribute decision making, etc., from which the optimal solution/s should be selected. The "top-down" process model is presented in Fig. 2.2.

In Fig. 2.2, the TME process is divided into four phases; in the third phase, traditional tech mining processes, methods, techniques and tools can be embedded into the framework of TME. The selected methods and tools used in different phases of TME are shown in Table 2.4.

Table 2.4 implies that TME is typically an interdisciplinary undertaking comprising strategy planning, innovation management, computer science, tech mining,

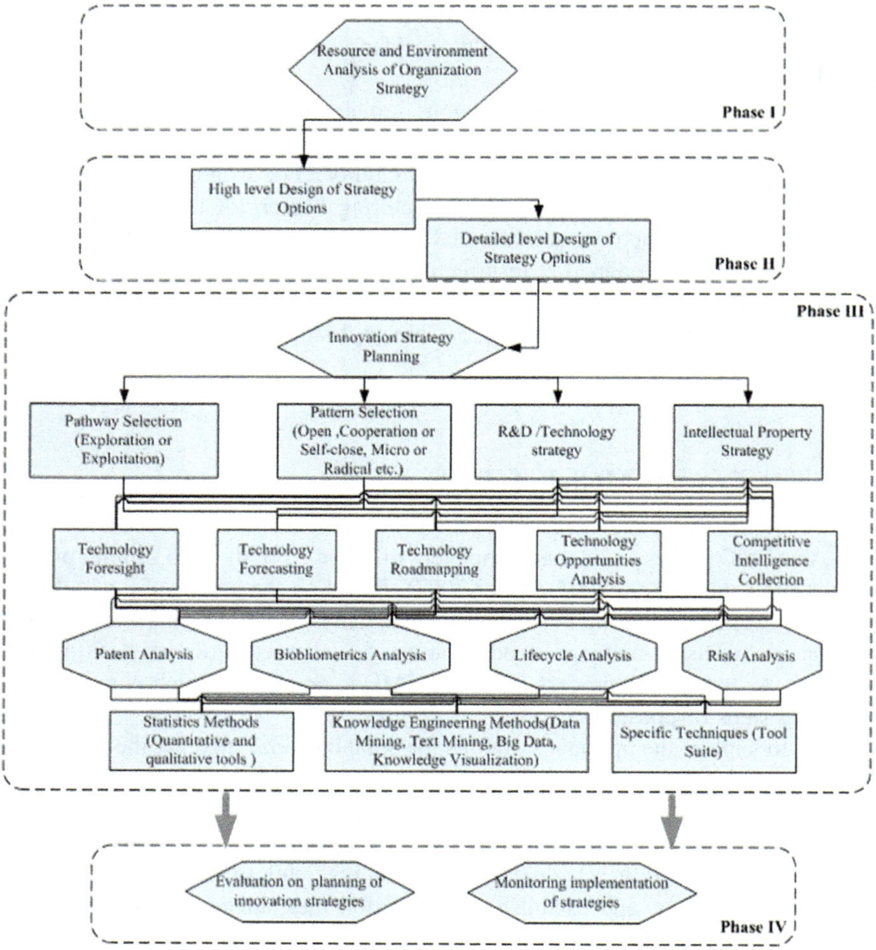


**Fig. 2.2** Top-down model of TME process

performance evaluation, quality control and management, among other disciplines. In addition, these four phases represent a prototype only; each phase can be extended to a more concrete process containing detailed steps under technical standards. In phase III, most of the techniques and tools of tech mining are deliberately separated into independent components, although they should be implemented as an integrated methods framework in the tech mining processes (Porter and Cunningham 2005). For example, collecting competitive intelligence is often facilitated by technology road mapping, technology foresight and technology opportunity analysis (Salles 2006; Trumbach et al. 2006; Roberta 2008; Shi et al. 2010; Shin and Lee 2013; Newman et al. 2013; Noh et al. 2015). However, these tech mining techniques can be implemented as components to bring the flexibility and scalability of the engineering framework to the analysis process, and the

**Table 2.4** Analytical methods and tools used in different phases of the TME framework

| | Description | Theory/methodology | Tools | Input | Output |
|---|---|---|---|---|---|
| Phase I | Requirement analysis | Strategy analysis theory | PEST, SWOT, etc. | Current organizational situations | Report of resources and environmental analysis |
| Phase II | Design of optional strategy solutions | Strategy planning and management | Five Forces Model, BCG Matrix, Mckinsey7S Model, etc. | Outputs of Phase I | Optional strategy solutions |
| Phase III | Innovation strategy planning | Innovation strategy management | Innovation management tools, computer science, tech mining techniques and tools | Outputs of Phase II | Planning solutions for organizational innovation strategy |
| Phase IV | Evaluation and monitoring | Performance evaluation and quality monitoring | Quality and performance evaluation and monitoring tools (Balanced Scored Card, Cause-effect Analysis) | Outputs of Phase III | Reports of evaluation and improvement policies |

practice of tech mining can utilize different combinations of techniques and tools and even different processes (Porter 2007) in phase III of TME.

In terms of the outputs of TME, technology road map (TRM) is a definitely crucial product because of the value of decision support for the planning of FIP, and the other strategic management activities (Yu et al. 2015). Based on the strategic decision-making on FIP, innovation pattern, the guidance for R&D and technology management, product and service innovation, marketing tactics and so on could be figured out.

## 2.6 The Quality Assurance Mechanism of TME

In terms of engineering, the QA (Quality Assurance) mechanism is a critical element. The QA mechanism is derived from the traditional philosophy of product quality. To improve and control product quality, researchers, managers and engineers designed a variety of frameworks, methods and tools for quality management, such as TQC (Total Quality Control), the PDCA (Plan-Do-Check-Action) Cycle, QFD (Quality Function Deployment), Six-Sigma (6σ) Management, and so on.

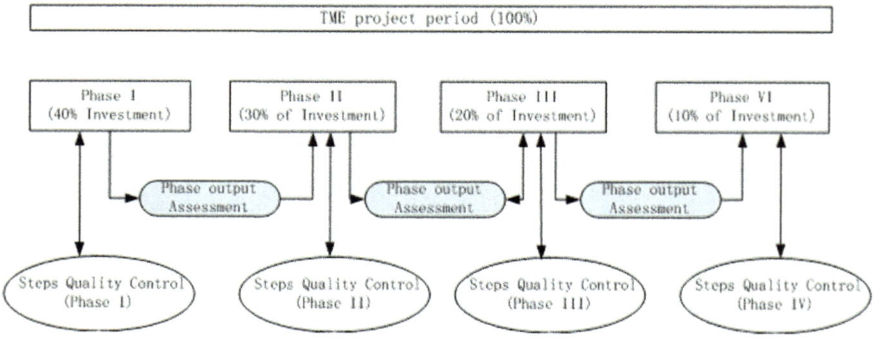


**Fig. 2.3** The quality control process for TME projects

Although the output of TME appears to be similar to a service product, the final strategic solution or suggestions should be taken as a concrete product; therefore, traditional theories of product quality management could be better references. Compared with consumer products such as electronics or even software, however, the quality of the TME product cannot be detected or revealed in the short run. In fact, most performance management tools focus on the implementation of strategy, e.g., CSP (Corporate Social Performance), EVC (Economic Value Added) or the BSC (Balanced Score Card). It is difficult to accurately measure strategic quality due to the lack of uniform and convincing metrics and methods. Generally, the basic philosophy and methodology of TQC, PDCA, and Lean Production/Management are very helpful in TME quality control. Based on the traditional methodology of quality management, the mechanism of phase quality control in TME is designated LTM (Lean Tech Mining). In LTM, the quality control mechanism is defined as embedded double chains, as shown in Fig. 2.3.

Figure 2.3 shows that LTM relies on the naïve quality control philosophy, i.e., the quality assurance of each step in each phase of the TME project is utilized to guarantee the total quality. Although allocating investment to the TME project would cause debate, TME is different from the tech mining process. In a TME project, the organization's strategic requirement analysis and the design of strategy options appear to be more important than the process of tech mining itself. LTM clearly references the core philosophy of software engineering.

## 2.7 A Suggestion to Improve the Architecture of Innovation Strategy Based on TME

Traditional strategy management, especially innovation strategy planning within organizations (nations, industries and enterprise) typically utilizes one or several tech mining techniques. At the level of architecture, however, there is an obvious gap between strategy management and tech mining. On the basis of TME, tech mining should be integrated into the architecture of innovation strategy, even strategy campaigns for the entire organization. In other words, if the organization strategy campaign is macro engineering (project), TME could be an important sub engineering method (project). The use of TME may be highly significant for the improvement of organizational strategy architecture.

Obviously, the conceptual framework of TME addressed in this paper is only a beginning of the related research and applications, especially for those interdisciplinary studies between innovation strategy and technology analysis and management.

## 2.8 Discussion

Obviously, because of the high uncertainty, the planning of future innovation pathway is a typically complicated engineering for any organization in macro, industrial and micro level, especially for those SMEs. Tech mining is an emerging tool of technology management. It integrates many techniques and methods of technology analysis; and technology road map is the critical important output of tech mining. To reduce the risk of the high uncertainty in future innovation development, and enhance the delivery performance of tech mining, a conception framework of tech mining engineering is proposed.

Based on the traditional tech mining process model, a new engineering framework named TME (Tech Mining Engineering) is advanced and illustrated in this paper. TME is a natural philosophical and methodological approach to engineering tech mining, which comprises many different techniques and tools deriving from computer science, information processing, competitive intelligence, strategy management, and so on. TME is not a wholly new concept in that there are many related activities and campaigns for innovation strategies, especially at the national and industry levels. These organizations (whether at the macro, industry or micro level) must use the techniques and tools of tech mining frequently in planning future innovation pathways and developing their innovation strategies; however, the conceptual framework of TME remains valuable and significant for several reasons.

First, TME could enhance the interactions between innovation strategy and tech mining techniques and tools.

Second, the proposed framework of TME could facilitate the creation of a new framework of engineering science by integrating several faculties.

Third, the TME framework describes a promising research area with enormous market potential. Although innovation strategy is important for organizations, the implementation of tech mining appears to be a sophisticated and complicated task, and many potential consumers, especially SMEs, would benefit from the guidance of a professional team.

Although entrepreneurs rarely know whether the macro or industrial strategies related to innovation and technology development are optimal when they are first implemented, R&D and technology development could be critical to the survival of enterprises, regardless of whether a firm is an industry leader or a well-known MNE. The failures of firms like Kodak, Motorola, and Nokia derived from many factors, one of which is technology innovation strategy. The stories of these firms show that even industry giants cannot guarantee the success of their technology strategies. SMEs therefore need professional tech mining services to aid in the planning and implementation of innovation and technology strategies. The interesting question is where and how can SMEs gain access to high-quality services related to tech mining. In the current cell phone market, inverse to the failure of Motorola and Nokia, Korea's Samsung created the "Samsung miracle," succeeding in becoming the biggest cell phone manufacturer in the global market by transcending Nokia, Ericsson, Motorola and other firms. This miracle derived from an enterprise that was bankrupt in the 1990s. In addition, it is questionable whether Apple's strategy of "micro innovation" is a good pattern for other companies, especially MNEs, or if it is only applicable to Apple. Seeking to address these questions on the innovation strategies of different organizations, the TME framework proposed in this paper may be unable to provide the right answer directly, but it is valuable and helpful for us to find the best approaches to improve the processes and activities of tech mining under the basic philosophy of engineering management.

The conceptual framework of TME discussed in this paper is only a skeleton and many details can be elaborated upon in future research. For example, the phases of TME are defined in this conceptual framework, but the concrete steps to be taken in each phase, the format and content of each step's outputs, the means of evaluating these outputs, the measurements and metrics of quality and the exploitation of the latest technologies all require further exploration. Even so, TME describes a novel and promising area of research and application.

**Acknowledgments** This work was supported by the Fundamental Research Funds for the Central Universities (No. 2015XZD15), Soft Science Research Project of Guangdong Province (No. 2013B070206020), Guangdong Province Key Laboratory Open Foundation (No. 2011A06090100101B), and Technology Trading System and Science and Technology Service Network Construction Project of Guangdong Province (No. 2014A040402003).

## References

Adner, R. (2006). Match your innovation strategy to your innovation ecosystem. *Harvard Business Review, 84*(4), 98–99.

Becker, H. A., & Sanders, K. (2006). Innovations in meta-analysis and social impact analysis relevant for tech mining. *Technological Forecasting and Social Change, 73*(8), 966–980.

Bowonder, B., Dambal, A., Kumar S. (2010). Innovation strategies for creating competitive advantage. *Research-Technology Management, 53*(3), 19–32.

Branzei, O., & Vertinsky, I. (2006). Strategic pathways to product innovation capabilities in SMEs. *Journal of Business Venturing, 21*(1), 75–105.

Chen, Y. X., & Turut, Q. (2013). Context-Dependent Preferences and Innovation Strategy. *Management Science, 59*(12), 2747–2765.

Choi, S. C., Kim, H. B., & Yoon, J. H. (2013). An SAO-based text-mining approach for technology roadmapping using patent. *R&D Management, 43*(1), 52–74.
Fauchart, E., & Keilbach, M. (2009). Testing a model of exploration and exploitation as innovation strategies. *Small Business Economics, 33*(3), 257–272.
Ghazinoory, S., Ameri, F., & Farnoodi, S. (2013). An application of the text mining approach to select technology centers of excellence. *Technological Forecasting and Social Change, 80*(5), 918–931.
Gizem, I., Erhan, B., & Tufan, K. (2013). The selection of technology forecasting method using a multi-criteria interval-valued intuitionistic fuzzy group decision making approach. *Computers & Industrial Engineering, 65*, 277–285.
Groenveld, P. (1997). Roadmapping integrates business and technology. *Research Technology Management, 40*(5), 48–55.
Guo, Y., Ma, T. T., & Porter, A. L. (2012). Text mining of information resources to inform forecasting Innovation pathways. *Technology Analysis and Strategic Management, 24*(8), 843–861.
Halme, M., & Korpela, M. (2014). Responsible innovation toward sustainable development in small and medium-sized enterprises: A resource perspective. *Business Strategy and the Environment, 23*(8), 547–566.
Harold, A. L. (2011). Three eras of technology foresight. *Technovation, 31*, 69–76.
Hopkins, M. M., & Siepel, J. (2013). Just how difficult can it be counting up R&D funding for emerging technologies (and is tech mining with proxy measures going to be any better)? *Technology Analysis and Strategic Management, 25*(6), 655–685.
Huang, L., Guo, Y., Porter, A. L., et al. (2012). Visualising potential innovation pathways in a workshop setting: The case of nano-enabled biosensors. *Technology Analysis and Strategic Management, 24*(5), 527–542.
Jin, G. M., Jeong, Y. J., & Yoon, B. G. (2015). Technology-driven roadmaps for identifying new product/market opportunities: Use of text mining and quality function deployment. *Advanced Engineering Informatics, 29*(1), 126–138.
Jose, M. V. G., & Fernando, P. M. (2013). Combining tech-mining and semantic-TRIZ for a faster and better technology analysis: A case in energy storage systems. *Technology Analysis and Strategic Management, 25*(6), 725–743.
Kostoff, R. N., & Schaller, R. R. (2001). Science and technology roadmaps. *IEEE Transaction on Engineering Management, 48*(2), 132–143.
Lee, S. J., Yoon, B. G., & Park, Y. T. (2009). An approach to discovering new technology opportunities: Keyword-based patent map approach. *Technovation, 29*, 481–497.
Li, M. N. (2015). A novel three-dimension perspective to explore technology evolution. *Scientometrics, 105*(3), 1679–1697.
Linstone, H. A. (2010). Three eras of technology foresight. *Technovation, 31*, 69–76.
Miles, I. (2010). The development of technology foresight: A review. *Technological Forecasting and Social Change, 77*, 1448–1456.
Mittra, J., Tait, J., Mastroeni, M., et al. (2015). Identifying viable regulatory and innovation pathways for regenerative medicine: a case study of cultured red blood cells. *New Biotechnology, 32*(1), 180–190.
Miyazaki, K., & Islam, N. (2007). Nanotechnology systems of innovation—An analysis of industry and academia research activities. *Technovation, 27*(11), 661–675.
Nazrul, I., & Kumiko, M. (2010). An empirical analysis of nanotechnology research domains. *Technovation, 30*(4), 229–237.
Newman, N. C., Porter, A. L., Newman, D., et al. (2013). Comparing methods to extract technical content for technological intelligence. *Journal of Engineering and Technology Management, 32*, 97–109.
Noh, H. Y., Jo, Y. G., & Lee, S. J. (2015). Keyword selection and processing strategy for applying text mining to patent analysis. *Expert Systems with Applications, 42*(9), 4348–4360.

Oliveira, M. G., & Rozenfeld, H. (2010). Integrating technology roadmapping and portfolio management at the front-end of new product development. *Technological Forecasting and Social Change, 77*, 1339–1354.

Park, H. S., Ree, J. J., & Kim, K. S. (2013a). Identification of promising patents for technology transfers using TRIZ evolution trends. *Expert Systems with Applications, 40*(2), 736–743.

Park, H. S., Yoon, J. H., & Kim, K. S. (2013b). Identification and evaluation of corporations for merger and acquisition strategies using patent information and text mining. *Scientometrics, 97*(3), 883–909.

Phasl, R., Farrukh, C. J. P., & Probert, D. R. (2004). Technology roadmapping—A planning framework for evolution and revolution. *Technological Forecasting and Social Change, 71*, 5–26.

Porter, A. L. (2007). How "Tech mining" can enhance R&D management. *Research-Technology Management, 50*(2), 15–20.

Porter, A. L., & Cunningham, S. W. (2005). *Tech mining: Exploiting technologies for competitive advantage*. New York: Wiley.

Porter, A. L., & Newman, N. C. (2011). Mining external R&D. *Technovation, 31*(4), 171–176.

Roberta, B. (2008). Issues in defining competitive intelligence: An exploration. *Journal of Competitive Intelligence and Management, 4*(3), 3–14.

Salles, M. (2006). Decision making in SMEs and information requirements for competitive intelligence. *Production planning and control, 17*(3), 229–237.

Shi, M. J., Liu, D. R., & Hsu, M. L. (2010). Discovering competitive intelligence by mining changes in patent trends. *Expert Systems with Applications, 37*, 2882–2890.

Shin, J. S., & Lee, H. K. (2013). Low-risk opportunity recognition from mature technologies for SMEs. *Journal of Engineering and Technology Management, 2013*(30), 402–418.

Trumbach, C. C., Payne, D., & Kongthon, A. (2006). Technology mining for small firms: Knowledge prospecting for competitive advantage. *Technological Forecasting and Social Change, 73*(8), 937–949.

Tseng, Y. H., Lin, C. J., & Lin, Y. I. (2007). Text mining techniques for patent analysis. *Information Processing and Management, 43*(5), 1216–1247.

Wang, Z. Y., Li, G., & Li, C. Y. (2012). Research on the semantic-based co-word analysis. *Scientometrics, 90*(3), 855–875.

Wong, M. K., Abidi, S. S. R., & Jonsen, I. D. (2014). A multi-phase correlation search framework for mining non-taxonomic relations from unstructured text. *Knowledge and Information Systems, 38*(3), 641–667.

Wood, M. S., & Williams, D. W. (2014). Opportunity evaluation as rule-based decision making. *Journal of Management Studies, 51*(4), 573–602.

Yoon, J. Y., & Kim, K. (2011). Identifying rapidly evolving technological trends for R&D planning using SAO-based semantic patent networks. *Scientometrics, 88*(1), 213–228.

Yoon, B. G., Park, I. C., & Coh, B. Y. (2014). Exploring technological opportunities by linking technology and products: Application of morphology analysis and text mining. *Technological Forecasting and Social Change, 86*, 287–303.

Yu, X. W., Hu, H., & Chen, X. P. (2015). Technology road mapping for innovation pathways of fibrates: A cross-database patent review. *Tropical Journal of Pharmaceutical Research, 14*(8), 1459–1467.

Zhang, Y., Zhou, X., Porter, A. L., et al. (2014). How to combine term clumping and technology roadmapping for newly emerging science & technology competitive intelligence: "problem & solution" pattern based semantic TRIZ tool and case study. *Scientometrics, 101*(2), 1375–1389.

# Chapter 3
# Profile and Trends of FTA and Foresight

**Per Dannemand Andersen and Lars Alkærsig**

**Abstract** This chapter presents the profile and trends of the academic discipline of Future-oriented Technology Assessment (FTA) and its approaches. This chapter presents the profile and trends of the academic discipline of FTA and its approaches. This is achieved through analyses of the development, focus and scientific impact of FTA. As such, this chapter contributes to the discussion on the conceptual development and academic positioning of FTA (or Foresight). The chapter is based on bibliometric analyses of the special issues of five international journals published after the FTA conferences in 2004, 2006, 2008 and 2011: Technological Forecasting and Social Change, Futures, Technology Analysis and Strategic Management, Science and Public Policy, and Foresight. Methodologically, the paper draws on the principles of Elsevier's Scopus and Thompson Reuter's Web of Science. The chapter concludes that the field of FTA seems to have remained remarkably stable over the last decade. As an academic field, FTA has targeted a small number of journals for its publications, which has helped to further define and focus the field. Finally, the chapter concludes that publications in special issues of international journals resulting from the FTA conferences have a similar level of quality and scientific impact as articles published in standard issues of these journals.

**Keywords** Foresight · FTA · Academic discipline · Trends · Publications

## 3.1 Introduction

Approaches and processes for dealing with the futures of science, technology and society have been around for several decades (Dalkey and Helmer 1962; Jantsch 1967; Irvine and Martin 1984). Along with the establishment of Future-oriented

P.D. Andersen (✉) · L. Alkærsig
DTU Management Engineering, Technical University of Denmark,
Diplomvej 372, 2800 Kongens Lyngby, Denmark
e-mail: pean@dtu.dk



T.U. Daim et al. (eds.), *Anticipating Future Innovation Pathways Through Large Data Analysis*, Innovation, Technology, and Knowledge Management,
DOI 10.1007/978-3-319-39056-7_3

Technology Assessment (FTA), or Foresight, as an area of practice in public policy development and in enterprises' strategic planning, FTA is also developing into an academic discipline. The publication *'Towards a new Knowledge Area'* from Norway's Research Council concludes that Foresight in Norway during the period of 2003–2009 has developed qualitatively from an application-oriented competence milieu to a more academically oriented knowledge area (Norwegian Research Council 2010). However, most academic Foresight literature is still descriptive or normative, relating to the practice of Foresight (Miles 2008), and it is generally acknowledged that there is a gap between its practice and theory (Barré and Keenan 2008; Hideg 2007). This has spurred a search for a more solid theoretical foundation and possible 'theoretical underpinnings' of the academic field of FTA or Foresight (Fuller and Loogma 2009; Weber et al. 2009; Öner 2010; Andersen and Andersen 2014).

In a contribution to science sociology, Andrew Abbott has suggested that '*...an academic discipline, or field of study, is a branch of knowledge that is taught and researched at the college or university level. Disciplines are defined (in part), and recognized by the academic journals in which research is published, and the learned societies and academic departments or faculties to which their practitioners belong*' (Abbott 2001). Following this definition, FTA and Foresight have many of the attributes of an academic discipline.

First, an increasing number of graduate studies in FTA, forecasting, Foresight and related areas are offered throughout the world—for example, in Australia, the USA, Canada, Finland and Norway. Courses for practitioners are found in many more countries, such as a course offered regularly by the Manchester Business School.

Second, academic departments whose names include terms such as Foresight, technology assessment or forecasting can be found at several universities and research centres. The most important research milieus are naturally found in countries such as Great Britain, Finland and Germany, which have large and long-lasting public Foresight programmes.

Third, several journals have FTA and Foresight as a core area. A complete search in SCOPUS on 'articles' published in 2014 or earlier that include the word 'Foresight' in their titles, abstracts or as a keyword yielded 2664 hits. Some of these articles have been published in economic journals such as the Journal of Economic Dynamics and Control, Journal of Economic Theory and Economics Letters. Excluding these economics-oriented journals, the top 10 journals from the search results are depicted in Fig. 3.1. From these findings, it is apparent that Foresight seems to be rapidly building its awareness in academic publishing, with more than a quadrupling of its academic output over the latest decade. Two non-English language journals are present in the Top-10 list: the French 'Futuribles Analyse Et Prospective' and the Russian 'Foresight Russia'. Furthermore, it is worth noting that the 'Journal of Futures Studies' is based in Taiwan. In addition to these publications, the leading journals in this field—the journals most frequently publishing articles

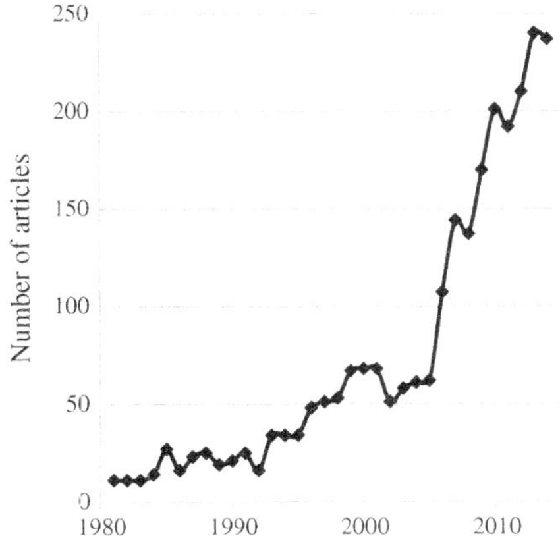


| Top-10 journals | |
|---|---|
| Futures | 142 |
| TFSC | 130 |
| Foresight | 117 |
| Futuribles | 54 |
| Journal of Futures Studies | 46 |
| Int. J. of Foresight&Innovation Policy | 43 |
| Foresight Russia | 36 |
| TASM | 28 |
| Science and Public Policy | 18 |
| Int. J. of Technology Management | 15 |

**Fig. 3.1** Development of in number of articles from 1980 to 2014 (incl.) with the word 'Foresight' in title, abstract or keyword and the top 10 journals in which these articles were published—including the number of articles in each journal. Journal in economics are excluded from the list. *Source* SCOPUS

with the term 'Foresight' in the title, abstract or keywords—are Futures, Technological Forecasting and Social Change (TFSC), Foresight, the International Journal of Foresight and Innovation Policy, Technology Analysis and Strategic Management (TASM), Science and Public Policy (SPP), and the International Journal of Technology Management.

Fourth, there appears to be no learned societies that specifically target FTA or Foresight.[1] However, international conferences within technology and innovation management have established special sessions on FTA and Foresight (e.g. the ISPIM conferences). A few international conferences specifically target FTA, such as the FTA conference, which has been organized by the European Commission five times since 2004. Hence, the FTA conferences have played an important role in formalizing this research community in the absence of a 'learned society'. According to the call for the 2014 version of the European Commission's FTA conference, these conferences have provided '*...a common platform for the closely related communities of Foresight, forecasting and technology assessment, where experts interact and help in guiding strategy, policy and decision-making to anticipate and shape future developments*'.

This chapter seeks to address these attributes by gaining an insight into the development and focus of this academic discipline. We achieve this by analysing the role of FTA conferences in shaping the academic platform for the closely related disciplines of Foresight, forecasting and technology assessment and to present a profile and trends in these disciplines.

---

[1]Organizations such as World Futures Studies Federation are focused on Futures Studies, which we consider somewhat distinct from FTA and Foresight.

### 3.1.1 Trends in FTA and Foresight

Several studies have described the historical development of Foresight (Georghiou 2001; Georghiou et al. 2008; Miles 2010). Many of these studies may have a tendency to focus on developments in English-speaking regions of the world (i.e. UK, USA, Canada, New Zealand and Australia) and less on the trends and developments elsewhere. However, despite national differences, there is no doubt that the practice of Foresight has developed well beyond the bounds of technology Foresight (Butter et al. 2008).

In this paper, we will focus on a few of the trends in FTA and Foresight.

First, FTA and Foresight have shifted from a focus on intra-organizational planning and forecasting in science and technology to a greater emphasis on open and inter-organizational 'strategizing', including external stakeholder opinions on alternative futures (Könnölä et al. 2009). This means fewer predictive methods and more user engagement or participative methods, as well as more 'alternative scenarios' methods in the practice of FTA and Foresight (Georghiou and Cassingena Harper 2011).

Second, FTA and Foresight seem to be experiencing a 'systems turn', taking a more systemic approach both in their practice and understanding of innovation (Oner and Saritas 2005; Saritas 2011; Andersen and Andersen 2014). This turn can be seen as implementing the systemic and evolutionary understanding of innovation that is predominant in the academic field of innovation studies (Martin 2014).

Third, FTA and Foresight are said to take a less national approach and more regional or sectoral approach (Georghiou et al. 2011). FTA has broadened its interest in policies for national innovation systems to comprise related concepts such as sectoral innovation systems (e.g. Abadie et al. 2010) and regional innovation systems and clusters (e.g. Roveda and Vecchiato 2008). The interest of the FTA community in the regional level is not new but has received increasing attention in recent years, both from the wider innovation system community (e.g. Asheim and Gertler 2005; Cooke 2012) and from policy-makers.

## 3.2 Data and Methodological Approach

The analyses are based on articles in five international journals published after the European Commission's FTA conferences in 2004, 2006, 2008 and 2011: Technological Forecasting and Social Change (TFSC), Technology Analysis & Strategic Management (TASM), Science and Public Policy (SPP), Futures and Foresight. In addition to the aforementioned special issues, the FTA seminar in 2006 also resulted in a book with 11 articles published by Springer Verlag. The latter is not included in this study. In total, 92 articles from the five journals were included in the analysis (see Table 3.1).

**Table 3.1** Overview of the special issues forming the empirical foundation for the analysis

| | 2004 | 2006 | 2008 | 2011 |
|---|---|---|---|---|
| Book on Springer Verlag[a] | | 12 articles, pp. 1–169 (2008) | | |
| TFSC | 8 articles in Vol. 72, No. 9, pp. 1059–1174 (2005) | 7 articles in Vol. 75, No. 4, pp. 457–582 (2008) | 9 articles in Vol. 76, No. 9, pp. 1135–1260 (2009) | 9 articles in Vol. 80, No. 3, pp. 379–470 (2013) |
| TASM | | | 6 articles in Vol. 21, No. 8, pp. 915–1001 (2009) | 8 articles in Vol. 24, No. 8, pp. 729–861 (2012) |
| SPP | | | 7 articles in Vol. 37, No. 1, pp. 3–78 (2010) | 12 articles in Vol. 39, No. 2, pp. 140–270 (2012) |
| Futures | | | 8 articles in Vol. 43, No. 3, pp. 229–356 (2011) | 7 articles in Vol. 59, pp. 1–72 (2014)[b] |
| Foresight | | | | 6 articles in Vol. 14, No. 4, pp. 279–351 (2012) |
| | | | | 5 articles in Vol. 15, No. 1, pp. 6–73 (2013) |
| Total number of articles analysed in this chapter | 8 | 7 | 30 | 47 |

[a]Not included in this analysis
[b]Futures has changed to a volume/issue system and does not use issue numbers

Methodologically, the paper draws on the facilities of Elsevier's Scopus and Thompson Reuter's Web of Science. Methodological support has been provided by Scopus' back-office user assistance.

## 3.3 Results

### *3.3.1 Trends in Academic Underpinning and Focus of FTA*

We cited Andrew Abbott above for his statement that academic fields are defined (in part) by the academic journals in which their research is published

(Abbott 2001). That leads to the conclusion that the underpinnings of an academic field are defined in part by the journals that are relevant to that field.

In this section, we analysed the references in the 12 special issues and compared them with the general reference profile of the journals as presented in the Journal Citation Report by Thompson Reuter's ISI Web of Science (see Table 3.2).

The right-hand column in Table 3.2 shows the number of times articles published in each of the referenced journals (over all years) was cited in 2013. The result is based on a search of the Journal Citation Report by Thompson Reuter's ISI Web of Science. We analysed data for four of the five journals in which special issues from the FTA conferences have been published. These include TFSC, TASM, SPP and Futures. The journal Foresight is not indexed by Web of Science.

The left and middle column in Table 3.2 shows how many times article-specific journals were referenced in articles in the FTA special issues. As only one special issue was published after the FTA conferences in 2004 and 2006, we merged the articles from these two conferences with the output from the 2008 conference. Hence, we generated two comparable populations with 45 articles from the FTA conferences in 2004, 2006 and 2008 and 47 articles from the FTA conference in 2011.

The result in Table 3.2 came from exporting all references in articles published in special issues and then sorting them according to a gross list of potential journals. This sorting sought to take into account differences in the notation of the individual journal in the reference lists. For example: 'Technological Forecasting and Social Change', 'TECHNOL FORECAST SOC' and 'TFSC' refer to the same journal. Only the top 10 most cited journals for each of the four journals are listed in Table 3.2. In Table 3.2, the right column shows that articles in the four journals generally reference high-quality (measured by their Journal Impact Factor) journals, such as Research Policy, Strategic Management Journal, Technovation, Science Technology and Human Values, Administrative Science Quarterly, and Nature.

Three observations can be made from Table 3.2. First, references to 'sector journals' such as Energy Policy, Environment Science Policy, Ecological Economics and Global Environmental Change indicate that energy and the environment are important domains for research published in these four journals.

Second, it appears that publications from the FTA conferences refer to a more coherent set of journals than the four journals in general. Articles in the three journals, Foresight, Futures and TFSC, are regularly the most cited across the four journals.

Third, the journal Research Policy is present in the top 10 list of most cited journals in all four FTA outlets. This likely reflects that articles in Research Policy are also among the top 2 most cited journals for three of the four FTA journals in general (except Futures). It might also highlight the fact that the theoretical rationale for policy-oriented Foresight exercises is supported by the perspective of evolutionary economics and the innovation systems approach, as Research Policy is acknowledged to be a leading journal in the field of innovation studies.

Fourth, it can be observed that the journal Nanotechnology is present in three of the top 10 most cited special issue journals from the 2004, 2006 and 2008 FTA conferences but only once in the special issues from the 2011 FTA conference. This could mean that the policy interest in nanotechnology was peaking in these years.

**Table 3.2** Analysis of journals in the field of FTA and Foresight

| | Special issues from FTA conferences 2004, 2006 and 2008 | | Special issues from FTA conferences 2011 | | Journal in general all years (WoS data) | |
|---|---|---|---|---|---|---|
| TFSC | Foresight | 64 | Futures | 30 | TFSC | 696 |
| | TFSC | 46 | TFSC | 26 | Research Policy | 233 |
| | Futures | 30 | Foresight | 21 | Futures | 141 |
| | Research Policy | 20 | Nanotechnology | 19 | Strategic Management J. | 136 |
| | Nature | 8 | Research Policy | 10 | Technovation | 128 |
| | Nanotechnology[a] | 7 | Nature | 3 | Energy Policy | 109 |
| | Eur. J. Oper. Res. | 3 | Eur. J. Oper. Res. | 2 | Adm. Science Quarterly | 76 |
| | Int. J. Foresight Innov. Policy | 3 | Int. J. Foresight Innov. Policy | 2 | Management Science | 74 |
| | Management Science | 3 | Technovation | 2 | Harvard Bus. Rev. | 72 |
| | Research Evaluation | 3 | Scientometrics | 1 | Acad. Management Rev. | 72 |
| TASM | Foresight | 44 | Foresight | 21 | Research Policy | 237 |
| | Futures | 12 | Futures | 21 | TASM | 185 |
| | TFSC | 10 | TFSC | 14 | Strategic Management J. | 106 |
| | Journal of Forecasting | 5 | Strategic Management | 8 | TFSC | 96 |
| | Research Policy | 4 | Research Policy | 6 | Technovation | 94 |
| | SPP | 3 | SPP | 5 | R&D Management | 69 |
| | TASM | 3 | Energy Policy | 4 | Organization Science | 51 |
| | Long Range Planning | 2 | Long Range Planning | 4 | Acad. Management Rev | 45 |
| | Futures Research Quarterly | 1 | Technovation | 4 | Ind. and Corp. Change | 41 |
| | Organization Science | 1 | Adm. Science Quarterly | 3 | Energy Policy | 40 |
| SPP | Foresight | 22 | Foresight | 43 | Research Policy | 137 |
| | Futures | 7 | Futures | 26 | SSP | 121 |
| | Research Policy | 6 | SPP | 20 | Sci. Tech. & Human Values | 35 |
| | SPP | 5 | TFSC | 19 | Environ. Science Policy | 30 |
| | TFSC | 2 | Research Policy | 18 | Public Understand. of Sci. | 27 |
| | Journal of Forecasting | 1 | Long Range Planning | 4 | Futures | 24 |
| | Management Science | 1 | Nature | 3 | Nature | 23 |
| | Nanotechnology | 1 | TASM | 3 | Social Studies of Science | 23 |
| | n.a. | | Organization Science | 2 | TFSC | 23 |
| | n.a. | | Sci. Tech. & Human Values | 1 | Science | 22 |

(continued)

**Table 3.2** (continued)

| | Special issues from FTA conferences 2004, 2006 and 2008 | | Special issues from FTA conferences 2011 | | Journal in general all years (WoS data) | |
|---|---|---|---|---|---|---|
| Futures | Foresight | 52 | Foresight | 28 | Futures | 299 |
| | TFSC | 21 | Futures | 18 | TFSC | 84 |
| | Futures | 17 | TFSC | 9 | Nature | 34 |
| | Research Policy | 6 | R&D Management | 6 | Foresight | 30 |
| | Journal of Forecasting | 4 | SPP | 6 | Energy Policy | 28 |
| | SPP | 4 | Management Decision | 4 | J Futures Studies | 27 |
| | Energy Policy | 2 | TASM | 4 | Long Range Planning | 24 |
| | Management Science | 2 | Management Science | 2 | Harvard Bus. Rev. | 23 |
| | Nanotechnology | 2 | Research Policy | 2 | Ecological Economics | 22 |
| | Industrial and Corp. Change | 1 | Adm. Science Quarterly | 1 | Global Environ. Change | 21 |

Numbers are only comparable within each square in the table
[a]'Nanotechnology' might cover more than one journal with the same name

## 3.3.2 *Trends in FTA Approaches*

In the following paragraph, we will strive to investigate the trends in the field of FTA and Foresight, as indicated in the first section of the chapter. However, a study on the trends in FTA and Foresight based on special issues from the FTA conferences of course has its biases. The articles in the special issues do primarily reflect the choices made by the special issues editors and the members of the scientific committees. As noted in the editorial note in one of the special issues: '*The papers ... were carefully selected and further nurtured to bring out three key themes*' (Könnölä et al. 2009). Hence, conclusions must be drawn with a good amount of caution.

The analysis of trends in FTA approaches is based on analyses of keywords and abstracts. Both keywords and abstracts of the 92 papers were extracted from an EXCEL-file exported from the search facility in SCOPUS. Unfortunately, SCOPUS was not able include keywords from SPP Vol. 37, No. 1 (2010) and Futures Vol. 43, No. 3 (2011). This also calls for cautiousness on drawing conclusions based on keyword analysis.

In the special issues from the FTA conferences in 2004, 2006 and 2008, 14 of the 45 articles have keywords identifiable as FTA methods. Nineteen of the 47 articles from the FTA conference in 2011 have keywords—and strings of keywords—identifiable as FTA methods (see Table 3.3). At first glance, scenario is clearly identifiable as the most popular FTA method. Scenarios are mentioned in more than half of the articles from the 2011 conference with identifiable FTA methods.

**Table 3.3** FTA methods appearing in keywords in alphabetic order

| Special issues from FTA conferences 2004, 2006 and 2008 | Special issues from FTA conferences 2011 |
|---|---|
| • Barometer, indicator | • Delphi method; Scenario planning |
| • Bibliometrics | • Delphi method; Scenarios |
| • Bibliometrics | • Early warning signals; Horizon scanning; Weak signals |
| • Delphi | • Exploratory modelling and analysis |
| • Roadmapping | • Horizon scanning |
| • Roadmapping | • Indicator; Patent |
| • Robust portfolio modelling | • Key enabling technologies |
| • Scanning process | • Megatrends |
| • Scenarios | • Modelling |
| • Scenario planning | • Roadmapping |
| • Scenario planning | • Roadmapping; Scenarios; Vision-building |
| • Text mining | • Scenarios |
| • Trends | • Scenario design |
| • Visioning | • Scenario planning |
| | • Scenario planning |
| | • Scenario practice |
| | • Scenario |
| | • Scenarios; Vision; Weak signals |
| | • Visioning |

If any trend is detectable from this keyword analysis, it is that scenarios seem to have become increasingly important. However, this is only partially confirmed by a similar analysis of terms in abstracts in the same papers (see Table 3.5). The term 'scenario' is actually less frequent in 2011 than in the earlier versions of the FTA conference but appears with twice the frequency in the abstracts submitted for the 2014 FTA conference.

We also analysed the content of the abstracts of the 92 FTA conference papers. Furthermore, as one of the authors of this chapter had access to abstracts submitted to the 2014 version of the FTA conference, these abstracts were also included in the analysis. Analysis of the abstracts was carried out using the free software, Text Analyzer, specifically its 'Unfiltered Wordcount' function. In the word count, we omitted common English words such as 'the', 'and', 'of' and 'to'. The 10 most frequent words (apart from these common English words) for the three data sets are listed in Table 3.4.

First, it can be observed that the term 'Foresight' is more frequently used than the term 'FTA'. That applies to both the edited special issues from the FTA conferences as well as also the non-edited abstracts for the FTA's 2014 conference. The term 'Foresight' is actually the most frequent word apart from common English words such as 'the' and 'and'.

**Table 3.4** Top 10 terms in abstracts

| Special issues from FTA conferences 2004, 2006 and 2008 | | Special issues from FTA conferences 2011 | | Abstracts submitted for the FTA 2014 conference | |
|---|---|---|---|---|---|
| Foresight | 1.06 | Foresight | 1.05 | Foresight | 0.68 |
| Research | 0.67 | Policy | 0.84 | Future | 0.58 |
| Future | 0.59 | FTA | 0.83 | Technology | 0.57 |
| Technology | 0.56 | Future | 0.73 | Research | 0.52 |
| Innovation | 0.47 | Technology | 0.68 | Innovation | 0.52 |
| Process | 0.44 | Innovation | 0.66 | Policy | 0.42 |
| Policy | 0.44 | Research | 0.64 | Analysis | 0.33 |
| Analysis | 0.37 | Challenges | 0.49 | Development | 0.31 |
| Development | 0.36 | Analysis | 0.45 | FTA | 0.29 |
| FTA | 0.32 | Methods | 0.42 | Process | 0.24 |

Numbers refer to the per cent of all abstracts that include each term

**Table 3.5** Selected terms in abstracts

| Abstracts in special issues from FTA conferences in 2004, 2006 and 2008 | | Abstracts in special issues from FTA conference in 2011 | | Abstracts submitted for the FTA 2014 conference | |
|---|---|---|---|---|---|
| System | 0.09 | Systems | 0.22 | Systems | 0.21 |
| | | System | 0.16 | System | 0.19 |
| Participatory | 0.05 | Participatory | 0.05 | Participatory | 0.06 |
| Participation | 0.09 | Participation | 0.02 | Participation | 0.04 |
| National | 0.21 | National | 0.22 | National | 0.13 |
| Region | 0.05 | Regions | 0.01 | Regional | 0.06 |
| Regional | 0.03 | | | Region | 0.03 |
| Scenario | 0.12 | Scenario | 0.09 | Scenarios | 0.22 |
| | | | | Scenario | 0.14 |

Numbers refer to fraction of each word in all abstracts combined

Second, the three samples have many words in common. In all three cases, eight words are among the most frequently used words: Foresight, research, future, technology, innovation, policy, analysis and FTA. Hence, based on these findings, no major development or trend in the field of FTA is detectable, which indicates a rather stable approach to the field.

In order to investigate the trends in the field of Foresight and FTA as described in Sect. 3.2 in this paper, we analysed the frequency of selected terms in the abstracts (see Table 3.5).

As mentioned in the introduction, FTA and Foresight are said to have taken a more systemic approach. This is confirmed by the abstract analysis. The frequency of the term 'system(s)' more than doubled from the 2004–2008 FTA conferences to the 2011 conference. The same frequency was observed for abstracts submitted for the FTA 2014 conference.

FTA and Foresight are also said to have become more focussed on user engagement and participatory methods. This trend is not confirmed by abstract analysis. The frequency of the term 'participatory' is actually quite stable in the three cases, and the term 'participation' decreased from the 2004–2008 FTA conferences to the 2011 FTA conference.

Finally, FTA and Foresight are said to take less of a national approach and more of a regional or sectoral approach. This trend can also not be supported by abstract analysis. No clear trend can be observed in the use of 'national', 'region' and related terms in the abstracts.

In conclusion, the attempt to use analyses of keywords and abstracts to verify generally acknowledged trends and developments within FTA and Foresight was not particularly successful. The clearest observation was that the field seems to have become stable—with stability in key journals and in the use of keywords.

### *3.3.3 Quality and Impact of Articles in the Special Issues*

As mentioned, the FTA conferences have strived to contribute to shaping the academic platform of the related disciplines of Foresight, forecasting and technology assessment. In academia, such an impact is often measured in the number of citations for a publication. Quite often in academia, the number of citations is also perceived as a proxy for quality. The logic is that the more citations an article has, the better the quality and impact of that article. However, traditions for publications and citations vary between academic fields, and comparisons must be made only between comparable articles.

To do this, we compared citations for articles from FTA conferences published in 12 special issues with all other articles published in the journals from the same year. Data were downloaded from SCOPUS through the end of October 2014. The analysis was based on a simple citation analysis using comparisons with standard issues of each journal (see Table 3.6).

The analysis was first conducted by calculating a simple average of citations for articles in each journal and comparing that number with the average citations for articles in all other articles in that journal in the same year (see Table 3.6). From this simple analysis, it is obvious that there is no significant difference or systematic pattern between FTA special issues and normal issues of the journals. Average articles in FTA special issues seem to have fewer citations (6.2) than articles in normal issues of the same journals (6.9). However, this difference is not significant. The only significant pattern is a relationship between the time since publication and the number of citations. Articles published in 2014 have very few citations during the same year.

The same data were exposed to a more thorough statistical analysis, specifically a $t$-test with equal variances. This method of analysis also showed no significant differences.

**Table 3.6** Comparisons between the average number of citations for articles published in FTA special issues and the similar number of citations for the rest of the articles published in the journals the same year

| Special issue | Average citations per article for the special issues | Average citations per article for the rest of the articles published in the journal that year |
|---|---|---|
| TFSC Vol. 72, No. 9 (2005) | 18.3 | 21.5 |
| TFSC Vol. 75, No. 4 (2008) | 19.4 | 18.8 |
| TFSC Vol. 76, No. 9 (2009) | 11.3 | 17.5 |
| TFSC Vol. 80, No. 3 (2013) | 4.7 | 2.4 |
| Foresight Vol. 14, No. 4 (2012) | 1.0 | 1.5 |
| Foresight Vol. 15, No. 1 (2013) | 0.0 | 0.3 |
| Futures Vol. 43, No. 3 (2011) | 5.3 | 3.8 |
| Futures Vol. 59 (2014) | 0.3 | 0.1 |
| TASM Vol. 21, No. 8 (2009) | 2.0 | 6.7 |
| TASM Vol. 24, No. 8 (2012) | 2.3 | 2.2 |
| SPP Vol. 37, No. 1 (2010) | 5.1 | 6.3 |
| SPP Vol. 39, No. 2 (2012) | 4.1 | 2.1 |
| Average | 6.2 | 6.9 |

In conclusion, the analysis showed that 92 articles published in the 12 special issues had the same quality and impact—measured in number of citations—as articles published in these journals in general.

## 3.4 Conclusions

The aim of this paper was to present the profile and trends of the academic discipline of FTA and its approaches. In this chapter, we used the terms FTA and Foresight as identical fields. We analysed trends in the academic underpinnings and focus of FTA, trends in FTA approaches, and the quality and impact of the FTA special issues.

The first conclusion is that the field of FTA and Foresight seems have been remarkably stable over the last decade. As an academic field, FTA has targeted a small number of journals for its publications, and there appears to be no clear trend or development in the most important issues dealt with in the academic publications from the FTA conferences. This is not necessarily a problem. Rather, it indicates that FTA has found its framing, which could lead to further defining and focusing the field.

Next, we investigated generally assumed trends and developments in the FTA field: that FTA has taken a more systemic approach; that FTA has become more focused on user engagement and participatory methods; and that FTA has taken less of a national approach and focussed more on regional or sectoral issues. We investigated these alleged trends by analysing keywords and abstracts from 92 articles in international journals from the four earlier FTA conferences. Only development affiliated with a more systemic approach can be confirmed by analysing the abstracts. The other developments cannot be confirmed. This can be due to the validity of such analyses or a bias resulting from the editing process of the special issues. This can also be caused by the fact that some of the generally assumed developments in FTA are based on qualitative studies and not on more quantitative analyses. This leads us to suggest more detailed quantitative analyses of the development of the field of FTA and Foresight.

Finally, we investigated the quality and impact of the FTA special issues measured by the number of citations for the articles in these special issues. This analysis concludes that publications in a special issue of an international journal resulting from the FTA conferences have the same level of quality and impact as articles published in standard issues of these journals.

## References


Abadie, F., Friedewald, M., & Weber, K. M. (2010). Adaptive Foresight in the creative content industries: anticipating value chain transformations and need for policy action. *Science and Public Policy, 37*(1), 19–30.

Abbott, A. (2001). *Chaos of disciplines*. Chicago: University of Chicago Press.

Andersen, A. D., & Andersen, P. D. (2014). Innovation system foresight. *Technological Forecasting and Social Change, 88*, 276–286.

Asheim, B. T., & Gertler, M. (2005). The geography of innovation: Regional innovation systems. In J. Fagerberg, D. Mowery, & R. Nelson. (Eds.), *The Oxford handbook of innovation*. Oxford, USA: Oxford University Press.

Barré, R., & Keenan, M. (2008). Revisiting foresight rationales: What lessons from the social sciences and humanities? In C. Cagnin, et al. (Eds.), *Future-oriented technology analysis—Strategic intelligence for an innovative economy*. Berlin: Springer.

Butter, M., et al. (2008). Editors' introduction to the European Foresight Monitoring Network. *Foresight, 10*(6), 3–15.

Cooke, P. (2012). *Complex adaptive innovation systems: Relatedness and transversality in the evolving region (Regions and Cities)* (1st ed.). London: Routledge.

Dalkey, N., & Helmer, O. (1962). *An experimental application of the Delphi method to the use of experts*. Santa Monica, CA: INFORMS.

Fuller, T., & Loogma, K. (2009). Constructing futures: A social constructionist perspective on foresight methodology. *Futures, 41*(2), 71–79.

Georghiou, L. (2001). Third generation foresight—Integrating the socio-economic dimension. *NISTEP Study Material*.

Georghiou, L., & Cassingena Harper, J. (2011). From priority-setting to articulation of demand: Foresight for research and innovation policy and strategy. *Futures, 43*(3), 243–251.

Georghiou, L., Harper, J. C., & Scapolo, F. (2011). From priority-setting to societal challenges in future-oriented technology analysis. *Futures, 43*(3), 229–231.

Georghiou, L., et al. (2008). *The handbook of technology foresight*. Cheltenham, UK: Edward Elgar Publishing.

Hideg, É. (2007). Theory and practice in the field of Foresight. *Foresight, 9*(6), 36–46.

Irvine, J., & Martin, B. R. (1984). *Foresight in science: Picking the winners*. London: Pinter Publishers.

Jantsch, E. (1967). *Technology forecasting in perspective*. Paris: OECD.

Könnölä, T., Smith, J., & Eerola, A. (2009). Introduction: Future-oriented technology analysis—Impacts and implications for policy and decision making. *Technological Forecasting and Social Change, 76*(9), 1135–1137.

Martin, B. R. (2014). *R&D policy instruments—A critical review of what we do and don't know*, Aalborg.

Miles, I. (2008). From futures to foresight. In *The handbook of technology foresight: Concepts and practice*.

Miles, I. (2010). The development of technology foresight: A review. *Technological Forecasting and Social Change, 77*(9), 1448–1456.

Norwegian Research Council. (2010). *Foresight i Norge 2009. Mot et nytt kunnskapsfelt*, Norwegian Research Council.

Öner, M. A. (2010). On theory building in foresight and futures studies: A discussion note. *Futures, 42*(9), 1019–1030.

Oner, M. A., & Saritas, O. (2005). A systems approach to policy analysis and development planning. *Technological Forecasting and Social Change, 72*(7), 886–911.

Roveda, C., & Vecchiato, R. (2008). Foresight and innovation in the context of industrial clusters: The case of some Italian districts. *Technological Forecasting and Social Change, 75*(6), 817–833.

Saritas, O. (2011). Systemic foresight methodology. In *Forth International Seville Conference on Future-Oriented Technology Analysis (FTA) FTA and Grand Societal Challenges—Shaping and Driving Structural and Systemic Transformations SEVILLE,* May 12–13, 2011.

Weber, M., Schaper-Rinkel, P., & Butter, M. (2009). *Sectoral innovation foresight.* Introduction/Interim Report, Brussels.

# Chapter 4
# Recent Trends in Technology Mining Approaches: Quantitative Analysis of GTM Conference Proceedings

**Nadezhda Mikova**

**Abstract** This paper performs a quantitative analysis of trends in technology mining (TM) approaches using 5 years (2011–2015) of Global TechMining (GTM) conference proceedings as a data source. These proceedings are processed with a help of Vantage Point software, providing an approach "tech mining for analyzing tech mining." Through quantitative data processing (bibliometric analysis, natural language processing, statistical analysis, principal component analysis (PCA)), this study presents an overview, explores dynamics and potentials for existing and advanced TM methodologies in three layers: related methods, data sources, and software tools. The main groups and combinations of TM and related methods are identified. Key trends and weak signals concerning the use of existing (natural language processing (NLP), mapping, network analysis, etc.) and emerging methods (web scraping, ontology modeling, advanced bibliometrics, semantic the theory of inventive problem solving (TRIZ), sentiment analysis, etc.) are detected. The results are considered to be taken as a guide for researchers, practitioners, or policy makers involved in foresight activity.

**Keywords** Technology mining · Foresight · Future-oriented technology analysis (FTA) · Conference proceedings · Vantage Point

## 4.1 Introduction

Quantitative methods are increasingly being used in studies devoted to future-oriented technology analysis (FTA). There is a need to validate expert assessments with empirical evidence by searching for implicit signs of technological change in large amounts of data. "Tech mining" is exploiting information about emerging technologies to inform technology management (Porter and Cunningham 2005). As a special form of "big data" analytics, TM is becoming especially popular in FTA studies. In the context of information overload and limited resources, the

N. Mikova (✉)
Higher School of Economics, Myasnitskaya St. 9/11, 101000 Moscow, Russia
e-mail: nmikova@hse.ru



T.U. Daim et al. (eds.), *Anticipating Future Innovation Pathways Through Large Data Analysis*, Innovation, Technology, and Knowledge Management,
DOI 10.1007/978-3-319-39056-7_4

question is how to use TM in combination with other related methods on different stages of technology monitoring, what data sources to select and how to automate this process in order to improve the results.

Approaches to technology mining have been developing over the past decades. Several journals consider TM a core research area. A search in Web of Science (WoS) database (core collection) on publications in this field [using keyword "Text mining AND Technology" in "topic" (title, abstract, author keywords, keywords plus)] gave 464 records. The growing interest in this theme can be seen from Fig. 4.1.

TM-related studies have been published in ~80 different journals including specialized and economics-oriented journals and such multidisciplinary ones as Technological Forecasting & Social Change (TF&SC), Scientometrics, Technological Analysis and Strategic Management (TASM), Journal of Technology Management & Innovation, PLoS ONE, and others. The top 10 multidisciplinary journals publishing TM papers are presented in Table 4.1.

Studies devoted to TM are conducted by several top organizations located in different countries: USA, UK, China, South Korea, Germany, and others (see Fig. 4.2).

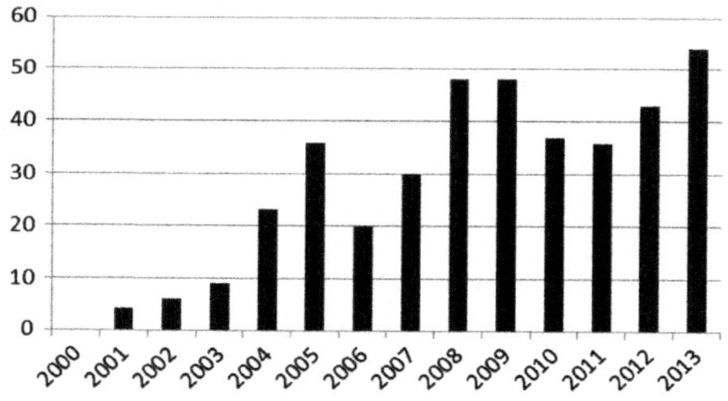


**Fig. 4.1** The amount of WoS publications devoted to TM in 2000–2013. *Source* WoS (core collection)

**Table 4.1** Top 10 multidisciplinary journals on TM topic in 2015

| Journal name | Amount of TM-related papers |
|---|---|
| Technological Forecasting and Social Change | 26 |
| Expert Systems with Applications | 21 |
| Scientometrics | 13 |
| Lecture Notes in Artificial Intelligence | 10 |
| Lecture Notes in Computer Science | 8 |
| Technology Analysis and Strategic Management | 7 |
| R&D Management | 4 |
| Journal of American Society for Information Science | 3 |
| Decision Support Systems | 3 |
| Technovation | 2 |

*Source* WoS (core collection)

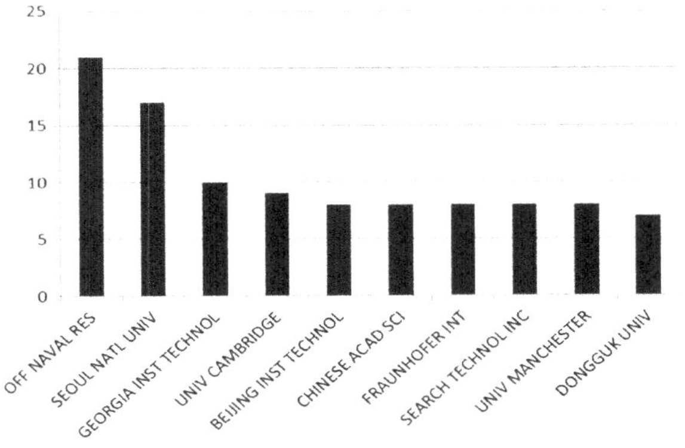


**Fig. 4.2** Top 10 organizations conducting TM studies in 2015. *Source* WoS (core collection)

Different information sources (publications, conference proceedings, research reports, etc.) can be used for exploring trends in TM approaches. In this paper, conference proceedings are considered a core information source for analyzing TM context, because in this case, the conferences serve as the platforms for exchanging experiences and discussing critical points of theory and practice with the participation of leading experts. The Global Technology Mining (GTM) conference was chosen for tracking the most promising TM areas as the most important forum for discussing TM-related issues. It has an aim "to engage cross-disciplinary networks of analysts, software specialists, researchers, policy makers, and managers to advance the use of textual information in multiple science, technology, and business development fields" (GTM Conference 2015).

This paper provides an overview of TM approaches which have been developing in the past five years (2011–2015) using a quantitative approach "tech mining for analyzing tech mining," which will enable the identification of key trends and weak signals concerning the use of existing and novel TM methods in three layers: related methods, data sources, and software tools.

Following the description of methodology, the paper will deliver the generated results. The conclusion section will briefly discuss the main trends in this area and indicate possible areas for future research.

The next section of the paper will present a methodology for analyzing TM and related approaches.

## 4.2 Methodology

In the framework of this study, the following tasks are resolved:

- Identifying key areas (mainstreams) of development in the TM field.
- Analyzing emerging trends and developments in TM approaches in three layers: related methods, data sources, and software tools.
- Discussing future TM trends.

Providing an approach "tech mining for analyzing tech mining," this study includes the following methodological stages: collecting data, processing data, and analyzing the results.

### 4.2.1 Collecting Data

For the purpose of this research, a collection of conference proceeding was created using the abstracts of GTM presentations for five years (2011–2015). This collection consisted of 188 proceedings downloaded from GTM official site (GTM Conference 2015).

Forming this collection covers the following procedures:

1. Finding the official programs of the conferences with information about the sessions, the participants' names, the titles of presentation, abstracts, and keywords (if presented).
2. Downloading the abstracts and saving them in *.xls format.
3. Converting *.xls files into *.xml format (smart-XML in Vantage Point).

As a result, five *.xml files (for GTM Conferences 2011–2015) were created using structured and unstructured data for the following fields: title, year, abstract, keywords, country, and organization.

### 4.2.2 Processing Data

In order to generate keywords for this study, the field "abstract" (rather than "keywords" or "title") was chosen for further data processing, because the field "title" does not contain comprehensive information about the core of the research; besides not all the presentations had authors' keywords, while keywords did not always fully characterize the study.

Using the methodology presented by Mikova and Sokolova (2014a), data processing included the following procedures:

1. Importing collections into Vantage Point software.
2. Prepreparing data.
3. Clustering keywords derived from abstracts (factor analysis).

The obtained smart-XML files were imported into Vantage Point software using customized import filters for *uploading collections* in *.xml format. In the framework of *prepreparing data*, the following NLP procedures were conducted: removing duplicate documents, carrying out a linguistic analysis and stemming of the text, and excluding stop words. After the collection was cleaned, the keywords had been further clustered (based on the keywords co-occurrence) using PCA. The analysis was conducted in 5 iterations through discussions with experts who participated in keywords selection and filtering, as well as in advising results.

### 4.2.3 Visualization

Keywords clustering based on factor analysis (PCA) helped to show only the most important clusters on the map (main components). Vantage Point software named the resulting clusters based on the most frequent keywords. Each cluster was characterized by a vector of keywords (descriptors), which it included and which were listed on the map in receding order according to the frequency in which they occurred. The number of clusters in Vantage Point can be preset (for instance, 5 factors or 10 factors), or be calculated by the software itself, based on the number of documents that are being processed. The collection of conference proceedings was divided into 50 clusters. Figure 4.3 shows the obtained cluster map.

Further analysis included a preliminary review of clusters, studying the links between them, and additional consultations with experts in order to identify key trends on this basis.

Next section of the paper will present the results of quantitative analysis and discuss the key findings.

## 4.3 Results and Discussion

The quantitative analysis of GTM conference proceedings for the period 2011–2015 helped identify the key groups (mainstreams) of research in the TM field.

Much of the research is connected with *the methodological issues of using TM* for analysis of technological development. The authors consider the stages of TM methodology, explore new methods and information sources, and propose new developments in TM analytics. The main goal of such research is to develop a systemic methodology for using TM for different types of technological analysis in various fields of knowledge. For example, it can be a methodology for discovering emerging technology trends with the help of TRIZ and technology roadmapping; a presentation of the approaches for finding patterns of emergence in S&T; the development of an integrated methodology for detecting emerging technologies using roadmaps, patents, and publications.

A number of studies are devoted to *the use of TM methods in FTA activity*. They employ such methods for the following goals:

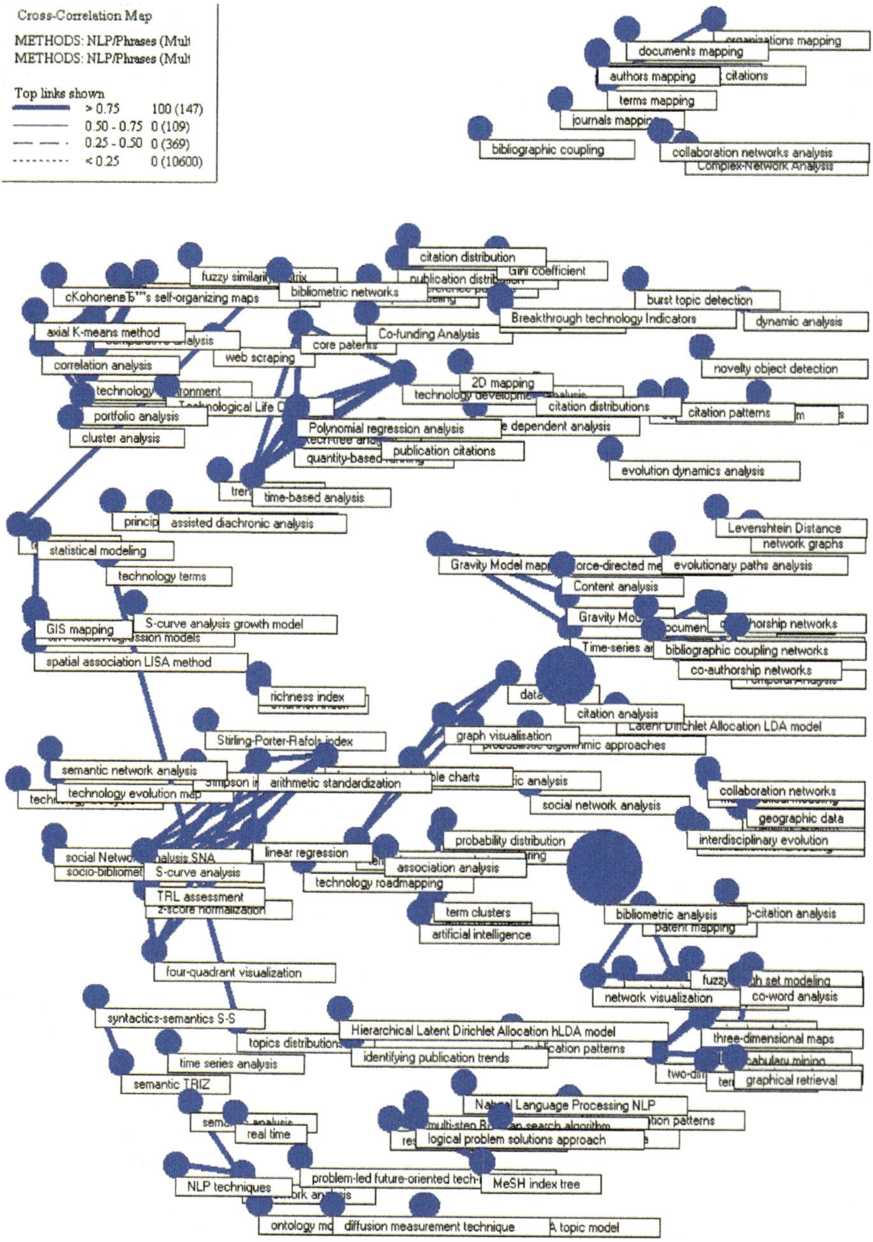


**Fig. 4.3** A cluster map produced in Vantage Point

- Identifying trends and emerging technologies (bibliometric analysis, citation analysis, and others).
- Detecting weak signals (citation dynamics analysis, LDA models, technology life cycle analysis, visualization, and others).
- Developing technology roadmaps (bibliometric analysis, clustering, graph visualization, statistical analysis, and others).
- Scenario building (bibliometric analysis, NLP, network analysis, and others).

Part of the studies proposes TM approaches for tracking technology evolution considering it from the *technology life cycle* perspective. Gartner's Hype Cycle (Gartner 2015) can be employed in such methodology, building an ontology of a technology trend and a system of indicators of its "presence" in documents of different genres. Herein, the indicators are interrelated with the ontology through linguistic markers. The research devoted to exploring technology life cycle stages may use such methods as bibliometric analysis, hierarchical clustering, semantic network analysis, mapping of technology evolution, mathematical modeling, statistical analysis, semantic TRIZ, PCA, subject-action-object (SAO) modeling, ontological modeling, and others.

A small number of the authors *elaborate and use indicators* for measuring technology development. It can be the indicators for improving the accuracy of outstanding papers prediction; for investigating S&T graduate training programs; for identifying patent collaborative patterns for emerging technologies; for developing the relevant criteria for selecting the project proposals by funding agencies; for detecting potentially disruptive technologies by opinion mining in social science networks.

A number of studies focus on using TM for *profiling* in technology areas: profiling science and innovation policy, profiling of research performance, using researcher profiling for identifying potential competitive or cooperative specialists.

*Automated tools* for assisting decision makers are in great demand. There are various techniques helping TM practitioners conduct technology analysis using TM and related techniques. As a rule, proposed software tools include such functions as data extraction and processing, cluster generation, topic identification, and information mapping.

There are also attempts to provide *visualization techniques* which are frequently combined with TM methods. These studies explore the methods of quick generation of helpful knowledge from text in the format of two-dimensional maps and networks. In some cases, these tools help not only present information, but also interpret and assess the results.

### 4.3.1 Methods

Some TM methods have been stabilized (quite intensively used in TM research), while the others can be categorized as emerging (actively involved in TM in the last 3–5 years).

In the framework of this research, TM and related methods are divided into two groups: main and supporting. The most popular combinations of them are studied, as well as key trends and weak signals concerning the use of existing and emerging methods are detected. The most frequently used (existing) main methods combined with TM are bibliometric and citation analysis. There are also various supporting techniques: trend analysis, clustering (K-means, hierarchical and nonhierarchical, etc.), and network analysis (social networks, ego networks, etc.). In most cases, visualization employs mapping (Kohonen's self-organizing maps (SOM), multidimensional scaling (MDS), force-directed placement (FDP), Kernel-based spectral clustering, local spatial autocorrelation (LISA) method, etc.) and network [for example, social network analysis (SNA)] analysis. Main and supporting methods can be mixed with other techniques, such as PCA, TRIZ, and probability distributions analysis.

As for emerging TM and related methods, they include web scraping; SAO modeling; semantic network analysis; dynamic clustering; analysis of evolutionary patterns and trend fitting; statistical modeling (for example, using agent-based modeling simulation); semantic TRIZ; probability latent-semantic analysis (using LDA model); real-time clustering, intelligence, and ongoing monitoring; ontology modeling; and sentiment analysis.

### *4.3.2 Data Sources*

While conducting TM research, the authors usually deal with the processing of data presented in a structured or unstructured format. An analysis of unstructured data has become possible with the use of electronic databases and the development of methods for processing large volumes of documents.

The most frequently used TM data sources are *patents* (USPTO, EPO, JPO, etc.) and *publications* (ISI WOS, Scopus, etc.). But the results of this study show that a wide variety of novel information sources have been involved in TM-related activities. *Web data* are used for mining topic trends, profiling science and innovation policies, understanding factors associated with growth of the number of companies, measuring technology transfer of universities, analyzing systemic evolution of technologies. *R&D (expenditures) data* help trace promising technologies, visualize the bridges between R&D and applications, identify the technology profiles of R&D performing firms, etc. *Social media (networks, blogs) data* are employed for detecting potentially disruptive technologies, evaluation of social media content on firm's marketing innovation, etc. *Geospatial data* are useful for spatial analyses of innovative excellence of the organizations over time. *Commercial, market, and business data* are helpful for gauging innovation pathways, supporting decision-making strategies in industrial processes, evaluating commercial potential of emerging technology, etc. *Internal (or strategic) documents and individual firm's annual reports* are used for profiling science and innovation policies, mining topic trends, analyzing publication activity in firms, etc. *News data* give information for identifying technology trends throughout the life cycle, predicting technology breakthroughs, tracing promising technologies, validating emerging science and technology topics, etc. *Funding and awards data* are important for investigating research sponsorship impacts, measuring interdisciplinarity of technology, linking federally funded research projects, evaluating the outcomes of government-funded research programs, validating emerging science and technology topics, comparing impacts of different funding modes, etc. *International projects and programs* provide information for designing research policy, finding relevant criteria for selection of project proposals, assessing patented technologies, etc. *Trainings programs data* are employed for studying emerging professions, etc. *Administrative and CV data* are useful for discovering and visualizing social networks of scientists, exploring emerging networks between different countries, profiling research network software, etc. *Presentations* (*SharePoint,* etc.) are used in developing decision-making strategies in industrial processes, etc. *Survey data and citizen requests* are applied for investigating graduate training programs in science and engineering, studying gender processes in science and advisor–advisee collaborations, measuring citizen participation in crowdsourced government, etc.

### *4.3.3 Software Tools*

As TM approaches deal with large volumes of information, many theoretical studies are devoted to developing and using automated software to process data, including linguistic and statistical analysis (Vantage, VOSviewer, etc.) and visualization tools (Gephi, UCINET, ClusterSuite, etc.). The faster information processing time significantly speeds up the sorting and filtering of data, analysis of trends and statistics, and the process of visualizing results. In such an analysis, both online (Carrot, PAS, etc.) and offline software applications (Vantage Point, CiteSpace, DIVA, Sci2, etc.) can be used. Many of the aforementioned tools have been developed by the authors themselves. Such applications, as a general rule, use information from electronic databases (publications, patents, news, etc.) and have a user interface to make queries, filter, and visualize the results (Mikova and Sokolova 2014b). Some programs offer special thesauri for NLP (WordNet), tools for creating ontologies (S&T Ontology), and possibilities for geospatial data processing (Google Earth, Google Maps, ArcGIS, I3Geo, etc.) and web scraping (automated web information monitoring systems).

A number of authors use software tools in combination. For example, Vantage Point can be applied together with the following applications: Gephi, VOSviewer, UCINET, Goldfire, ClusterSuite, MALLET, and others. There are also studies devoted to the issue of integrating and synchronizing different software solutions (for example, Vantage Point and Gephi).

Next section of the paper will present the conclusions formulated on the basis of analysis and the areas for further research.

## 4.4 Conclusion

The approach "tech mining for analyzing tech mining" using GTM conference proceedings as a data source helped identify and explore the main trends in TM approaches for the period 2011–2015. The results show that TM-related studies are conducted in different countries (USA, UK, China, Germany, South Korea, The Netherlands, Spain, Italy, Brazil, Colombia, Russia, Turkey, etc.) and cover a lot of subject areas. These studies can be divided into the following groups: focusing on methodological issues of using TM, devoted to the use of TM methods in FTA activity, considering technology evolution from the technology life cycle perspective, developing and using indicators for measuring technology progress, applying TM for profiling in technology areas, developing automated tools for assisting decision makers, providing visualization techniques which can be combined with TM methods.

Some TM-related methods are rather stabilized and are frequently used (bibliometric and citation analysis, trend analysis, clustering, network analysis, mapping, PCA, TRIZ, probability distributions analysis, etc.), while others just emerge and represent signals of change in the TM area (web scraping; SAO modeling; semantic network analysis; dynamic clustering; analysis of evolutionary patterns and trend fitting; statistical modeling; semantic TRIZ; probability latent-semantic analysis; real-time clustering, intelligence, and ongoing monitoring; ontology modeling; sentiment analysis). The methods have been becoming more and more sophisticated and may integrate several main approaches under one roof (for example, in a hybrid linguistic-semantic methodology). There are also attempts to cross TM methods with FTA approaches, using them in trend analysis, roadmapping, scenarios, etc.

In general, there is a tendency of expansion of pool of methodologies and techniques and involving well-known (publications, patents) and novel data sources (web data; R&D expenditure data; social media (networks, blogs) data; geospatial data; commercial, market, and business data; internal or strategic documents and individual firm's annual reports; news data; funding and awards data; international projects and programs; trainings programs data; administrative and CV data; presentations; survey data; and citizen requests) in technology monitoring. Some information sources can also be used in combination (patent-paper twins, etc.).

An increasing variety of web information can be processed using different kinds of software tools, among which the most popular are Vantage Point (and Thomson Data Analyzer), Gephi, VOSviewer, UCINET, Sci2, and Netdraw. However, in recent years, the authors have been using increasingly sophisticated software tools which support novel methods (WordNet, ClusterSuite, MALLET, ArcGIS, etc.). There is a trend of moving to quick processing of web information (social networks, blogs, etc.), applying dynamics methods (dynamic clustering, technology evolution analysis, etc.), using unsupervised TM algorithms, and real-time processing. Besides, more attention is paid to visualization methods (mapping and network analysis), which are more comprehensive inside, but serve to further simplify understanding and the interpretation of results.

In the future, it will be possible to analyze in detail the evolution of tech mining approaches to technology monitoring in terms of subject areas, countries, and centers of excellence.

**Acknowledgments** The chapter was prepared within the framework of the Basic Research Program at the National Research University Higher School of Economics (HSE) and supported within the framework of a subsidy by the Russian Academic Excellence Project '5–100'.

## References

GTM Conference. (2011–2015). Available via http://www.gtmconference.org

Mikova, N., & Sokolova, A. (2014a). Selection of information sources for identifying technology trends: A comparative analysis. HSE working papers. WP BRP 25/STI/2014. Available via http://www.hse.ru/pubs/lib/data/access/ram/ticket/6/1441928651196c35c49bf8abbded26a330de880864/25STI2014.pdf

Mikova, N., & Sokolova, A. (2014b). Global technology trends monitoring: Theoretical frameworks and best practices. *Foresight-Russia*, *8*(4), 64–83. Available via http://www.hse.ru/pubs/lib/data/access/ram/ticket/9/14419288231b11ad79b7424c01265dcb5772dc6a0d/Mikova-Sokolova_Global%20technology%20trends%20monitoring_2014.pdf

Porter, A., & Cunningham, S. (2005). *Tech mining: Exploiting new technologies for competitive advantage*. Hoboken: Wiley.

Gartner. (2015). Gartner Hype Cycle. Available via http://www.gartner.com/technology/research/methodologies/hype-cycle.jsp

# Chapter 5
# Anticipating Future Pathways of Science, Technologies, and Innovations: (Map of Science)$^2$ Approach

**Irina V. Efimenko, Vladimir F. Khoroshevsky and Ed. C.M. Noyons**

**Abstract** Anticipating future pathways of Science, Technologies, and Innovations is a complex task in any R&D field and is even more challenging for the complex landscape of promising R&D directions in multiple fields. As a solution, this study analyzes research papers in Scientometrics and Technology mining. It presents an approach and text mining tools for building maps of science of a special kind which is called the Map of Science Squared. Nodes of maps corresponding to R&D fields and locations (e.g., as centers of excellence) are created, weighted, and coupled whenever possible based on processing full texts or abstracts of research papers. The questions to answer with this are as follows: (1) Do Scientometrics and Technology mining cover the full range of topics both in terms of breadth and depth? (2) Do research papers appear "at the right time," i.e., just or soon after emergence of a topic? (3) Do researchers link R&D fields in non-traditional ways through their studies? (4) What fields are locally bound? (5) What conclusions on future pathways of Science, Technologies, and Innovations can be drawn on the basis of the analysis of the Scientometrics and Technology mining agenda?

**Keywords** Map of science · Emerging R&D fields · Scientometrics · Technology mining · Text mining · Semantic technologies · Ontology

---

I.V. Efimenko (✉)
Higher School of Economics, National Research University,
Myasnitskya Str. 20, 101000 Moscow, Russia
e-mail: veassi@mail.ru

V.F. Khoroshevsky
Dorodnitsyn Computing Center, Federal Research Center Computer Science and Control of RAS, Vavilov Str. 40, 119333 Moscow, Russia
e-mail: khor@ccas.ru

Ed.C.M. Noyons
Centre for Science and Technology Studies, Leiden University, 2300 AX Leiden, The Netherlands
e-mail: noyons@cwts.leidenuniv.nl



T.U. Daim et al. (eds.), *Anticipating Future Innovation Pathways Through Large Data Analysis*, Innovation, Technology, and Knowledge Management,
DOI 10.1007/978-3-319-39056-7_5

## 5.1 Introduction

Anticipating future pathways of Science, Technologies, and Innovations (ST&I) is a complex task in any R&D field. It becomes even more challenging if the task is to build a complex landscape of promising R&D directions in general, i.e., in multiple fields of science and technology or in the case when only a high-level specification of such fields is available at the beginning (e.g., in foresight projects in the interests of major state institutions, or corporations which provide solutions for diverse industries, or a whole country). In such cases, it is difficult to define a starting point for research to be carried out, both in terms of information sources and specific topics for an in-depth analysis.

As a solution, this study analyzes research papers in Scientometrics and Technology mining in order to find the "points of departure." In many cases, R&D fields extracted from such papers can suggest promising directions, both due to the nature of the business tasks which are often behind the studies in Scientometrics and Technology mining and due to a flair for innovation which characterizes the authors of these studies. Thus, the Scientometrics and Technology mining agenda can help analysts and researchers to draw useful conclusions on future pathways of ST&I, since it plays the role of a superstructure, similar to the one of intelligent meta search engines in information retrieval.

In order to prove the "predictive power" of Scientometrics and Technology mining, we show that they cover a full range of topics both in terms of breadth and depth and that they appear "at the right time," i.e., just or soon after emergence of a topic (e.g., an emerging technology or a new problem to be solved such as an epidemic outbreak). Discerning interdisciplinary connections which emerge between R&D fields is also important.

In addition to R&D fields, it is of interest to identify topics which are associated with a specific location, i.e., studied together with some locations or by authors belonging to various countries, regions, or cities. Topics and locations provide evidence for focuses of interest in the world and can be used, e.g., for benchmarking, as well as for the identification of centers of excellence being objects of studies, and potential partners.

A complete set of relevant questions includes the following:

1. Do papers in Scientometrics and Technology mining provide coverage for a "representative sample" of R&D fields in a variety of ST&I areas?
2. Is the analysis deep enough to focus on specific topics including emerging ones, down to the names of specific technologies, and objects of research (as opposed to an analysis on a level of disciplines or broad areas of research only)?
3. Do researchers in Scientometrics and Technology mining focus their attention on topics which are just emerging or, at least, still on the rise (as opposed to technologies and research areas being de facto standards already)?
4. Do researchers link R&D fields in non-traditional ways through their studies, do such links reflect emerging interdisciplinary connections between R&D fields, or can they be considered to be weak signals of future pathways of ST&I?

5. What fields are locally bound in terms of authors and their focuses of interest?
6. What fields are locally bound in terms of the objects of studies and potential centers of excellence?
7. What R&D fields are in focus of Scientometrics and Technology mining today, which fields are old-timers and newcomers, what are the trends? Do both disciplines have the true potential for anticipating future pathways of ST&I, what are the differences between them?

The approach presented in this chapter which we call the Map of Science Squared is intended to answer all these questions. When describing the basic principles, methods, and tools behind this approach, we refer more or less to all these aspects, since the main idea of this study is to develop the approach itself and the corresponding framework with text mining, business intelligence (BI), and other tools. For the case study and the discussion of text processing results, we focus mostly on questions (1)–(4), and (7), with an emphasis on questions (1) and (2). Such limitations are due both to the lack of space, which is relevant mostly for questions (5)–(7), and to the need for further elaboration in the case of questions (3) and (4). The examples presented show the validity of the approach in terms of questions (3) and (4); however, further implementation of complex methods for expert evaluation is needed which is in our short-range plans.

The chapter proceeds with a short discussion of related work in Sects. 5.2 and 5.3, which is followed by a description of the approach (Sect. 5.4). Sections 5.5 and 5.6 present the results of processing text collections in Scientometrics and Technology mining. They provide interesting findings and a short discussion of results in the context of anticipating future pathways of ST&I. They are followed by the conclusions in Sect. 5.7.

## 5.2 Map of Science Squared

Maps of Science (MoS) are well known as a useful and captivating tool and are widely used both by researchers and policy makers in ST&I with recent efforts addressing planning-related problems (Börner et al. 2012a, b; Klavans and Boyack 2011; Boyack and Klavans 2014; Chen and Leydesdorff 2014). We present an approach and new methods and tools for building maps of science of a special kind (MoS Squared). These are maps of ST&I through the lenses of Scientometrics and Technology mining. In a sense, it is a "scientometric analysis of Scientometrics" and "technology mining in Technology mining." Mapping ST&I is one of the key tasks in science and technology studies, and we build maps of Scientometrics and Technology mining themselves, so we get it "squared."

Methodologically, the representations we build are not maps of science in the strict sense of the word. Specific "nodes" of maps are created, weighted, and coupled whenever possible based on mentions in full texts or abstracts of research papers in Scientometrics and Technology mining. Our approach is based on

Information Extraction techniques (Khoroshevsky 2009; Wimalasuriya and Dou 2010; Piskorski and Yangarber 2013). The text mining and mapping tools which were developed in this study make use of hybrid statistical and linguistic analysis (Efimenko and Khoroshevsky 2014), widely accepted thesauri and classifications, as well as of the publication classification scheme proposed by Leiden University (Waltman et al. 2010).

Our science mapping approach deals not with the papers themselves but with the names of R&D fields and locations, which were extracted from those documents at earlier stages of document processing. Thus, we couple not only documents, but also directly R&D fields, being aware of their semantics at the moment of coupling.

## 5.3 Background

Researchers in Scientometrics have developed a variety of science mapping approaches based on citation, textual similarity, or a hybrid (Small 1973; Marshakova-Shaikevich 1973; Rip and Courtial 1984; Leydesdorff 1987; Garfield 2001; Boyack et al. 2013). Co-citation analysis, bibliographic coupling, and direct citation are among the most popular approaches (Robinson et al. 2013; Porter and Cunningham 2010). Our study exploits general principles of some of those approaches, but transforms them taking into account the semantics-based nature of the proposed algorithm and the character of input data.

Text mining approaches which make use of mapping techniques in light of analyzing a wide range of R&D fields go a long way in Technology mining (Porter and Newman 2011). They are often used to identify promising topics in external R&D and to draw conclusions on the community engaged in a particular R&D domain, as well as to provide support in science and technology policies (Huang et al. 2015). As opposed to studies in Scientometrics, which can be oriented either toward narrow R&D fields, or to various maps of sciences as a whole (Leydesdorff 1987; Boyack and Klavans 2014), research and developments in technology mining are more often focused on specific themes of ST&I (e.g., *renewable energy, rice studies*) or even specific products and technologies (e.g., *LED, semiconductors, specific types of solar cells, nanocomposite coatings*). If to consider our study as the one in Technology mining, its most important distinctive feature is in the fact that we mine papers not *in*, but *on* R&D fields.

Both papers representing science mapping approaches for identification of emerging topics and those oriented toward technology intelligence and technology management issues are naturally related to future-oriented ST&I analyses (Small et al. 2014; Porter and Zhang 2015). However, in our case, we do not regard papers in Scientometrics and Technology mining as sources of "ready answers," but as input data for the further analysis. A similar logic is behind the methodologies which are based on the text content of documents being results of foresight studies, e.g., technology roadmaps. Corresponding papers have appeared in last few years (Ye and Feng 2013).

## 5.4 Methods and Tools

The main constituents of the proposed approach include ontology-based information extraction (OBIE) (Wimalasuriya and Dou 2010), coupling, and BI techniques. In turn, ontology makes use of a number of thesauri and classification schemes. Key objects extracted from documents are names of R&D fields and Locations.

### 5.4.1 *Classification of R&D Fields*

It is well known that providing an adequate classification for R&D fields, especially for multidisciplinary and rapidly evolving ones, is among the greatest challenges in science and technology studies, bibliometrics, and cognate sciences. Maps of science help to solve this problem when used in many practical applications since they allow users to discover emerging R&D fields. However, some classification is needed as a starting point to build a map itself.

A number of widely used thesauri and classifications were considered as a basis for the ontology of R&D fields in this study. They include Field of Science and Technology (FOS) Classification in the Frascati Manual,[1] Web of Science Categories and Subject Areas, Scopus Subject Areas, International Standard Industrial Classification of All Economic Activities,[2] and a number of specialized thesauri such as the INIS Thesaurus which is a major tool for describing nuclear information and knowledge in a structured form,[3] as an example. The NOWT-WoS classification system[4] was selected as a basic one. Most differences between NOWT-WoS classification and our results are in subareas of the third or lower levels and are due to two major factors: the flexibility achieved as a result of the semantic analysis of input data and the employment of several additional classifications for selected R&D fields.

For example, medicine was the most diverse in terms of diversity at the stage of preliminary shallow semantic analysis: several hundreds of subfields and specific subjects such as diagnoses, medical specialties appeared in results (84 subareas and several hundreds of unique subjects of research appeared in the final results). It was decided to pay special attention to the healthcare domain. Several classifications allowing users to examine medicine-related topics from different points of view were used and partially integrated into our ontology. They include International

[1]http://www.oecd.org/sti/inno/frascati-manual-revision.htm.

[2]http://unstats.un.org/unsd/publication/seriesM/seriesm_4rev4e.pdf.

[3]https://www.iaea.org/inis/products-services/INIS-Thesaurus/index.html.

[4]http://www.cwts.nl/pdf/nowt_classification_sc.pdf.

Classification of Diseases ICD-10 and classifications of medical specialties that are common worldwide or recognized in Europe, North America, and some other countries.[5] For Locations, a number of generally accepted thesauri were used as well.

However, commonly used thesauri and classifications alone turned out to be insufficient for solving our tasks. To provide a real data-driven approach, Information Extraction is used. For example, we use a hybrid approach based on statistics and linguistic rules to extract information on emerging fields and complex, multiword names, low-level concepts which are not provided in standard classifications (e.g., *soil erosion, RFID, dye-sensitized solar cells*), parataxis (e.g., *complementary and alternative medicine*), names of professions (e.g., *Chemists*), descriptive names (e.g., *Italian Regions, developed, and developing countries*).

A number of stop gazetteers are used to avoid false triggering (e.g., *Oslo Manual*). Special rules are provided for ambiguous cases, where specific terms can refer both to a subject and to an object or various aspects of scientometric and technology mining studies (e.g., *Economics* as a discipline and *Economic impact of S&T*. Other examples refer to *Psychology, Social sciences, Geography, Ecology, Management*).

In this chapter, given the context of this book, we focus on hard sciences and technologies providing a few examples in Social Sciences and Humanities for illustrative purposes only. Industries are considered on a shallow level as well. R&D fields are attributed to industries in several cases only, i.e., in the case of key industries in current technological change (e.g., *the automotive industry*) and when it seems difficult to attribute a topic to an area of research due to cross-disciplinary and/or the service-oriented nature of a field (e.g., *logistics*). Otherwise, the extracted topics are considered in terms of areas and subareas of research, but not industries.

## 5.4.2 Ontology

Ontologies of Locations and the one of R&D fields were built. These two ontologies have similar structure. They specify hierarchies (*regions of the world–countries–states, provinces,* etc.—*cities* for Locations; *fields and disciplines* of various levels down to specific *subjects of study* for R&D fields). They also include knowledge on the status of the object to be extracted, which is mostly relevant in the case of full text papers. Thus, we have Locations related to authors (authors' affiliation), Locations being objects of studies (extracted mostly from "Data," "Methods," or similar sections), and other Locations mentioned in a paper (e.g., for

---

[5] http://www.gmc-uk.org/educaion/standards.asp
http://www.webmd.com/a-to-z-guides/medical-specialists-medical-specialists
http://apps.who.int/classifications/icd10/browse/2015/en
http://www.who.int/hrh/statistics/Health_workers_classification.pdf.

comparison purposes or in an overview). A similar scheme is used for R&D fields. The main task is to distinguish between R&D fields extracted from different parts of a paper and to attribute an extracted subject of study to an ontology node based on its semantics.

Extraction is provided for Locations which are mentioned explicitly and for implicit ones based on names of councils, foundations, and other institutions (e.g., *NSF* → *USA*, *NISTEP* → *Japan*). At the moment, only major institutions are taken into consideration. Such expressions as *G7*, *BRICs*, etc. are also considered to be Locations.

The hierarchy of R&D fields has 5 levels at most. At the moment, we extract information mostly on those fields or subjects which are explicitly mentioned. Some of them are presented implicitly through the names of institutions, but it is not a simple task to process such names in order to make correct inference on R&D fields, and we plan to solve it later. At the same time, some implicit markers are already used. For example, mentioning PubMed or MeSH indicates relation to a biomedical domain which is considered in the ontology and in rules for text processing.

We consider it useful to include not only names of the major R&D fields, but also the specific subjects of study into the map (real examples include *superconductivity, organometalloidal compounds, augmented reality, wastewater, Bose–Einstein condensate, bionanofluidics, transnational organized crime, vehicle-to-grid technologies, bioinspired mechanisms, oil shale, underwater connection, oenology, cochinchina momordica seed, tissue plasminogen, ebola, sleep disorders*, etc.). For the purpose of uniformity, they are called R&D fields as well. For Technology mining, it is natural to choose not a wide field but a specific business task or a technology as a starting point of the analysis. For Scientometrics and Bibliometrics, it seems to be a trend nowadays, too, to focus on specific topics (to a greater extent than on a discipline or the area of research as a whole). It may be caused by increasing degree of interdisciplinarity and the multidisciplinary nature of many important phenomena on one hand, which in turn leads to a problem-oriented rather than a discipline-oriented approach, and by the research diversification, on the other hand.

### 5.4.3 Information Extraction Models and Tools

The objects of interest (i.e., R&D fields and Locations) are extracted taking into account their position within a text. At the moment, an object position is considered in terms of semantics, but not in terms of importance. Directions of future research might include advancement of the algorithm based on assigning weights to the names of R&D fields and Locations depending on the part of a paper where they appear.

In coupling techniques, bibliographic fields Abstract, or a full text where available, Keywords, and Title are considered as a whole and are called "Body." Other important fields include References and Authors' information.

The names of R&D fields are extracted from Body and References, but not from Authors' information since it could lead to wrong inferences. For example, not all the researchers in Higher School of Economics, Moscow, conduct research in Economics. Locations, just the other way round, are not extracted from References, due to the risks of mistakes resulting, for example, from processing contact details of a publisher. They are extracted from Body and Authors' information (with different status).

The overall approach implemented in our software tools is as follows.

OBIE is performed (Efimenko and Khoroshevsky 2014). A hybrid linguistic and statistical approach is applied. GATE environment provided by the University of Sheffield[6] is used as a platform for development of our own software tools (Efimenko et al. 2014). For each node of the ontology, there are a number of ontological gazetteers and rules which drive the extraction process. Ontological gazetteers and rules were built in advance in a semi-automated way based on linguistic patterns, thesauri, and the statistical processing of the large collections of abstracts and the full texts of research papers (60,000+ articles and reviews in various fields of ST&I). The hybrid approach is needed, since the most commonly used methods of clustering and classification would not allow us to solve the task of this study. In general, all the analyzed papers are related to two major topics: Scientometrics and Technology mining. So, the papers often look very similar from a statistical point of view, while the difference is in details requiring deeper linguistic analysis.

Thus, mono- or multiword terms (noun groups, NGs) corresponding to R&D fields, as well as to the names of Locations, are extracted from each document and are considered to be "Named Entities" in terms of Information Extraction. Each extracted term is associated with a node or several nodes of the ontology. For example, *traffic medicine* is related to Healthcare and to Transportation, *digital health*-related topics can be associated with Healthcare and IT.

### 5.4.4 Coupling Techniques and Analytics

For coupling, several types of co-occurrence or citing-cited matrices are built for each document and then for the whole collection. They involve the following pairs:

- R&D fields extracted from the Body of a paper compared to each other (R&D field/Body + R&D field/Body);
- R&D fields extracted from the Body of a paper compared to R&D fields extracted from References (R&D field/Body + R&D field/Ref);
- R&D fields extracted from the Body of a paper compared to Locations extracted from the Body (R&D field/Body + Locations/Body);

[6]https://gate.ac.uk.

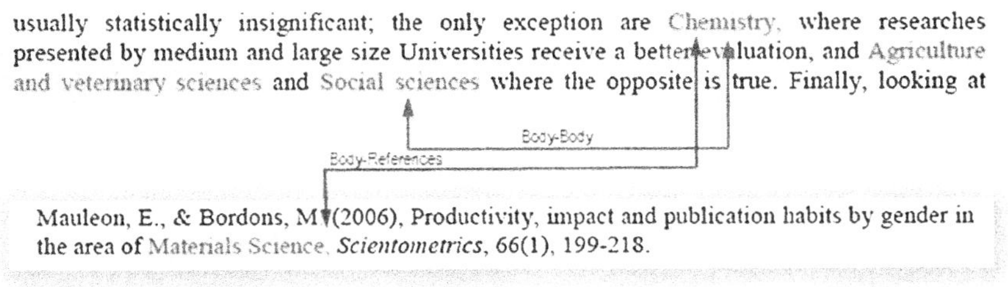


**Fig. 5.1** Types of coupling

- R&D fields extracted from the Body of a paper compared to Locations extracted from Authors' information (R&D field/Body + Locations/Author).

The process of coupling is shown in Fig. 5.1 (paper[7] is taken as an example of an input document).

Coupling R&D fields within a body of a document is methodologically close to co-citation, since we consider cases when the document "co-cites" a pair of R&D fields by mentioning, more or less explicitly, both of them. We can regard such cases as citing, since we do not process papers, e.g., *in* Physics, but scientometric and technology mining papers *on (about)* Physics.

Coupling R&D fields extracted from the body of a paper with those extracted from references is methodologically close to direct citation model. In some cases, it is considered to be the least accurate mapping approach (Boyack and Klavans 2010). However, the MoS Squared methodology allows us to overcome some shortcomings of direct citation by involving semantics, as well as statistics of co-occurrence of any two R&D fields, while in the "traditional" direct citation model, each pair of documents is unique. Such approach can be considered as a hybrid one, since it involves text-based methodology. The hybrid approaches are recognized as the most powerful ones.

Other types of analysis provided in MoS Squared approach include various rankings (e.g., top R&D fields in Scientometrics and/or Technology mining), time series (for analyzing old-timers, newcomers, and trends), and other classical BI tools. Figure 5.2 shows the general pipeline of paper processing in MoS Squared approach.

[7]Cicero T, Malgarini M, Benedetto S (2014) Research quality, characteristics of publications and sociodemographic features of Universities and Researchers: evidence from the Italian VQR 2004-2010 evaluation exercise. In: Proceedings of the science and technology indicators conference 2014, Leiden, the Netherlands.

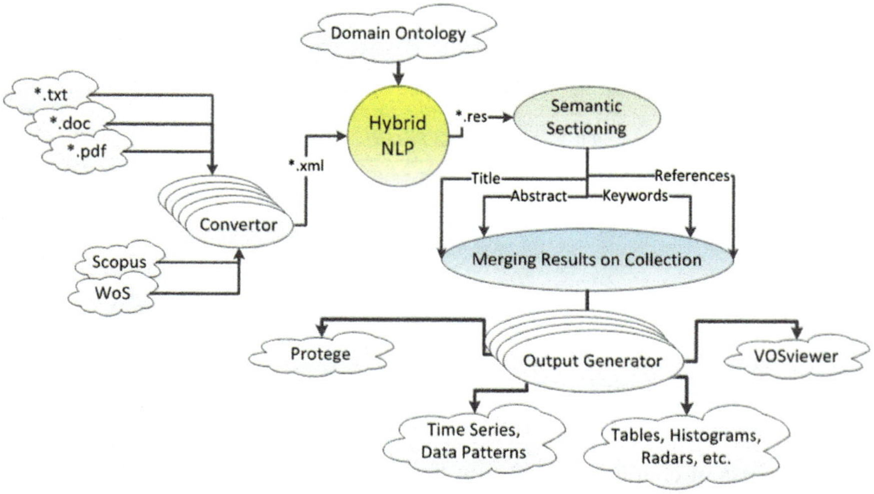


**Fig. 5.2** Map of science squared approach: pipeline

## 5.5 Data

Two corpora were created as follows: one for Scientometrics (including Bibliometrics) and one for Technology mining. These two sets intersect. However, it is of interest to consider them separately. Complex queries to Web of Science and Scopus were created which specified the scope of these two disciplines.

For Scientometrics, the query was *bibliometric OR scientometric OR bibliometrics OR scientometrics OR "STI indicator" OR "STI indicators" OR "science indicator" OR "science indicators" OR "technology indicator" OR "technology indicators" OR "innovation indicator" OR "innovation indicators."* Web of Science hits were completed by the papers from the STI Conference (2000–2014). The final collection in Scientometrics included 7950 abstracts and/or full text papers. The earliest papers were those published in 1969, which was good for defining old-timers. In addition to the abstracts and the full texts, WoS categories and research areas were processed as text input. Some field names being potential sources of errors, such as Library science and similar ones, were excluded from consideration.

For Technology mining, the initial idea was to include names of typical tasks into a query (e.g., *patent analysis, trend monitoring, processing roadmaps,* etc.), as well as to consider relevant technology management and IT fields (*NLP, semantic technologies, emerging technologies,* etc.). However, it would blur the focus. Only basic terms were used which included those for technology intelligence as an essential component of the Technology mining (Porter and Zhang 2015). In future, similar maps can be built through the prism of specific business tasks (*technology roadmapping, trend watch,* etc.).

The final query was *"technical intelligence" OR "technology intelligence" OR "technological intelligence" OR "tech mining" OR "technology mining" OR "technology scout" OR "technology scouting."* Web of Science brought 300 results. Scopus turned out to be a better source of information, possibly due to the fact that Technology mining is more conference-oriented as a newer and more IT-related field, in comparison with Scientometrics. At the same time, there are no journals at the moment dedicated to Technology mining similar to "Scientometrics" or the "Journal of Informetrics" as examples, though several journals in technological forecasting and innovation management do publish articles in Technology mining on a regular basis in recent years. For Scientometrics and Bibliometrics, in addition to journals focused on these disciplines, it is also easy to find lots of case studies, for instance a bibliometric analysis of a specific research field. Such papers are published in journals related to the corresponding fields and are available in Web of Science database.

Hits were refined in order to exclude mining industry, "technical intelligence" in animal behavior, forensics, etc. Papers on "pure" technology management, on one hand, and data mining or the NLP algorithms not related to technology intelligence, on the other hand, which still appeared in query results were also excluded. At the same time, it is worth mentioning that Technology mining as an all-sufficient R&D field took shape much later than Scientometrics. In order to provide data for earlier years, papers in several fields which could be considered as "methodological precursors" of Technology mining were included into the collection. These were primarily papers in technology intelligence and technological decision making or, more precisely, those of them which were related to gathering, storing and processing the large volumes of unstructured information on new technologies, centers of excellence, etc. even if it was supported by the means of methodology and not by the automation tools.

Additional texts were taken from the proceedings of the GTM Conference and VantagePoint library. In the end, 1000+ texts were included into the Technology mining collection. The collection included a mixture of abstracts and full texts (where available). The earliest paper was published in 1980.

At the processing stage, some papers were excluded, for example, those without abstracts. The data on the final collections are provided in Table 5.1.

## 5.6 Results and Discussion

### *5.6.1 The Scope of the Research in Scientometrics and Technology Mining. Co-occurrence of R&D Fields*

Results show that researchers in Scientometrics and Technology mining pay their attention to a huge variety of subjects and, at the same time, often provide an

**Table 5.1** Processed collections

| Scientometrics | | Technology mining | |
|---|---|---|---|
| WoS DB | STI proceedings | Scopus DB | GTM proceedings |
| 1969–1978 (50 rec.) | | | |
| 1979–1988 (239 rec.) | | 1980–1998 (30 rec.) | |
| 1989–1998 (759 rec.) | | | |
| 1999–2003 (899 rec.) | 2000 (67 pap.) | 1999–2003 (65 rec.) | |
| | 2002 (58 pap.) | | |
| 2004–2008 (1000 rec.) | 2004 (100 pap.) | 2004–2008 (261 rec.) | |
| | 2006 (103 pap.) | | |
| | 2008 (135 pap.) | | |
| 2009–2010 (1077 rec.) | 2010 (123 pap.) | 2009–2010 (157 rec.) | |
| 2011–2012 (1414 rec.) | 2012 (105 pap.) | 2011–2012 (192 rec.) | |
| 2013–2015 (1492 rec.) | 2014 (101 pap.) | 2013–2015 (241 rec.) | 2014 (43 pap.) |
| Total: 6930 rec. | Total: 792 pap. | Total: 946 rec. | Total: 43 pap. |
| Total: 8711 units | | | |

in-depth analysis of specific emerging topics, e.g., promising technologies. The list of categories, which were identified in the processed collections, i.e., the nodes of maps or the types of objects in terms of ontology not including the level of specific instances, is provided in Appendix.

Not only the names of R&D fields, but also the links between them were considered during the mapping process. The maps were built with VOSviewer tools (van Eck and Waltman 2010). The dynamic representation of the maps for different time periods clearly shows the diversification of the objects of interest in terms of R&D fields. An example for Technology mining papers at the level of research areas and subareas is provided in Fig. 5.3.[8]

Each object type (R&D field) contains the specific instances extracted from texts (e.g., *carbon nanotube field emission display*). Such instances cannot be listed in this chapter because of a vast number of them, but they can be placed on the maps, too. Often, they are the most interesting nodes in terms of anticipating future pathways of ST&I. Some examples are briefly discussed below. The flexible algorithms based on linguistic analysis helped us to make use of a problem (challenge)-driven approach, i.e., to discern R&D fields, often of a multidisciplinary nature, which correspond to the most important challenges of today (e.g., various types of *waste or water management*).

Figure 5.4 illustrates the role of Medicine and its subfields in Scientometrics throughout all periods, as well as the connections of Medicine and Healthcare with other R&D fields in various periods (e.g., *Physics—for Nuclear medicine, Neuroimaging,* etc., *Environmental studies*, etc.). The examples given in this chapter are based on the co-occurrence of R&D fields in the bodies of papers. Tasks

[8]The label “country” is related to the topics which are represented as locally bound ones.

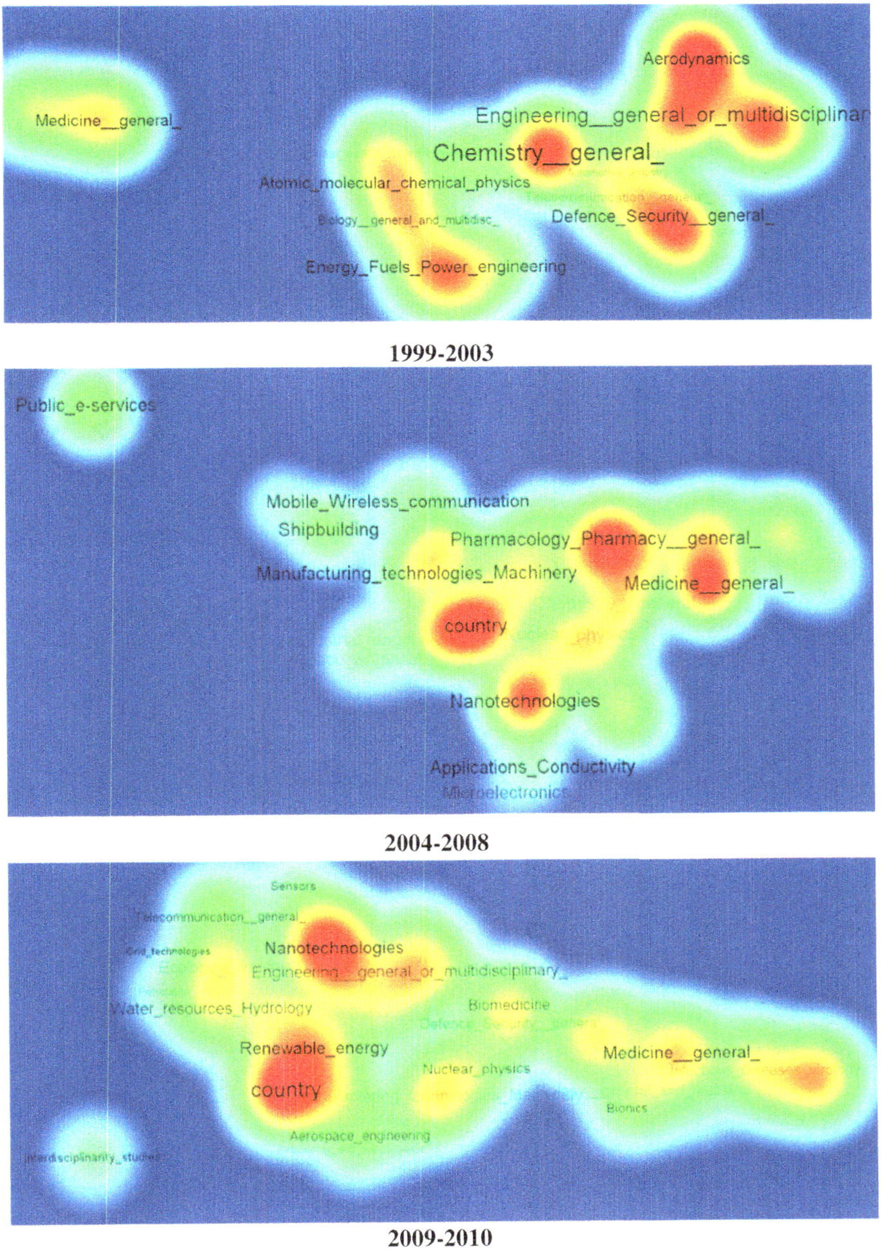


**Fig. 5.3** R&D fields in technology mining papers

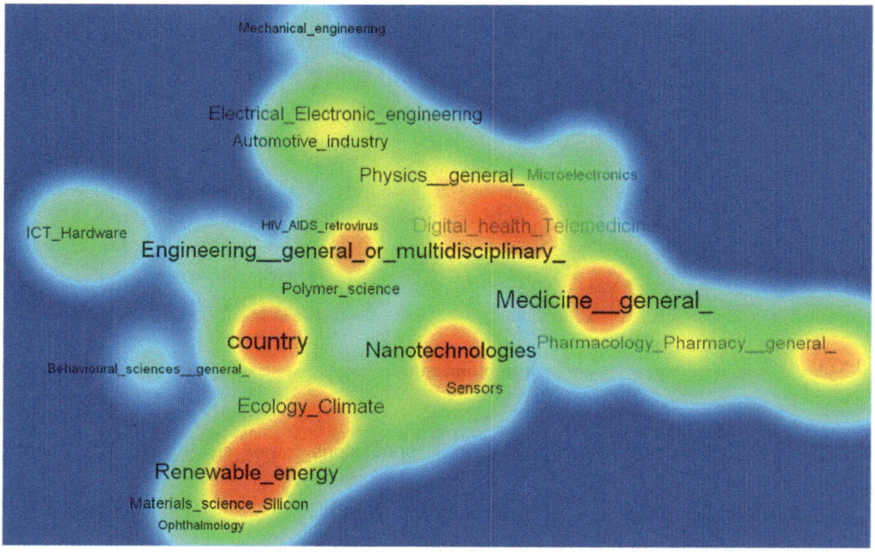


2011-2012

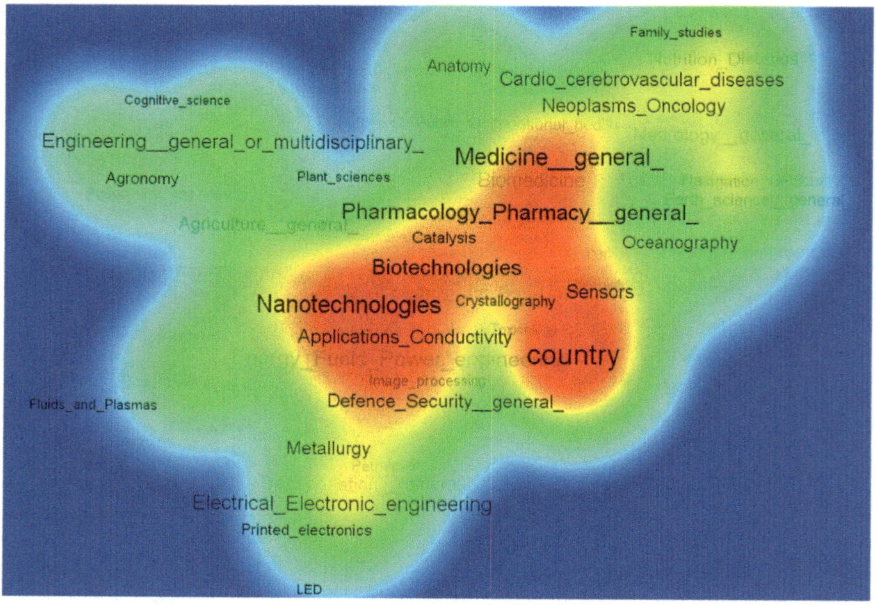


2013-2015

**Fig. 5.3** (continued)

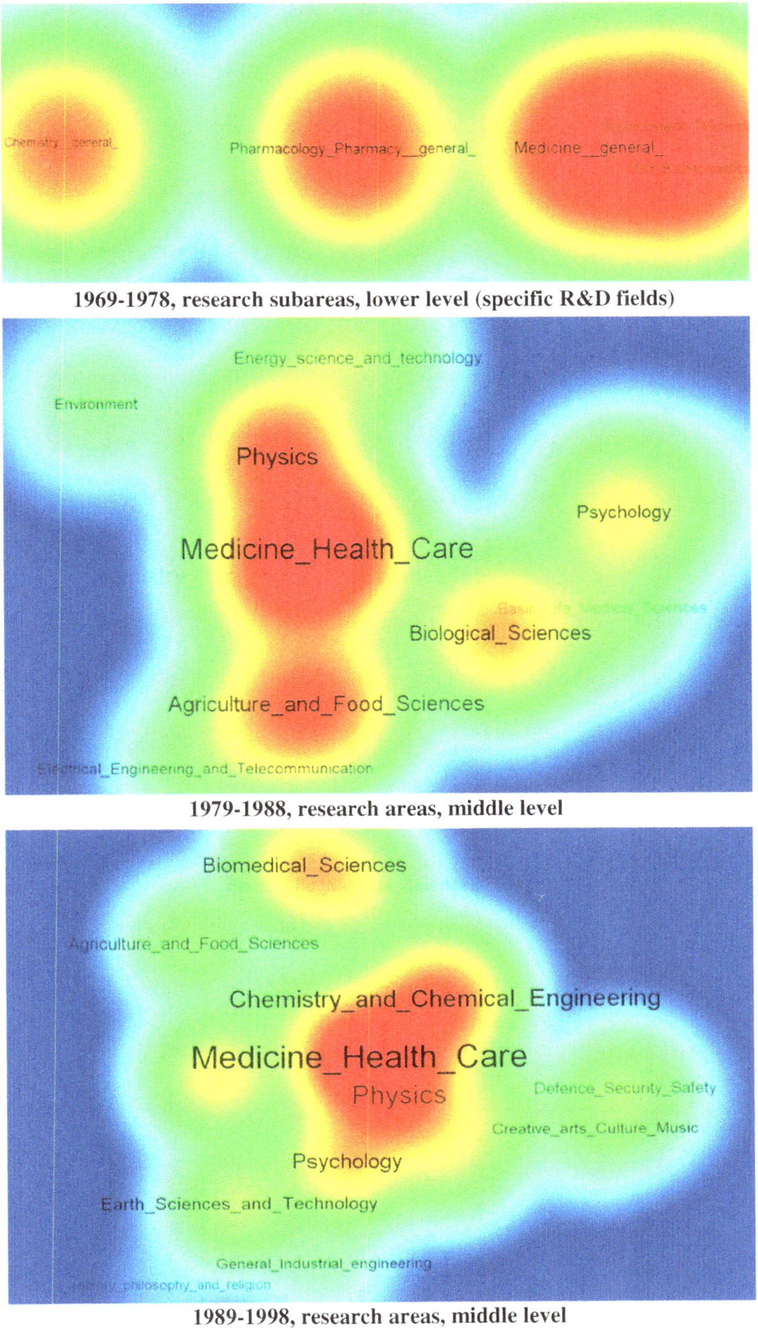


**Fig. 5.4** Medicine as a core topic in Scientometrics

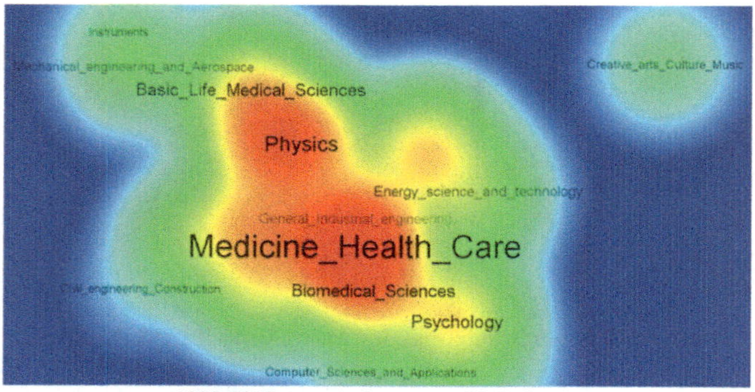


1999-2003, research areas, middle level

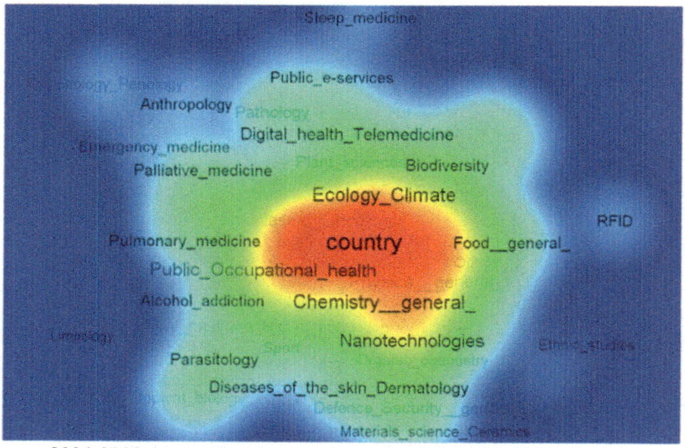


2004-2008, research subareas, lower level (specific R&D fields)

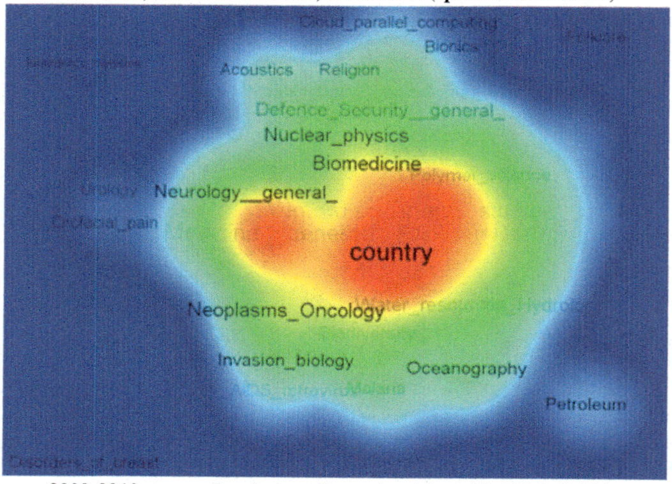


2009-2010, research subareas, lower level (specific R&D fields)

**Fig. 5.4** (continued)

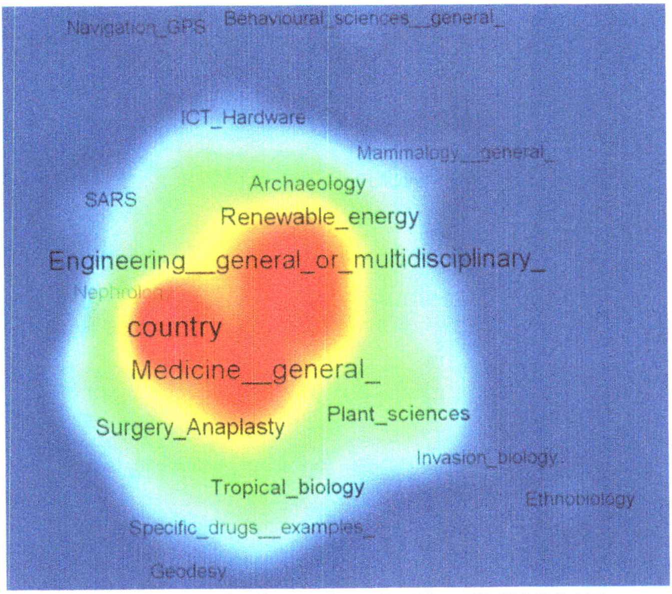


2011-2012, research subareas, lower level (specific R&D fields)

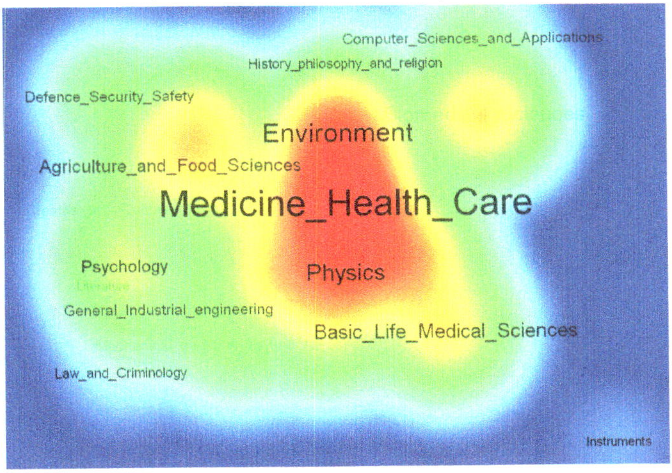


2013-2014, research areas, middle level

**Fig. 5.4** (continued)

for future research include the detailed evaluation of the MoS Squared approach potential in the assessment of emerging interdisciplinary connections between R&D fields. Besides, they include the validation of relevance of comparing R&D fields which were extracted from the Body of a paper with those extracted from References.

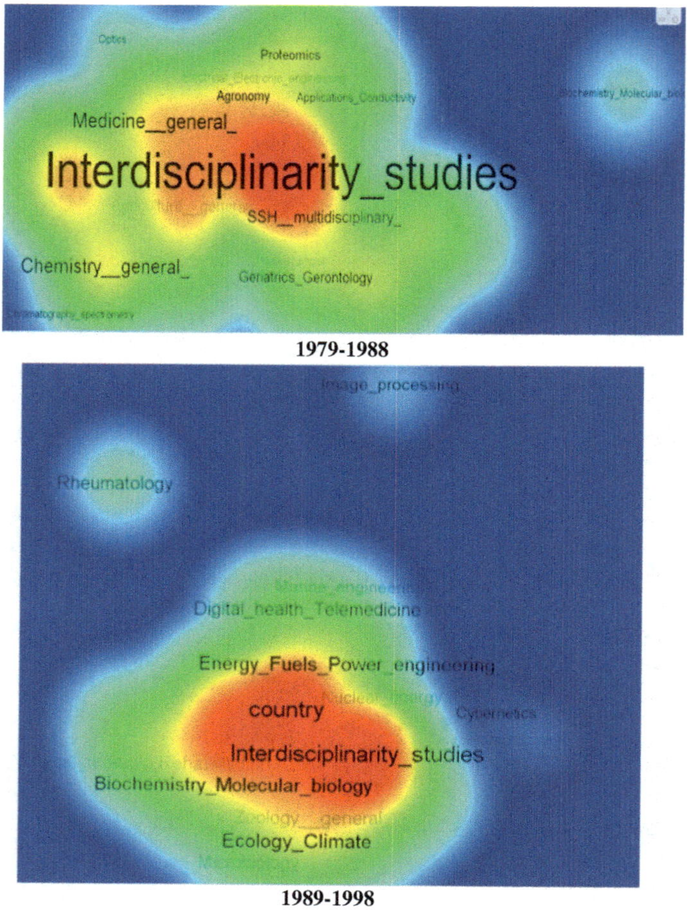


**Fig. 5.5** Interdisciplinarity in focus of Scientometrics

During this study, special attention was also paid to the cases where interdisciplinarity itself becomes the object of interest (see Fig. 5.5 for Scientometrics).

#### 5.6.1.1 Emerging Topics

As it was stated before, time-related issues are of high importance. Some examples which come into notice in terms of emerging ST&I topics were already provided (see Fig. 5.4 for Digital health in 1969–1978).

In the MoS Squared approach, time series are combined with the statistical analysis of results and the rankings of R&D fields. This makes it possible to extract knowledge on both emerging and mature trends. Entities on the top of the rankings represent a current agenda or a "big picture," while those taking lower positions are

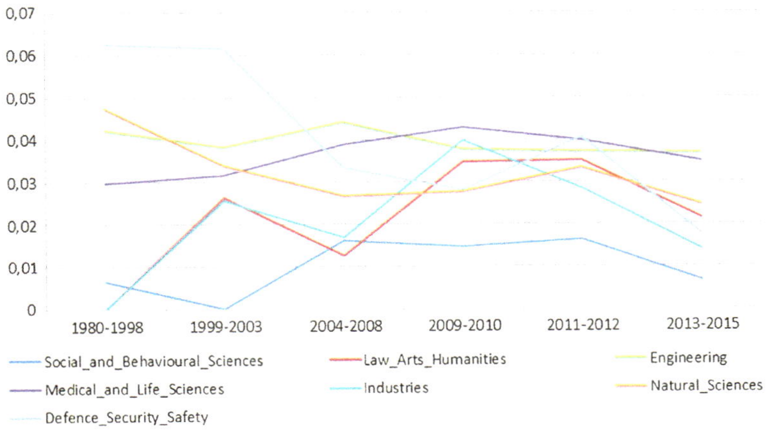


**Fig. 5.6** Major disciplines in Technology mining papers, upper level in hierarchy: time series

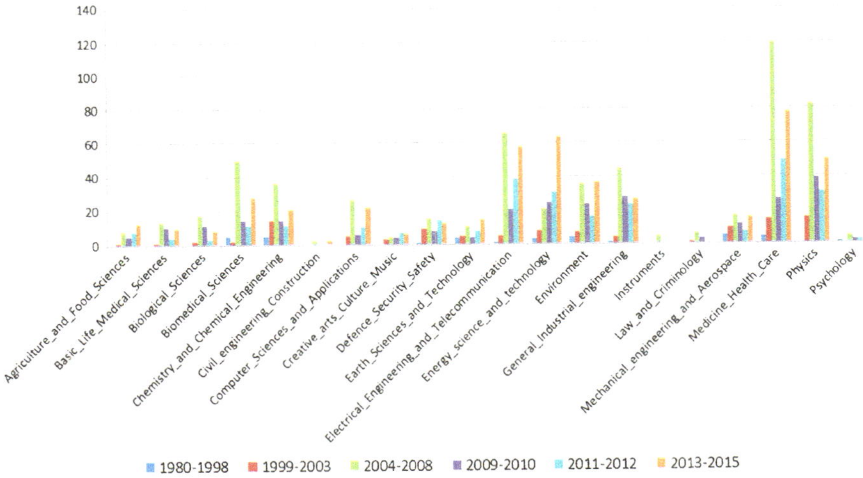


**Fig. 5.7** Top research areas in Technology mining papers, middle level in hierarchy

possibly related to “weak signals.” Some examples for the Technology mining collection are provided in Figs. 5.6, 5.7, 5.8 and 5.9.

For anticipating future pathways of ST&I, one should consider specific topics appearing in Information Extraction results. To test our hypotheses on the predictive power of scientometric and technology mining papers as input data, we have examined a number of cases. A couple of them are briefly discussed below.

One of the examples is related to *dye-sensitized solar cells* (“Renewable energy” node). The first developments related to these cells took place in the late 1960s and 1970s. However, it was after 2006 when a number of major discoveries were made. Important results were obtained in 2009 and 2010 followed by the developments of

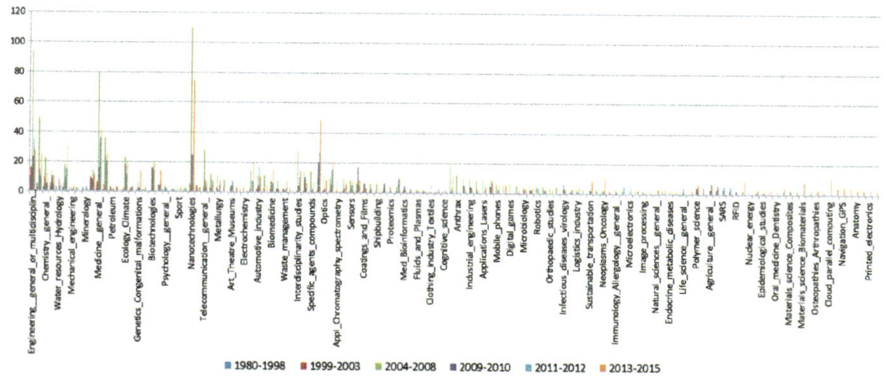


**Fig. 5.8** Top subareas of research and development in Technology mining papers, lower levels in hierarchy

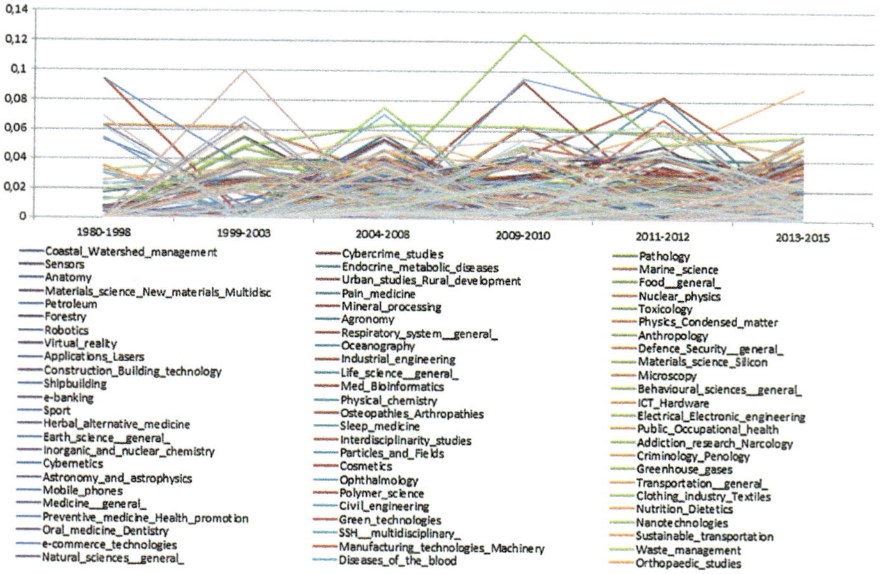


**Fig. 5.9** Subareas of research and development in Technology mining papers: time series

practical applications. The first papers on the dye-sensitized solar cells in the processed collections appear in 2009 in Technology mining collection (Guo et al. 2009) and in 2010 in Scientometrics collection (Shibata et al. 2010) with more papers appearing in later years. Since the technologies related to the dye-sensitized solar cells are still on the rise, we can state that the researchers in Technology mining and Scientometrics have "caught" evidence for promising technology at the stage of emergence.

Some pieces of evidence are provided by unexpected examples, such as the research on the R&D fields related to the Olympic torch which was published by Chinese authors in 2008.[9]

Another inspiring example is related to *Theranostics* which is considered to be a promising therapeutic paradigm in recent years. The first paper on Theranostics in Scientometrics collection appears in 2013 (Ho et al. 2013). In this year, Theranostics was not a brand-new concept already, but was, and still is, "climbing the peak." Such examples demonstrate the fact that papers in Scientometrics and Technology mining provide users with information on promising technologies not from the very conception (which would be impossible since it takes time for original papers to appear before such papers can be analyzed by scientometricians or Technology mining specialists), but still at early stages. Many examples provided in our results show that Scientometrics and Technology mining agenda includes technologies which are already more mature than those at the stage of a technology trigger, but are still on the rise—just what is of interest for anticipating future pathways of ST&I.

Some fields show interesting patterns in the dynamics of appearance, which may help users draw conclusions on external "motive forces" (Small et al. 2014). Examples include (for the collection in Scientometrics): *Terrorism research* (it appears in 1992 and then demonstrates upward trend and diversification, e.g., *Bioterrorism*), various types of *Influenza* (2009, 2011, 2012, 2013), *Whales* and *Whaling industry* (1982, then only starting from 2011), *Entomology* (in the context of diseases and Forensic Entomology: 1992, 2007, 2009, 2012, 2013), and many others. It could be an object of future research to provide an in-depth analysis of the drivers of studies in Scientometrics and Technology mining. It should help us understand what reasons are behind the fact that some fields are more often studied than the other ones, if it is caused by the availability of input data, "market pull" effect, researcher background, external events (see the example on the Olympic torch above) or other factors.

## 5.7 Conclusions

The study presented an approach and software for building the maps of science of a special kind as tools for anticipating future pathways of Science, Technologies, and Innovations. Research papers in Scientometrics and Technology mining were used as input data. The approach was called Map of Science Squared (the maps of ST&I through the lenses of Scientometrics and Technology mining). To test the approach, it was necessary to prove the "predictive power" of Scientometrics and Technology

[9]Huang L, Guo Y, Zhu DH (2008) Research of Olympic torch based on the methods of technology monitoring. *Journal of Beijing Institute of Technology (English Edition)*, 17 (SUPPL.), pp. 196–201.

mining papers by answering a number of questions. The most important ones were related to the breadth and the depth of Scientometrics and Technology mining agenda, as well as to timeliness with regard to the appearance of research papers. Additional issues covered locally bound topics and the co-occurrences of R&D fields.

The results presented show the validity of the approach. They show that both Scientometrics and Technology mining agendas have true potential for anticipating future pathways of ST&I. Some interesting findings on differences between two disciplines were not discussed in the chapter due to the lack of space. The most important task for future research is the one of further implementation of expert evaluation methods, primarily with respect to time models and the concept of emergence, motive forces which drive research in Scientometrics and Technology mining, as well as to the meaning of statistically significant, emerging and non-traditional links between R&D fields in scientometric and technology mining papers.

**Acknowledgments** This research is partially supported by the Russian Foundation for Basic Research, grant № 15-01-06819, "Research and Development of Ontological Models of the Centers of Excellence/Competence in the Emerging Areas of Science and Technology and their Identification Based on Heterogeneous Information Sources Processing."

## Appendix: Major R&D Fields

**Engineering sciences**:

- *Civil engineering and Construction*: Civil engineering; Construction and Building technology;
- *Electrical engineering and Telecommunication*:
  - *Telecommunication*: Telecommunication (general); Mobile phones; Mobile and wireless communication; Remote control; Underwater connection; Navigation/GPS;
  - *Transportation*: Transportation (general); Sustainable transportation; Shipbuilding;
  - *Other fields*: Automation and Control Systems; Electrical and electronic engineering; LED; Microelectronics; Printed electronics; RFID; Robotics;
- *Energy science and technology*: Energy and Fuels/Power engineering; Mineral processing; Nuclear energy; Petroleum; Renewable energy;
- *General and industrial engineering*: Industrial engineering; Manufacturing technologies and Machinery; Rock engineering;
- *Instruments*: Microscopy;
- *Mechanical engineering and Aerospace*: Acoustics; Aerospace engineering;
- *Other fields*: Engineering (general or multidisciplinary); Biometrics; Bionics; Biotechnologies; Grid technologies; Hydraulics_Pneumatics; Metrology; Nanotechnologies;

**Industries** (*examples*): *Automotive industry*; *Clothing industry, Textiles*; *Insurance industry*; *Logistics industry*;

**Language, Information and Communication** (*examples*): Literary studies;

**Law, Arts and Humanities** (*examples*):

- *Creative arts, Culture and Music*: Art, Theatre, Museums; Folklore; Musicology;
- *History, Philosophy and Religion*: Archaeology; Religion;
- *Law and Criminology*: Criminology and Penology; Cybercrime studies;

**Medical and Life Sciences**:

- *Agriculture and Food sciences*: Agriculture (general); Agricultural engineering; Agronomy; Dairy and Animal science; Food (general); Oenology_Wine making; Oils; Rice studies; Seed industry;
- *Basic Life/Medical sciences*: Life science (general); Biochemistry/Molecular biology; Biophysics; Cell biology; Materials science: Biomaterials; Medical/Bioinformatics; Microbiology;
- *Biological sciences*:
  - *Zoology*: Zoology (general); Mammalogy (general); Specific mammals (examples);
  - *Other fields*: Biology (general and multidisciplinary); Entomology; Ethnobiology; Evolutionary biology; Horticulture; Ichthyology/Fisheries; Invasion biology; Mycology; Ornithology; Plant sciences; Tropical biology;
- *Biomedical Sciences*:
  - *Pharmacology and Pharmacy*: Pharmacology and Pharmacy (general); Cosmetics; Theranostics; Specific drugs;
- *Medicine/Health Care*:
  - *Allergy and Immunology*: Allergy and Immunology (general); Vaccination;
  - *Diseases of the circulatory system*: Cardio- and cerebrovascular diseases; Raynauds phenomenon;
  - *Diseases of the genitourinary system*: Andrology; Disorders of breast; Gynecology; Nephrology; Urology;
  - *Diseases of the musculoskeletal system and connective tissue*: Disorders of muscles; Orthopaedic studies; Osteopathies/Arthropathies; Rheumatology; Soft tissues;
  - *Diseases of the respiratory system*: Respiratory system (general); Otorhinolaryngology; Pulmonary medicine; Smoking research;
  - *Endocrine, nutritional and metabolic diseases*: Diabetes; Endocrine and metabolic diseases; Nutrition and Dietetics; Obesity research;
  - *Gastroenterology*: Gastroenterology (general); Hepatology;
  - *Infectious and parasitic diseases*: Infectious diseases/Virology (general); Epidemiological studies; Parasitology; Tropical medicine; African sleeping

sickness; Anthrax; Chagas disease; Ebola; e-coli; HIV/AIDS/retrovirus; Influenza; Leishmaniasis; Malaria; SARS; Yellow fever; Zoonoses;
- *Mental health and behavioural disorders*: Addiction research and Narcology; Alcohol addiction; Mental health; Sexology;
- *Neurology*: Neurology (general); Pain medicine; Sleep medicine;
- *Paediatrics*: Paediatrics (general); Adolescent Medicine;
- *Other fields*: Medicine (general); Allied health and Nursing; Anesthesiology; Biomedicine; Digital health and Telemedicine; Disability; Diseases of the blood; Diseases of the skin, Dermatology; Emergency medicine; Forensic medicine; Genetics and Congenital malformations; Geriatrics and Gerontology; Herbal and Alternative medicine; Hospital service; Neoplasms and Oncology; Non-invasive treatment; Nuclear medicine and Medical imaging; Ophthalmology; Oral medicine and Dentistry; Orofacial pain; Palliative medicine; Pathology; Physical therapy and Rehabilitation; Preventive medicine, Health promotion; Primary care, General practice, Family medicine; Proteomics; Public and Occupational health; Regenerative medicine/Synthetic biology; Reproductive health, Childbirth and Neonatology; Serology; Surgery, Anaplasty; Traffic medicine and Traumatology; Transplantation, Prostheses, Implantable devices; Venereology; Veterinary sciences;

- *Cognitive science*;

**Natural Sciences**:

- *Natural sciences (general)*;
- *Astronomy and Astrophysics*;
- *Chemistry and Chemical Engineering*: Chemistry (general); Aerosol; Analytical chemistry; Catalysis; Chemometrics/Chemoinformatics; Electrochemistry; Inorganic and nuclear chemistry; Materials science: Textiles; Organic chemistry; Physical chemistry; Polymer science; Specific agents/compounds;
- *Computer Sciences and Applications*: Cloud and parallel computing; Cybernetics; Data storages; Digital games; e-banking; e-commerce technologies; ICT, Hardware; Image processing; Public e-services; Speech technologies; Virtual reality;
- *Earth Sciences and Technology*: Earth science (general); Geochemistry and Geophysics; Geodesy; Geography, Geosciences (multidisciplinary); Geology; Marine engineering; Meteorology and Atmospheric sciences; Mineralogy; Mountain science; Oceanography; Paleontology; Sensors; Soil science;
- *Environment*: Aquatic sciences (general); Arctic/Antarctic studies; Biodiversity; Coastal and Watershed management; Disaster reduction (Tsunami, Earthquake, Famine, Fires, etc.); Ecology, Climate; Forestry; Green technologies; Greenhouse gases; Limnology; Marine science; Ozone layer studies; Urban studies and Rural development; Waste management; Water resources and Hydrology;

- *Physics*: Physics (general); Aerodynamics; Applications: Chromatography/Spectrometry; Applications: Conductivity; Applications: Lasers; Atomic, molecular, chemical physics; Coatings and Films; Crystallography; Fluids and Plasmas; Materials science: Ceramics; Materials science: Composites; Materials science: New materials, Multidisciplinary; Materials science: Silicon; Metallurgy; Nuclear physics; Optics; Particles and Fields; Physics, Condensed matter;

**Social and Behavioral Sciences** (*examples*):

- *Social sciences and Humanities (multidisciplinary);*
- *Social and Behavioral sciences (multidisciplinary)*: Behavioural sciences (general); Violence studies;
- Educational sciences;
- *Economics and Business*: Family business;
- *Political science and Public administration*: Postwar studies;
- *Psychology*: Psychology (general); Evolutionary psychology—Altruism studies; Life-threatening behavior; Suicidology;
- *Sociology and Anthropology*: Anthropology; Ethnic studies; Family studies; Hospitality, Leisure, Tourism; Sport;

**Defence, Security, Safety**: *Defence, Security (general)*; *Biosecurity*; *Nuclear safety*;

**Inter-_transdisciplinarity studies**.

## References

Börner, K., Boyack, K. W., Milojević, S., & Morris, S. (2012a). An introduction to modeling science: Basic model types, key definitions, and a general framework for comparison of process models. In A. Scharnhorst, K. Börner, & P. van den Besselaar (Eds.), *Models of science dynamics: Encounters between complexity theory and information science*. Springer.

Börner, K., Klavans, R., Patek, M., Zoss, A. M., Biberstine, J. R., Light, R. P., et al. (2012b). Design and update of a classification system: The UCSD map of science. *PLoS ONE, 7*(7), e39464. doi:10.1371/journal.pone.0039464

Boyack, K. W., & Klavans, R. (2010). Co-citation analysis, bibliographic coupling, and direct citation: Which citation approach represents the research front most accurately? *Journal of the American Society for Information Science and Technology, 61*(12), 2389–2404.

Boyack, K. W., & Klavans, R. (2014). Creation of a highly detailed, dynamic, global model and map of science. *Journal of the Association for Information Science and Technology, 65*(4), 670–685.

Boyack, K. W., Small, H., & Klavans, R. (2013). Improving the accuracy of co-citation clustering using full text. *Journal of the American Society for Information Science and Technology, 64*(9), 1759–1767.

Chen, C. M., & Leydesdorff, L. (2014). Patterns of connections and movements in dual-map overlays: A new method of publication portfolio analysis. *Journal of the Association for Information Science and Technology, 65*(2), 334–351.

Efimenko, I., Khoroshevsky, V., & Ena, O. (2014). Peaks, slopes, canyons, plateaus: Identifying technology trends throughout the life cycle. In *Proceedings of 4th Annual Global TechMining Conference, GTM-2014*, Leiden, Netherlands, 2014

Efimenko, I. V., & Khoroshevsky, V. F. (2014). *New technology trends watch: An approach and case study* (No. 8722, pp. 170–177). *Lecture Notes in Computer Science (subseries Lecture Notes in Artificial Intelligence).*

Garfield, E. (2001). From bibliographic coupling to co-citation analysis via algorithmic historio-bibliography: A citationist's tribute to Belver C. Griffith. *A paper presented at the Drexel University*, Philadelphia, PA, November 27, 2001.

Guo, Y., Huang, L., & Porter, A. L. (2009). Profiling research patterns for a new and emerging science and technology: Dye-sensitized solar cells. In *Atlanta Conference on Science and Innovation Policy, ACSIP 2009.*

Ho, D. N., Choi, K. Y., & Lee, S. J. (2013). Bibliometric analysis of theranostics: Two years in the making. *Theranostics, 3*(7), 527–531.

Huang, Y., Zhang, Y., Porter, A., Youtie, J., Kay, L., & Zhu, D. (2015). Funding proposal overlap mapping: A tool for science and technology management. In *Proceedings of the GTM 2015.*

Khoroshevsky, V. (2009). Ontology driven multilingual information extraction and intelligent analytics. In *Proceedings of NATO Advanced Research Workshop on Web Intelligence and Security*, November 18–20, 2009 in Ein-Bokek, Israel

Klavans, R., & Boyack, K. W. (2011). Using global mapping to create more accurate document-level maps of research fields. *Journal of the American Society for Information Science and Technology, 62*(1), 1–18.

Leydesdorff, L. (1987). Various methods for the mapping of science. *Scientometrics, 11*, 291–320.

Marshakova-Shaikevich, I. (1973). System of document connections based on references. *Scientific and Technical Information Serial of VINITI, 6*(2), 3–8.

Piskorski, J., & Yangarber, R. (2013). Information extraction: Past, present and future. In T. Poibeau et al. (Eds.), *Multi-source, multilingual information extraction and summarization, theory and applications of natural language processing* (pp. 23–49). Heidelberg: Springer. doi:10.1007/978-3-642-28569-1_2

Porter, A. L, & Cunningham, S. W. (2010). *Tech mining: Exploiting new technologies for competitive advantage* (Chinese edition 2010). New York: Wiley, 2005.

Porter, A. L., & Newman, N. C. (2011). Mining external R&D. *Technovation, 31*(4), 171–176.

Porter, A. L., & Zhang, Y. (2015). Tech mining of science & technology information resources for future-oriented technology analyses. In J. C. Glenn & T. J. Gordon (Eds.), *Futures research methodology version 3.1. The millennium project.* Washington, DC

Rip, A., & Courtial, J. P. (1984). Co-word maps of biotechnology: An example of cognitive scientometrics. *Scientometrics, 6*(6)

Robinson, D. K. R., Huang, L., Guo, Y., & Porter, A. L. (2013). Forecasting innovation pathways for new and emerging science and technologies. *Technological Forecasting and Social Change, 80*(2), 267–285.

Shibata, N., Kajikawa, Y., & Sakata, I. (2010). Extracting the commercialization gap between science and technology—Case study of a solar cell. *Technological Forecasting and Social Change, 77*(7), 1147–1155.

Small, H. (1973). Co-citation in the scientific literature: A new measure of the relationship between two documents. *Journal of the American Society for Information Science, 24*, 265–269. doi:10.1002/asi.4630240406

Small, H., Boyack, K. W., & Klavans, R. (2014). Identifying emerging topics in science and technology. *Research Policy, 43*(8), 1450–1467.

van Eck, N. J., & Waltman, L. (2010). Software survey: VOSviewer, a computer program for bibliometric mapping. *Scientometrics, 84*(2), 523–538.

Waltman, L., van Eck, N. J., & Noyons, E. C. M. (2010). A unified approach to mapping and clustering of bibliometric networks. *Journal of Informetrics, 4*(4), 629–635.

Wimalasuriya, D. C., & Dou, D. (2010). Ontology-based information extraction: An introduction and a survey of current approaches. *Journal of Information Science, 36*(3), 306–323.

Ye, C., & Feng, L. (2013). Future-oriented technology analysis of technology roadmap based on text mining (pp. 1126–1130). In *Proceedings of the 10th International Conference on Fuzzy Systems and Knowledge Discovery (FSKD)* Shenyang, Peoples Republic of China.

# Part II
# Text Analytic Methods

# Chapter 6
# Towards Foresight 3.0: The HCSS Metafore Approach—A Multilingual Approach for Exploring Global Foresights

**Stephan De Spiegeleire, Freija van Duijne and Eline Chivot**

**Abstract** The policymaking community puts ever more emphasis on basing policy on rigorously collected and curated objective evidence ('evidence-based policy'). But what is the equivalent for the future of what 'evidence' is for the past and present? This paper presents some examples of what we call 'Foresight 3.0': an attempt to distil more insights about the entire futurespace by systematically collecting, parsing, visualizing and analysing a large database with elements of the future (*futuribles*) as they are perceived across the globe. This global 'futurebase' allows policy analysts and policymakers to gauge the bandwidth of views on these *futuribles* across different constituencies (and different languages and cultures). The hope is that such an approach will allow us to transcend some of the widely acknowledged bias problems with many current approaches to foresight. This paper introduces one of the approaches *The Hague* Centre for Strategic Studies pursuing to move the current debate about foresight for strategic planning in more of a 3.0 direction: the HCSS *Metafore* approach. The paper briefly presents the main steps in the Metafore research protocol that was used in some recent HCSS work for the Dutch government's 'Strategic Monitor', which tries to anticipate the future in the area of foreign, security and defence policy. The essence of the protocol is that we try to collect a much larger corpus of foresight studies in a particular field than has hitherto been possible and then code these with both manual (for the smaller sets) and semi-automated (for the larger sets) coding tools. The results are then visualized in different ways and analysed. We describe the protocol and provide some

---

Our fascination with the future is only surpassed by our inability to grasp it.

---

S. De Spiegeleire (✉) · E. Chivot
The Hague Centre for Strategic Studies (HCSS), The Hague, The Netherlands
e-mail: sdspieg@gmail.com

E. Chivot
e-mail: e.c.m.chivot@gmail.com

F. van Duijne
Ministry of Economic Affairs, The Hague, The Netherlands
e-mail: f.h.vanduijne@minez.nl



T.U. Daim et al. (eds.), *Anticipating Future Innovation Pathways Through Large Data Analysis*, Innovation, Technology, and Knowledge Management,
DOI 10.1007/978-3-319-39056-7_6

illustrative examples. The paper concludes by discussing some of the strengths and weaknesses of such a 'Metafore' approach for strategic policymaking.

**Keywords** Foresight · Text mining · Metafore protocol

## 6.1 Introduction

Foresight is an essential aspect of strategy development. Foresight helps organizations in anticipating change, in navigating their often turbulent environments and in making decisions that offer a better chance at keeping them ahead of others. Many public and private organizations are aware that new insights can be gained from looking at the future in a more systematic way. They see value both in the deliverables and in the very process of engaging groups of people, widening their views, having discussions among decision-makers and influencing the thinking processes within the organization which may spread out to actors in their transactional context.

But while the value of foresight is widely (and—in our experience—increasingly) acknowledged, there are also a number of equally widely accepted limitations. Many of those have to do with the subjectivity of the process. In the past, foresight was often done by individual experts or by prestigious institutions ('Foresight 1.0'). The purpose of these foresight studies was typically to produce an 'authoritative' view of 'the' future. And their (presumed) quality was derived from the singular authority the expert(s) wielded in their field.

In the last quarter of the previous century (and in some fields, like defence and security, even earlier than that), the field gravitated to what could be called Foresight 2.0—in an analogy to 'Web 2.0', which became more interactive than the original static 'Web 1.0'. Foresight 2.0 exercises (or specialized foresight businesses like GBN or IFTF) typically bring together a diverse group of people from different backgrounds and with different points of view in order to try to make sense of the future through some Delphi-like techniques (Rowe and Wright 2011). The output of these studies is often a small set of scenarios that provided not a single authoritative view of the future as in Foresight 1.0, but a small number of plausible futures based on few key dimensions (e.g. the still very popular 2 × 2 matrices). The quality of such foresight exercises is presumed to result not from the singular authority of a person or an institute, but from the diversity of high-quality inputs by different experts.

In reality, it is extremely difficult for most organizations (both logistically and mentally) to replicate genuine diversity in such Foresight 2.0 exercises. The diversity that is ultimately drawn upon tends to be 'nearby diversity' (that subset of the broader diversity that organizations know or can invite) or 'simulated diversity' (with 'token' outsiders). This paper proposes an approach that fits in what we call 'Foresight 3.0'. Again in analogy to Web 3.0 (the semantic Web), Foresight 3.0 takes advantage of our increasing ability to tap into the enormous global amount of text-based multilingual information—also about various aspects of the future—that is available on the Internet, as well as of our increasing ability to extract semantic meaning from such large corpora of texts. The output of this approach is an even richer insight across not just a few scenarios as in Foresight 2.0, but across a much broader scenario space along a larger set of dimensions for each of which the approach tries to sketch (and analyse) the bandwidth of available views.

*The Hague* Centre for Strategic Studies (HCSS) have been working on such a 'Foresight 3.0' approach. This paper will start by introducing the idea and the rationale behind this approach. It will then describe the method used by HCSS and illustrate it with some examples from foresight studies published in 2012 in the field of security. It concludes by discussing the potential of metaforesights for strategic policymaking.

## 6.2 Metaforesight for Evidence-Based Policymaking

The idea that policymaking should become increasingly 'evidence-based' is becoming ever more engrained in the minds of policymakers across the globe.[1] The main ambition behind this is to help 'people make well informed decisions about policies, programmes and projects by putting the best available evidence from research at the heart of policy development and implementation' (Davies 2004, p. 3).

But facts and evidence are by definition about the past or—at best—the present. So what about the future? There are no 'data', no 'facts' and no 'evidence' about the future. We are bombarded daily with various diverging views, visions and opinions about the future; with numerous quantitative attempts to extrapolate data about the future from data about the past (an endeavour that has proved quite perilous in many policy domains); and with various methodologies that the foresight community applies to 'vision' different futures. Although many of those are inspiring, challenging, threatening or amusing, we have no ways of empirically validating their reliability ex ante. The few serious studies that have tried to assess their reliability ex post have arrived at quite bleak findings.[2] How then can we attempt to deal with the future in as dispassionate and rigorous a way as possible? And—most importantly—in a way that is actually useful for decision-makers. What is the equivalent for the future of what 'evidence' is for the past and present?

*The Hague* Centre for Strategic Studies have been looking for ways to present decision-makers with a balanced and informative overview of different insights about the future. We try not to take sides in the many substantive, ideological, methodological and political debates that permeate discussions about the future and the 'futures' field. We are constantly and painfully reminded of the various

[1]But note the caveats by (Pawson 2006).

[2]Gardner (2011), Tetlock (2005), see also Banerjee (1992), Batchelor (2007, 2011), Dovern and Weisser (2011), Franses et al. (2012), Henry (1989), Lamont (1995) and Stekler 2007).

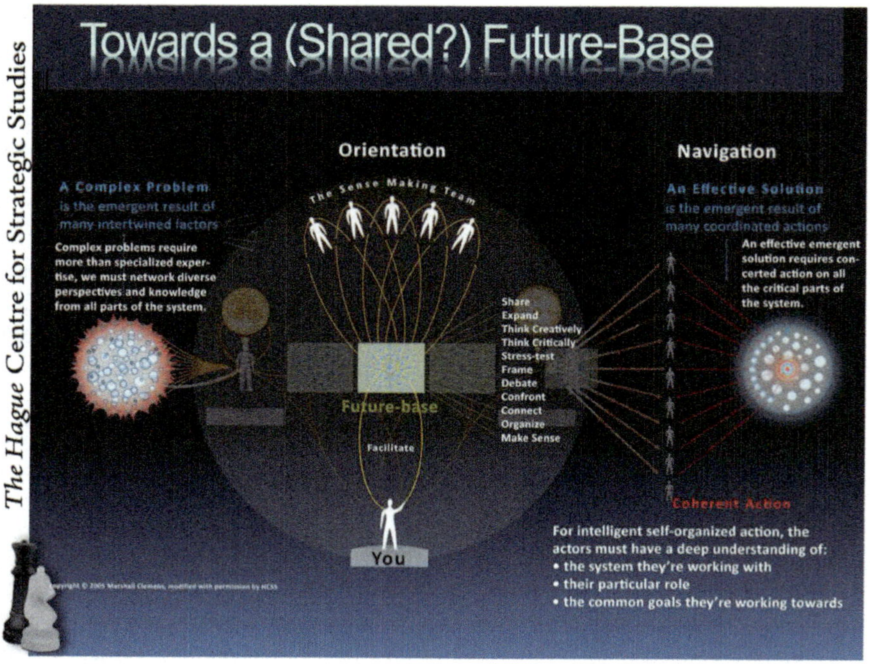


**Fig. 6.1** Towards a futurebase

well-known pathologies that we humans encounter when we try to wrap our minds around the future.[3] We have therefore developed a method to collect, process and visualize a large number of foresight studies in multiple language domains with the aim of mapping the bandwidth of views about the future in various policy areas. We have called this approach 'Metafore': 'moving beyond' (Greek: meta-ferein) trying to predict the future (forecasting) towards critically but constructively developing and curating a more intellectually modest and honest database of various diverse insights about the future culled from a variety of different methodological approaches, academic disciplines, ideological schools and cultural or linguistic backgrounds ((meta)foresighting). We call such a 'database' about the future a 'futurebase' (Fig. 6.1).

The main ideas behind the development of a broader shared database of foresight studies (futurebase) are related to recent insights from the field of complexity theory.[4] Many of today's most burning policy challenges are complex in the sense

[3]Examples of these pathologies include presentism and recentism; lack of imagination; herd mentality; overconfidence in data sets; underestimating 'framing' effects; reductionism and underappreciating system effects; dubious timing; stovepiping; systematic normative bias (e.g. in some areas in favour of 'gloom and doom'; in others in favour of unbridled optimism); analytical hybris; and 'masking' uncertainty (De Spiegeleire et al. 2010).

[4]For an introduction to these insights, see Mitchell (2009) and Page (2011).

that they are the emergent result of the (often poorly or at least incompletely understood[5]) interaction between many intertwined actors and factors. This then also implies that the solutions to these challenges are likely to be the emergent result of the actions of a variety of different actors that are directly or indirectly involved. These actors are bound to have their own idiosyncratic views and perspectives on what the future might bring and what their role might be in it. Although many (none more than governments or international organizations) may perceive themselves as being at the centre of that constantly self-reconfiguring complex ecosystem, very few—if any—are likely to end up there. But some key actors might still be able to position themselves more centrally by mapping different elements of the future (we call these elements *futuribles*[6]), by depositing and curating them in a futurebase, by putting this futurebase at the disposal of a broader audience and by thus nudging (Thaler and Sunstein 2009) their own ecosystem to become more future-oriented and possibly even more future-proof.

To illustrate what such a futurebase might look like, the remainder of this paper will document one of the ways in which HCSS has been using the Metafore approach in one particular context: (multilingual) metaforesight in 'national security'. This work is based on one of the contributions HCSS made to the Dutch government's 2012 Strategic Monitor (Bekkers et al. 2013), a yearly public interdepartmental effort by the Ministry of Defence, the Ministry of Foreign Affairs and the Ministry of Security and Justice to sketch some of the key new future developments in the area of 'national security'. This paper just presents some of the main methodological steps with some illustrations; readers who are interested in the full analysis are invited to take a look at the full HCSS Strategic Monitor (Bekkers et al. 2013).

## 6.3 The HCSS 'Metafore' Protocol

HCSS maintains a database with security-related foresight studies in different language domains. In 2013, the Dutch Ministry of Defence asked HCSS to augment this database with security-relevant foresight studies published in 2012 in a number of different language domains and to systematically compare the views they contain about the future of the international system from a security perspective. The languages selected included English, French, German, Russian, Chinese and Turkish in order to not only provide an overview of 'Western' perspectives, but also from other major regions that already play and/or are likely to play an important role in the future global security environment.

[5]And—according to some scholars—even fundamentally defying (a priori) comprehension.

[6]We were inspired in some of this work by French futurist Bertrand De Jouvenel's concept of '*futuribles*' (de Jouvenel 1964). See also De Jouvenel (1965) and Malaska and Virtanen (2005, 2009).

### 6.3.1 The Team

HCSS assembled a multilingual team to conduct this analysis. Our team members were located in the Netherlands (English, French, German, Romanian languages), Ukraine (Russian language), the USA and Singapore (Chinese) and Ankara (Turkish). The team coordinated its activities entirely online through various communication methods, in particular Rizzoma, a Web-based collaboration environment.

### 6.3.2 The Collection Process

The HCSS team collected a broad set of publicly available foresight studies on the future of the international security environment for the various aforementioned language domains. These studies were published between 2011 and 2012 by diverse sources, including governmental ones (national military doctrines, security strategies, defence reviews, foresight), international organizations, academia (books and periodicals), research institutes, substantive statements by public intellectuals, industry representatives and religious leaders. Each team member carried out research on the Internet using search engines with an analogous search algorithm that was defined and translated for each language domain. The algorithm consisted of four main 'semantic baskets':

- '**Security**' as a central search term; 'defence'; 'security environment'; 'national security'; 'international security';
- '**The future**': words such as 'foresight', 'forecast', 'scenarios', 'trends', 'drivers', 'in the future', 'twenty-first century', etc.;
- '**International**': the study or document had to deal with the broader international security environment, for which we used the world 'global';
- '**Policy**': to ensure the studies were also policy-relevant, we also included the word policy.

Although in subsequent search iterations some adjustments were required to obtain more or better results, the objective always remained to yield comparable results across language domains and to avoid substantive biases. Google (including Google Scholar) was the main search engine that was used for all language domains, but on top of it, the team also used Yandex and Rambler for Russian and Baidu for Chinese studies. In addition, the team searched the full-text academic databases and the websites of some key research institutes, think tanks or governmental agencies resources.

In addition to the search queries, the types of sources and the specified time horizon, we also applied some additional criteria to make sure the final selection contained fairly 'serious' studies: their length had to be at least over ten pages and they had to contain an explicit (or at least implicit) methodology and a clear analytical line of argumentation. Documents that fulfilled all of these criteria were

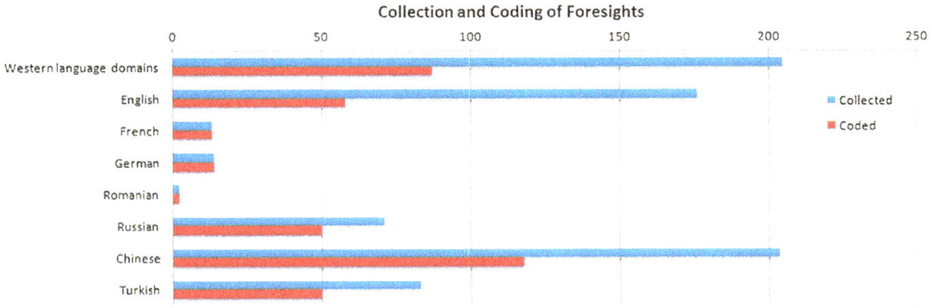


**Fig. 6.2** Collection and coding of foresights by language domain

downloaded and added to the HCSS Metafore Futurebase. The number of studies to be downloaded was about a hundred per language domain.

As shown in Fig. 6.2, the 205 studies selected for the 'Western' perspectives on security (i.e. English, French, German and Romanian) included 176 English studies, 13 French studies, 14 German studies and 2 Romanian studies. The Russian set contained 71 studies, the Chinese 204 studies and the Turkish 83 ones.

HCSS worked hard to avoid systematic selection biases. But there are still a number of 'short cuts' we use that make it hard to ascertain the actual representativeness of the final set of documents. First of all, the sources were limited to a few languages[7] and to public domain sources. Second, various heuristics and biases, such as faddishness, presentism and recentism, or cultural and other 'framing effects', are extremely hard to control for in the search queries.[8] Finally, any Internet-based search today depends to large degree on the ranking algorithms that are used by the main search engines (like Google).

The HCSS team also affixed some metadata to every study in the futurebase. One of those metadata is the organization that conducted the study. Publications from universities were labelled as 'Academia'. 'International/multilateral organizations' include EU institutions and the aforementioned organizations such as NATO or the UN. 'Private sector' includes reports published by a broad diversity of firms. Think tanks and knowledge institutes were included in the category 'research institutes'. Figure 6.3 shows the breakdown of these sources by who authored the study.

Another type of metadata the team manually added to the database for each study is what substantive area it focused in. Although we were primarily looking for 'general' security foresight studies, our search queries also yielded a number of foresights on more specific topics that still contained at least one passage dealing with security. Previous HCSS research has shown us that such more oblique

[7]The Turkish search effort, for instance, posed specific challenges, as scenarios with a 2020–2050 time horizon are rare in the Turkish language domain, especially in the security field. We have encountered similar problems in an anterior study on the Contours of Future Conflict for the Arabic language domains. See De Spiegeleire et al. (2011).

[8]See footnote 5.

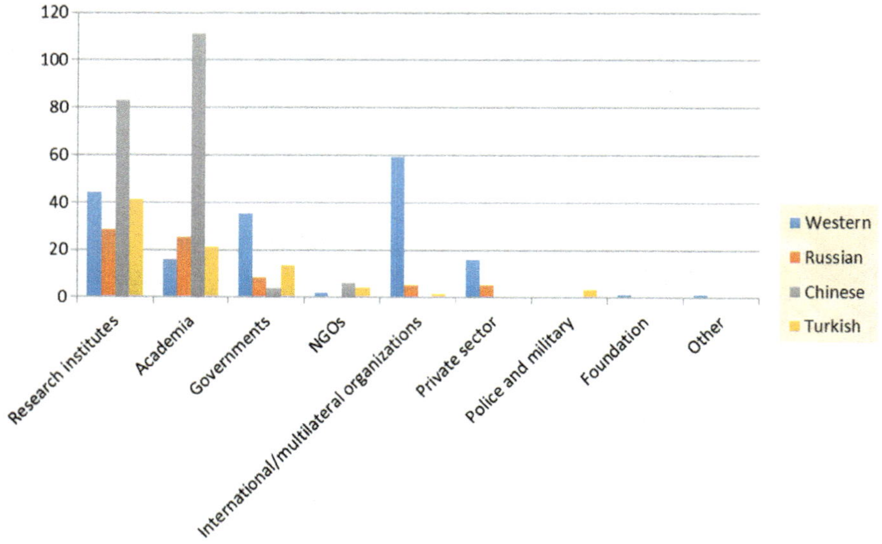


**Fig. 6.3** Western, Russian, Chinese and Turkish foresights by type of organization

references are often quite 'interesting' and often enrich the insights from the more general studies. Figure 6.4 shows the studies broken down by topics across the different language domains based on the metadata entered by our team in the Metafore database.

### 6.3.3 *The Coding Process*

The identified foresights were then subjected to both a manual and a semi-automated coding effort.

### 6.3.4 *Manual Coding*

For the manual coding, the team used a collaborative coding software program called Dedoose. The objective was to carefully tag the relevant passages from these texts based on a single coding scheme that contained some important categories of *futuribles* (elements of the future that we are interested in—e.g. a certain region or a certain theme). This allowed us then to systematically compare the relative importance of such *futuribles* across languages.

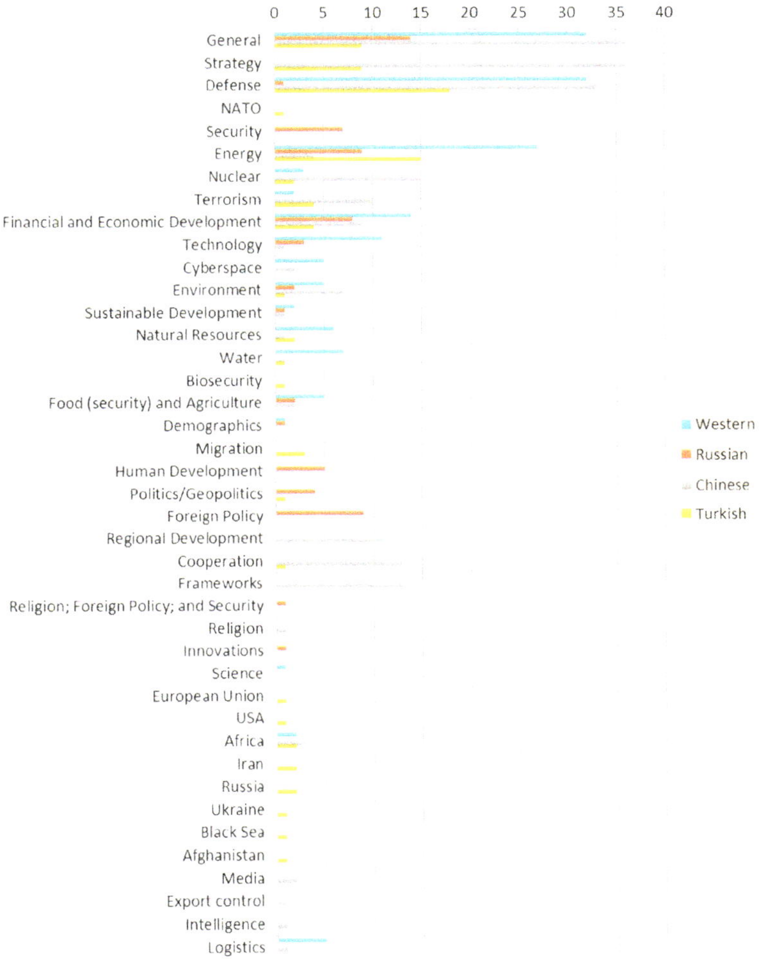


**Fig. 6.4** Western, Russian, Chinese and Turkish foresights by topic

#### 6.3.4.1 Coding Scheme

All team members labelled ('coded') all relevant specific passages describing key future security developments across the selected literature in Dedoose based on a singe coding scheme, the highest level of which is presented in Table 6.1.

'**Key developments**' are the full selected text fragments that were tagged in Dedoose. They had to be future-oriented and could be of a negative (security dangers, threats, risks), positive (opportunities, positive trends) or neutral nature. Past and strictly current events were not coded. Key developments were labelled according to their main topics—for instance: 'the costs of environmental degradation', 'the rise of new and/or major powers' or 'efforts and investments in renewable resources'. This yielded a long list of 'key developments'.

Table 6.1 The coding scheme

| |
|---|
| Key developments |
| Actors |
| Regions |
| Drivers |
| Domains |

'**Actors**' are the main parties mentioned in various text fragments that may get involved in future security developments. To facilitate the analysis, we coded either 'state actors' or 'non-state actors'. Wherever applicable, the team also labelled the anticipated level of cooperation between these actors as either 'more cooperative' or 'less cooperative'.

'**Regions**' are the geographical areas where key developments occur and which are thought to be strategically important. If a foresight *only* dealt with Japan or Kazakhstan or was very much focused on a particular region of strategic importance (e.g. the Black Sea region), the team coded and labelled these regions as such.

'**Drivers**' are forces that are likely to trigger changes in the future and generate elements of the key development. As we were coding, we defined a number of high-level drivers with a number of subdrivers. To give an example, for the 'economic' driver, subdrivers included 'unemployment', 'economic growth', 'economic infrastructure', 'financial/debt crisis', 'austerity', 'poverty' or 'market, monetary and prices (in)stability'.

'**Domains**' are the primary (substantive, not geographical) arenas in which the key development may manifest itself. We started out with the categorization we also used in the HCSS study *Contours of Conflict in the 21st Century*: (De Spiegeleire et al. 2011):

- Security dimension, related to military conflicts, including the following:
  - Traditional military dimension (land, sea, air);
  - Modern military dimension (space, cyberspace).
- Political dimension,
- Economic dimension,
- Human terrain (societal, social, mental, moral, psychological dimension. For example, national identity).

In the process, the following additional types of domains were added:

- Legal (laws, regulations, policies, institutions),
- Environment
- Energy (oil, nuclear, renewable, solar),
- Global logistics and trade,
- Technology (cyber),

- Culture/history,
- Vital elements (food, water, health),
- Nuclear weapons as a subdomain of the 'security dimension',
- Other security and conflict domain (e.g. terrorism).

The results of the coding were then consolidated and classified by Dedoose in a tabular format that allowed us to generate a synoptic visualization of the findings.

#### 6.3.4.2 Textual Example

To illustrate this on the basis of a concrete example, let us take a closer look at one particular excerpt and at how it was coded in Dedoose:

> Retirement ages are already rising in Western economies as people are living longer and funding public pensions proves too much of a burden on fiscal positions. There could also be government incentives to try and raise the fertility rate. Russia has recently announced a scheme whereby couples producing three children or more are entitled to a certain amount of land. This again highlights the uncertainty around forecasting this far into the future (Ward 2011).

- This entire text fragment is coded as the '**key development**'. The main topic analysed here is labelled as 'demographic outlook'.
- **Driver(s)**: Demographic (high-level driver), with 2 subdrivers: (1) ageing and (2) fertility rates. Ageing is coded because '*people **are living** longer*' (see the use of tense = to be + ing: suggests an ongoing trend).
- Actor(s):
  1. Which ones? state actor ('*government*') or non-state actor ('*people*').
  2. Are they adopting a more or less cooperative attitude? This is not shown in this text fragment here.
- **Region(s)**: Developed world ('*Western economies*'), Russia.
- **Domains(s)**: Economic ('funding public pensions').

#### 6.3.4.3 Visual Example

To give just one example, Fig. 6.5 shows the breakdown across the different language domains for the key actors in international security and whether these actors are more likely to be cooperative or non-cooperative. We observe, for instance, that state actors are more frequently mentioned than non-state actors and that cooperative attitudes are expected to prevail over non-cooperative ones, regardless of the type of actor. These *two key overall findings* also appear to be remarkably *consistent across all language domains.*

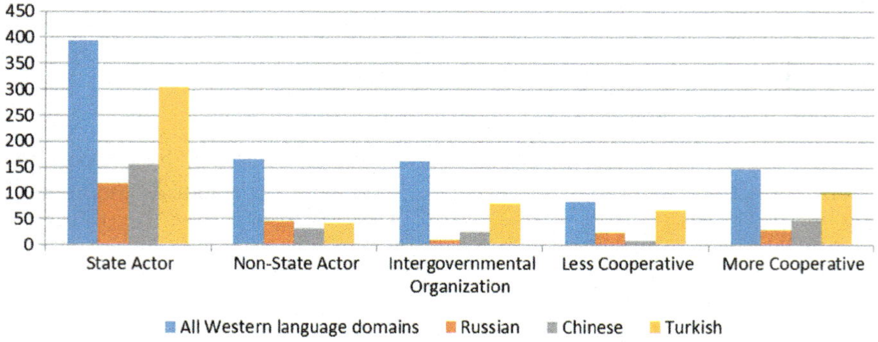


**Fig. 6.5** Actors and cooperation: manual coding results by language domain

### *6.3.5 Automatic Coding with Text Mining*

Alongside the manual coding of the identified security foresight studies, the team also used a few automated tools to text mine these studies.[9]

#### 6.3.5.1 Leximancer

Leximancer is a commercial text mining tool ('Leximancer', n.d.; Smith 2000, 2003; Smith and Humphreys 2006) that in essence builds a semantic network out of the co-occurrences of terms in certain sets of documents.[10] These themes are clusters of concepts that tend to travel together through this entire set of texts. In a sense, Leximancer does with words what social network sites such as Facebook or LinkedIn do with people. When people befriend each other, Facebook's graph draws a line between each of them. As people start connecting to more additional contacts, Facebook starts building up a denser network of associations between them all.

After some initial cleaning of texts (e.g. by identifying sentences; by ignoring frequently used words such as 'the' or 'it'; or by merging various grammatical versions of the same word such as 'link', 'links', 'linked' and 'linking' into the 'concept' 'link'), Leximancer starts drawing lines between words that co-occur within two sentences. The main underlying intuition is that the same 'line of thought' is close to each other. Leximancer uses this intuition to build up an

[9]The one exception here is the Chinese subset, which was only coded manually. HCSS is currently working on including a Chinese text mining tool into its Metafore tool suite.

[10]We want to emphasize that Leximancer does NOT use any predefined ontologies or taxonomies in building this semantic network (e.g. things like "we 'KNOW' that security is mostly 'military', and therefore we link 'security' automatically to that 'known' association"). Whereas taxonomies can certainly be useful for certain purposes, we intentionally opted for a fully 'open' approach like Leximancer's (without any preconceived notions).

ever-richer semantic 'network' as it systematically goes—in a number of iterations—through all of these documents sentence by sentence. When it finishes this analysis, it is able to identify the main 'themes' within a set of documents and adds a numerical value for the relative 'weight' of each theme. Since we only collected articles that deal with future security issues, these themes essentially summarize the main substantive issues that the texts collectively focus on.

Leximancer also generates lists of (ranked) concepts with some numerical values reflecting their centrality in the texts (not only how often they occur in the text, but also—and even more importantly—how often they co-occur with other words). This provides the ability to zoom in on all concepts that co-occur within two sentences of the word 'security'. These are also numerically ranked, allowing our analysts to quickly (and without any preconceptions or other biases) get a sense of what security-associated concepts tend to be more important throughout the set. Leximancer allows the user to study the actual snippets of text and take into account the broader context to 'make sense' of the purely numerical data.

#### 6.3.5.2 Visual Example

Table 6.2 shows the key themes that were automatically generated by Leximancer. The left column indicates the discovered themes, clusters of concepts that tend to travel together throughout the entire set of documents, in declining order of importance. The second column shows the concepts themselves that are contained in this theme and the third column the theme's relative importance in the set of studies, with the most important theme always receiving 100 % and the other ones scaled accordingly. We observe, for instance, that the most dominant theme in that year's security foresight crop can best be labelled '**resources**'.

Leximancer then also allows to select some key terms and to find out which other terms tend to co-occur with that selected term throughout the text. Figure 6.6, for instance, shows which words tend to be associated with 'security' across the different types of sources.

#### 6.3.5.3 Paper Machines for Zotero

We also use an automated topic modelling tool that is part of the free, open-source 'Paper Machines' text mining software. Paper Machines is a plugin to Zotero, one of the most popular free bibliographical management programs currently available. Zotero allows analysts to automatically download the bibliographical information including the actual underlying document from various full-text databases (such as Google Scholar, or EBSCO) in database format that can subsequently be used for formatting footnotes and bibliographies. Paper Machines can then use this bibliographical information to text mine the underlying documents and visualize trends over time in the main themes from a set of documents. Paper Machines thus

**Table 6.2** Themes and concepts generated by Leximancer for the English language foresights

| Theme | Concepts | Connectivity (%) |
|---|---|---|
| **Resources** | Energy, production, resources, water, sector, gas, supply, price, costs, reduce, investment, improve, case, current, significant, result, demand, power, natural, oil, example, potential, emissions, produce, sources, land, chain, competition, estimated, standards, consumers, reduction, lower, carbon, compared | 100 |
| **Countries** | Countries, change, global, climate, level, region, important, economic, related, world, China, continue, social, become, remain, financial, scenarios, emerging, trends, tion, problems, events, strong, crisis, education | 94 |
| **System** | System, including, development, required, provide, support, management, process, based, activities, operations, needs, planning, range, integration, innovation, local, opportunities, building, ensure, control, implementation, specific, defined, funding | 89 |
| **Environmental** | Environmental, effects, time, risk, major, sustainable, terms, lead, key, further, direct, recent, structure, human, place, several, efforts, general, main, value, conditions, reform, making, during, given, ing | 82 |
| **Policy** | Policy, government, international, national, challenges, different, play, political, role, take, states, institutions, Eu, concerns, groups, interests, following, United States, society, clear | 78 |
| **Increase** | Increase, market, growth, areas, food, economy, impact, large, share, expected, growing, agricultural, decades, higher, urban, account, factors, health, domestic, due, low, capital, affect, rapid, export, period, flows | 74 |
| **Security** | Security, future, public, work, issues, response, force, focus, report, community, actions, strategic, present, threats, discussed, individual, cooperative | 51 |
| **Possible** | Possible, approach, address, order, strategy, critical, study, form, initiatives, established, framework, decisions, practices | 51 |
| **Use** | Use, technology, project, industry, services, likely, infrastructure, greater, existing, data, model, disposal, business, companies, design | 50 |
| **Considered** | Considered, limited, measures, contribute, benefits, protection, create, advanced, private, combined, allow, particular, target, analysis, meet, better, long-term, associated | 48 |
| **Fuels** | Fuels, efficiency, capacity, access, generation, electricity, addition, available, transport, renewable | 32 |
| **Nuclear** | Nuclear, research, information, capability, network, involved, assessment | 20 |
| **Population** | Population, rate, people, rising, trade, age | 19 |
| **Consumption** | Consumption, year, percent, total, million | 11 |

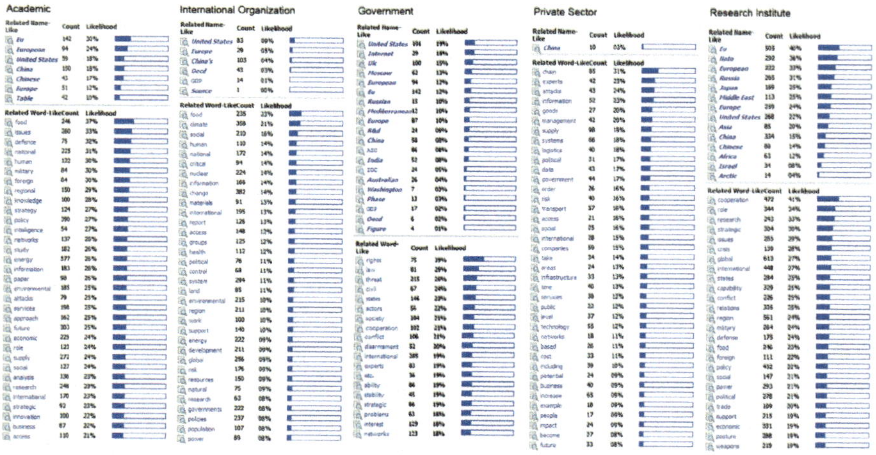


**Fig. 6.6** Associations of 'security' with other concepts by publishing organization

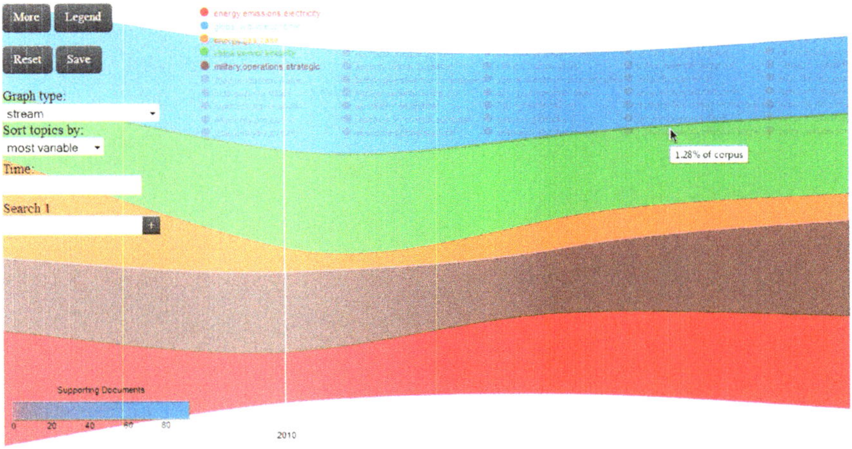


**Fig. 6.7** Example of a streamgraph (standard output)

provides a 'picture' of a Zotero library's contents, allowing to compare collections, identify patterns and assess on which topics the material is mostly focused.

The output of the automated coding through text mining is reflected in visuals such as streamgraphs (see figure below), showing the evolution of the importance of themes over time.

Figure 6.7, for instance, shows that the 'red' topic, the 3 key terms of which are energy, emissions and electricity, is the most important topic in this subset of the futurebase. We also see that its importance declined a bit around the beginning of 2010, but then increased in importance again throughout 2011. The same output can also be visualized through word clouds as shown in Fig. 6.8.

Figure 6.8 shows the breakdown of another subset of the futurebase by source. It shows, for instance, that the red topic in this case (which consists of the words in the red word cloud, which seems to refer to military capabilities) is the most dominant in the private sector and in the research world, but is much less important in the other 3 sources. We point all that all of these data and visuals are generated fully automatically by the clustering algorithm contained in Paper Machines.[11]

## 6.4 The Potential of Metaforesight Studies for Strategic Policymaking

This paper has illustrated the potential and the workings of a multilingual metaforesight protocol. Working with a global team, a substantive collection of foresight studies in different 'language domains' can be harvested, coded,

[11]In this case, a basic latent Dirichlet allocation topic modelling algorithm.

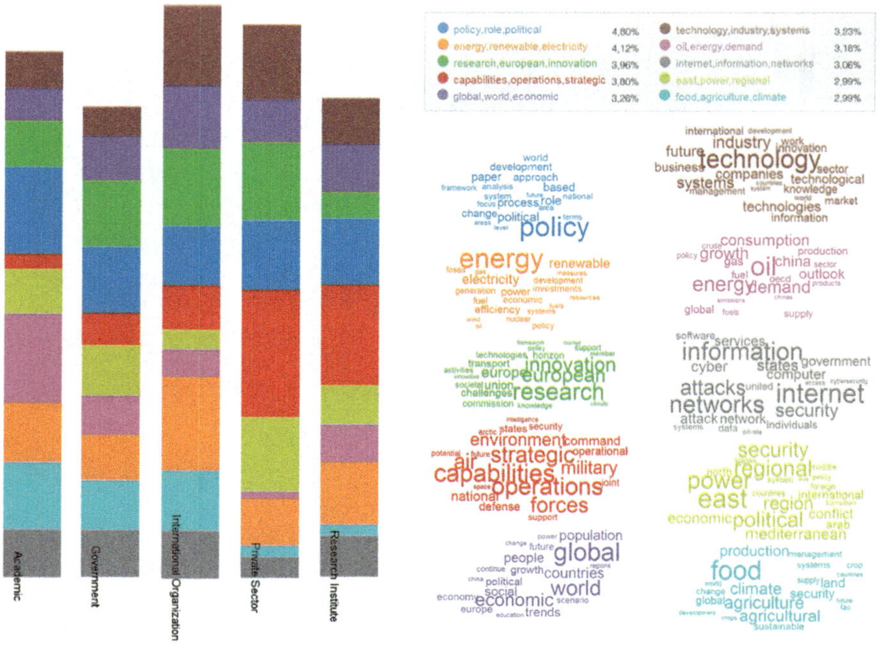


**Fig. 6.8** Word clouds and topics (based on a streamgraph)

visualized and analysed. The coding process combines the qualities of manual, human coding and automated coding through text mining. At the heart of the protocol is the constant reflection and cross-checking across the languages (and sources) of the sense making that lies beneath the coding. The coding itself is not intended to reduce but to synthesize the data, and the original source documents can always be reaccessed to refine the coding process and build insights as a result of the analysis.

The starting point for the Metafore protocol has been to overcome the subjectivity and normative aspects of foresight studies. By analysing a pool of foresight studies globally, the metaforesight approach tries to capture a broader bandwidth of views than is possible when one works within one, implicitly biased viewpoint. The coding process of weighting the different types of content that are mentioned in the reports helps to reveal the dominant themes in the foresight studies. It shows the way certain drivers of change are discussed, which stakeholders are involved and many other aspects of meaning that are important in these foresight studies. It also reveals the blind spots of themes that are discussed more in-depth in some language zones, but more or less ignored in other parts of the world. In this way, the metaforesight study gives a multiperspective approach. The insights can be truly refreshing and may trigger deeper drill-downs into the riches of these multiple futures.

The importance of this to policymakers cannot be underestimated. We submit that 'Foresight 3.0' approaches such as the one that was presented and illustrated in this paper represent the 'futures' equivalent of the unbiased 'evidence' that is increasingly becoming the basis of policymaking. Any prudent strategic decision-maker would be well-advised to carefully stress-test the robustness of his strategic decisions against as broad a segment of the entire (theoretical) futurespace as she can absorb. Multilingual approaches will increasingly play a key role in this.

Despite these unique advantages with respect to strategic policymaking, multilingual meta-analysis also has some inherent drawbacks. Policymakers often miss the linear 'narrative' quality of scenarios that they can fully absorb. What we (Chivot et al. 2013; De Spiegeleire et al. 2005 ) have therefore done in other work[12] is to use the distribution of views on various parts of the future ('*futuribles*') and the factors and clusters that tend to emerge from it to construct scenario ensembles that can then be 'narrated' in a more conventional way.[13]

Foresight 2.0—typically based on collaborative scenario exercises—is particularly valued by public and private actors for its ability to trigger a strategic discussion about the future or about how decisions in the present can or should be inspired, hedged, etc. by some stress-testing against a small number of 'chewable' futures (van der Heijden 2005). In the defence and security world, it is even used more 'hard'—analytically—than 'soft'—discursively to generate and prioritize robust options across a broader scenario space (Davis 2007, 2008; De Spiegeleire 2011). In comparison with the discursive process, Foresight 3.0 confronts decision-maker with a lot more uncertainty, which is both useful (as it may prepare them better) and also more challenging as most humans feel distinctly uncomfortable with uncertainty—both at a cognitive and at an emotional level.

Finally, it should be noted that the methodology of Foresight 3.0 is recent and still in development. Currently, manual coding is a large component of the data analysis, but has limitations in time constrains, human error, subjectivity and cognitive fallacies. However, this could be a temporary issue: we are just entering the age of big data analysis, which holds a strong promise for the analysis of large pools of documents. New tools from this field are being introduced to the Metafore protocol, which not only opens up possibilities for enhanced automated coding, but also helps to develop techniques for displaying insights from foresight studies in a way that supports strategic policymaking.

[12]On the same idea, see also van Asselt (2010).

[13]Our own thinking about the danger of 'narratives' has been profoundly influenced by what Nassim Taleb calls the 'narrative fallacy', which we strongly recommend all people engaged in this type of work to take a very close look at (Taleb 2007).

## References


Banerjee, A. V. (1992). A simple model of herd behavior. *The Quarterly Journal of Economics, 107*, 797. doi:Article.

Batchelor, R. (2007). Bias in macroeconomic forecasts. *International Journal of Forecasting, 23*, 189–203. doi:10.1016/j.ijforecast.2007.01.004

Batchelor, R. (2011). Accuracy versus Profitability. *Foresight: The International Journal of Applied Forecasting*, 10–15.

Bekkers, F., De Spiegeleire, S., Van Esch, J., Gehem, M., Sweijs, T., & Wijninga, P. (2013). De Toekomst in Alle Staten [The Future in All Its States]. HCSS strategic monitor 2013, HCSS Report. HCSS, The Hague.

Chivot, E., De Spiegeleire, S., Bullinga, R., & Jacobs, L. (2013). European capabilities assessment game—Towards scenario ensembles (HCSS Report for the European Defence Agency). The Hague Centre for Strategic Studies.

Davies, P. (2004). Is evidence-based government possible?

Davis, P. K. (2007). *Enhancing strategic planning with massive scenario generation: Theory and experiments*. Santa Monica, CA: RAND National Security Research Division.

Davis, P. K. (2008). Defense planning and risk management in the presence of deep uncertainty. In P. Bracken, I. Bremmer, & D. Gordon (Eds.), *Managing strategic surprise: Lessons from risk management and risk assessment*. Cambridge, UK; New York: Cambridge University Press.

de Jouvenel, B. (1964). L'art de la conjecture. Éditions du Rocher.

De Jouvenel, B. (1965). *Futuribles*. Santa Monica, CA: Rand.

De Spiegeleire, S. (2011). Ten trends in capability planning for defence and security. *The RUSI Journal, 156*, 20–28. doi:10.1080/03071847.2011.626270

De Spiegeleire, S., Boeke, S., Mans, U., Rademaker, M., & Toxopeus, R. (2005). Future worlds. an input into the NATO long-term requirements study (CCSS report for NATO consultation, command and control agency (NC3A)'s long-term requirement's study).

De Spiegeleire, S., Sweijs, T., Kooroshy, J., & i Novosejt, A. B. (2010). *STRONG in the 21st century. Strategic orientation and navigation under deep uncertainty*. The Hague: The Hague Centre for Strategic Studies.

De Spiegeleire, S., Sweijs, T., & Zhao, T. (2011). *Contours of conflict in the 21st century. A cross-language analysis of Arabic, Chinese, English and Russian perspectives on the future nature of conflict, HCSS security foresight programme*. The Hague: The Hague Centre for Strategic Studies.

Dovern, J., & Weisser, J. (2011). Accuracy, unbiasedness and efficiency of professional macroeconomic forecasts: An empirical comparison for the G7. *International Journal of Forecasting, 27*, 452–465. doi:10.1016/j.ijforecast.2010.05.016

Franses, P. H., McAleer, M., & Legerstee, R. (2012). Evaluating macroeconomic forecasts: A concise review of some recent developments. *Journal of Economic Surveys, 28*(2), 195–208. doi:10.1111/joes.12000

Gardner, D. (2011). *Future babble: Why expert predictions are next to worthless, and you can do better* (1st ed). Dutton Adult.

Henry, G. B. (1989). Wall street economists: Are they worth their salt? *Business Economics, 24*, 44+.

Lamont, O. (1995). *Macroeconomics forecasts and microeconomic forecasters* (Working Paper No. 5284). National Bureau of Economic Research.

Leximancer [WWW Document]. (n.d.). url:https://www.leximancer.com

Malaska, P., & Virtanen, I. (2005). Theory of futuribles.

Malaska, P., & Virtanen, I. (2009). Theory of futuribles and historibles. Futura 28.

Mitchell, M. (2009). *Complexity: A guided tour*. USA: Oxford University Press.

Page, S. E. (2011). *Diversity and complexity*. Princeton University Press.

Pawson, D. R. (2006). *Evidence-based policy: A realist perspective*. SAGE.

Rowe, G., & Wright, G. (2011). The Delphi technique: Past, present, and future prospects—Introduction to the special issue. *Technological Forecasting and Social Change, The Delphi technique: Past, present, and Future Prospects, 78*, 1487–1490. doi:10.1016/j.techfore.2011.09.002

Smith, A. E. (2000). Machine mapping of document collections: The Leximancer system. In *Proceedings of the fifth Australasian document computing symposium*. pp. 39–43.

Smith, A. E. (2003). Automatic extraction of semantic networks from text using Leximancer, In *Proceedings of the 2003 conference of the North American chapter of the association for computational linguistics on human language technology: Demonstrations*. pp. 23–24.

Smith, A. E., & Humphreys, M. S. (2006). Evaluation of unsupervised semantic mapping of natural language with Leximancer concept mapping. *Behavior Research Methods, 38*, 262–279.

Stekler, H. O. (2007). The future of macroeconomic forecasting: Understanding the forecasting process. *International Journal of Forecasting, 23*, 237–248. doi:10.1016/j.ijforecast.2007.01.002

Taleb, N. (2007). *The black swan: The impact of the highly improbable* (1st ed.). New York, NY: Random House.

Tetlock, P. E. (2005). *Expert political judgment: How good is it? How can we know?* Princeton University Press.

Thaler, R. H., & Sunstein, C. R. (2009). *Nudge: Improving decisions about health, wealth, and happiness*. New York: Penguin Books.

van Asselt, M. B. A. (Ed.). (2010). *Foresight in action: Developing policy-oriented scenarios, Earthscan risk in society series*. London; Washington, DC: Earthscan.

van der Heijden, K. (2005). *Scenarios: The art of strategic conversation*. Wiley.

Ward, K. (2011). *The world in 2050*. HSBC.

# Chapter 7
# Using Enhanced Patent Data for Future-Oriented Technology Analysis

**Christopher L. Benson and Christopher L. Magee**

**Abstract** Patents represent one of the most complete sources of information related to technological change, and they also contain much detailed information not available anywhere else. Thus, patents are the 'big data' source most closely related to future-oriented technology analysis (FTA). Not surprisingly, therefore, there is very significant practical and academic use of the patent database for understanding past technical change and attempting to forecast future change. This paper summarizes several new methods and demonstrates their combined effectiveness in establishing a cutting-edge capability for patent study not previously available. This capability can be stated as *a link between the information in patents and the dynamics of technological change*. The demonstrated capability relies upon the use of a database containing the rates of improvement for various technologies. We also specify the term we use for the analysed units of technology: a technological domain is *a set of artefacts that meets a specific generic function while utilizing a specific set of engineering and scientific knowledge*. This definition is unambiguous enough so technological domains can be linked with progress rates and are sufficiently flexible to accommodate the large scale and complexity of the patent database. The existence of an improvement rate database and its quality is a critical foundation for this paper. Establishing the overall capability also involves relating the rate of improvement of a technological domain to the patents in *that* domain. We show that a recently developed method called the classification overlap method (COM) provides a reliable and largely automated way to break the patent database into understandable technological domains where progress can be measured. In this paper, we show how this method overcomes the third limitation of the patent database. The major conclusion of the paper is that there is now an overall objective method named Patent Technology Rate Indicator (PTRI) for using *just* patent data to reliably estimate the rate of technological progress in a technological domain. Thus, the first link between the patent database information and the

C.L. Benson (✉) · C.L. Magee
Massachusetts Institute of Technology, 77 Massachusetts Ave, Cambridge, MA, USA
e-mail: cbenson@mit.edu

C.L. Magee
e-mail: cmagee@mit.edu



T.U. Daim et al. (eds.), *Anticipating Future Innovation Pathways Through Large Data Analysis*, Innovation, Technology, and Knowledge Management,
DOI 10.1007/978-3-319-39056-7_7

dynamics of technological change is now firmly established; robustness and back-casting tests have shown that the assertion of reliability is meaningful and that the estimate has predictive value. We demonstrate the key methodology of new elements (use of COM and rate estimation from the selected patent sets) for 15 technologies that some have thought have possible future importance. The 15 cases also demonstrate the usefulness of the overall method by estimating technological improvement rates that are significantly different for this group of technologies.

**Keywords** Future-oriented technology analysis · Patent analysis · Comparing emerging technologies

## 7.1 Introduction

This paper introduces the results of a new forecasting method called the Patent Technology Rate Indicator (PTRI) method that uses patent data to better predict time-based performance improvement rates of technologies whose performance trend is otherwise unknown. While the focus of the research is on quantitative performance trends, we do not want to suggest that such results will be all one desires for technological forecasting. Haegeman et al. (2013) explain the various focuses of several different disciplines within the FTA community:

> It is acknowledged that, within the FTA community (which comprises Foresight, Forecasting and Technology Assessment),[1] foresight practitioners have traditionally concentrated on participatory methods based on qualitative data, on the grounds that quantitative extrapolation from past data is not sufficient to address the uncertainties of the future and that emerging changes in the socio-economic and technological landscapes need to be taken into account. Another part of the FTA community, constituted by Forecasting and Technology Assessment practitioners, holds an opposite standpoint, considering qualitative and participatory approaches as a second best option, to which we are somehow compelled to refer until adequate quantitative methods arise. (Haegeman et al. 2013).

Our viewpoint is that both qualitative and quantitative approaches are needed for this complex issue and improvement of both is needed. Rosenberg's analysis done more than 20 years ago (Rosenberg 1982) categorized four areas of difficulty in any technological forecasting which includes the socio-economic aspect; these are as follows:

1. At emergence, the focal (or new technology) is not very capable;
2. Vital complementary technologies are potentially underdeveloped;
3. System design/evolution that may be necessary for large impact has not occurred;
4. The human user ingenuity that will greatly impact the technology and its impact has great diversity and is unknown at the early stages.

While we believe that a focus on quantitative performance improvement prediction can contribute to items (1) and (2) in Rosenberg's analysis, we believe that qualitative approaches will also be valuable not only in items (3) and (4) but also in (1) and (2).

Gao et al. (2013) introduced an important aspect of the quantitative approach by exploring technological performance over time using FTA techniques. In our analysis, we predict time-based technological improvements rates similar to the type made famous by Moore's law, where a specific technical metric (transistors/die) is measured over a period of time and is found to improve at a relatively constant percentage per year. This is not the first attempt at using technological improvement rates as part of forecasting, but most predecessors have done so by attempting to utilize learning rates, which compare the improvement of a technical metric with production (Nemet 2006) rather than with time. In particular, we are interested in estimating the yearly technical improvement rate of a technology, represented by the variable 'k' in Eq. 7.1.

$$q = q_0 \exp(k(t - t_0)) \tag{7.1}$$

While Sahal (1979) and Nagy et al. (2013) showed that the actual practical implications of the time-based and production-based improvement rates are very similar, this paper will focus solely on the time-based rates due to the evidence that they are more fundamental (Magee et al. 2014). Additionally, in performing this analysis we are building off of the strongly established results that show long-term time-based technical improvement rate stability (Magee et al. 2014), that is, that the improvement rate of a technology does not change considerably over time or at the very least changes considerably less between times than the rates change between technologies. This same argument is appropriate for the different complete technical metrics that can be used to measure the performance of a technology (i.e. $W_p$/\$ or kW hr/\$ for measuring solar PV output) (Benson and Magee 2014). Thus, we will focus almost entirely on the rate differences between technologies and not the differences within a technological domain between metrics.

The PTRI can estimate nearly any time-based technological improvement rate using a set of patents that represents each technological domain. The use of patents in FTA is given precedence by Gao et al. (2013) when they used patent indicators to estimate the level of technological maturity for a given domain. In a very similar way, the PTRI uses patent indicators as correlation factors for forecasting technological improvement rates of a domain and is based upon an extensive study reported in Benson and Magee (2015b). The results of the PTRI method can project relative improvement rates of technologies—which may be useful for investment decisions by private parties or governments. Additionally, the data can be used to aid in uncertainty analyses for future technological capabilities of a specific domain, which is often used in long-term product planning by large companies and the military. Both of these uses can aid in influencing both private and public policies, which has been the outcome of several FTA techniques in the past (Schaper-Rinkel 2013).

## 7.2 Methodology

The PTRI is based upon the finding by Benson and Magee (2015b) that the information contained in patents is sufficient for understanding differences in technical improvement rates between different domains. A number of patent metrics were studied by Benson and Magee (2015b) and were combined with multivariate regression tools to create a model for forecasting technological improvement rates. The resulting regression was found to be accurate for 12 years into the future. The PTRI method is summarized in Fig. 7.1.

The PTRI method begins with the identification of a technology of interest, and then, a technology needs to be converted to an appropriate technological domain. In order to convert from a technology to a technological domain, it is useful to think not just about the embodiment of an invention, but rather the use it fulfils and the underlying scientific principles that it makes use of. The intention behind this is to specifically clarify the unit of analysis by using a standardized definition of technological domain:

*A technological domain can be defined as follows: the set of artefacts that fulfil a specific generic function utilizing a particular, recognizable body of knowledge.*

Once a technological domain has been defined, the next step is the selection of a set of patents that represent the domain, and this step is very important because the set of patents that are selected will be the input data source for the method. The sets of patents can be selected by using a methodology called the classification overlap method (COM) that relies upon the different types of patent classification systems used by the US and International Patent Offices (UPC and IPC) (Benson and Magee 2013, 2015a). The input to the COM is a set of keywords related to the technological domain as well as potentially some supporting information such as key companies or inventors in the domain. These inputs can then be used to select a set

**Fig. 7.1** PTRI Method

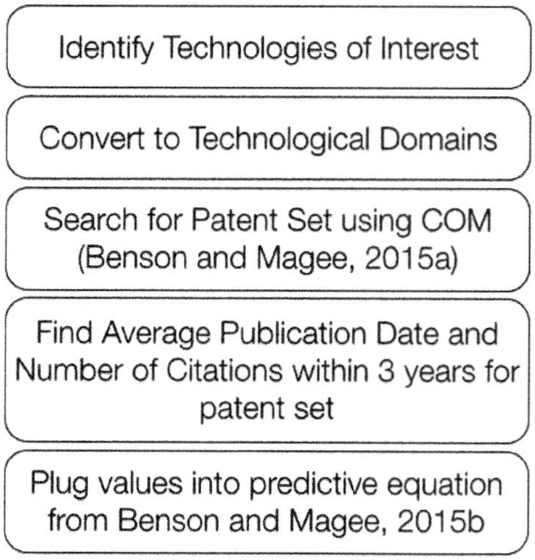

of patents contained within the overlap between the most appropriate UPC and IPC based upon the COM. All of the patent set searches for this paper were done using Patsnap (www.patsnap.com) and included only issues US patents from 1 January 1976 to 1 July 2013, and these dates were chosen so they could be compared with prior results from Benson and Magee (2015b).

Due to the importance of the patent sets for the PTRI method, it is important to ensure that the patents in the data set accurately represent the technological domain of interest. This is done manually by reading representative sampling of the data sets and qualitatively assigning each read patent a value of '1' for 'relevant or '0' for 'not relevant' to the domain of interest. The average relevancy score can then be added by summing the total relevant determinations and dividing by the number of total patents in the data set. In general, an acceptable value for relevance is greater than 0.65; however, a good patent set will have relevancy above 0.8.

Once the data set has been verified for relevancy, the patent indicators can be calculated using the metadata included in the patents. The PTRI uses two indicators for calculating the estimated technological improvement rate: average publication data and average number of forward citations within 3 years of publication as described in Benson and Magee (2015b).

The *average number of forward citations within 3 years of publication* is the average number of forward citations that each patent received within 3 years of publication for patents in a technological domain. The metric is calculated using Eq. 7.2 where SPC is the simple patent count, $FC_i$ is the number of forward citations for patent $i$, $t_{ij_{\mathrm{pub}}}$ is the publication year of patent $i$, $t_{ij_{\mathrm{pub}}}$ is the publication date of forward citation $j$ of patent $i$, and the function IF(arg) only counts the values if the argument is satisfied.

$$\sum_{i=1}^{\mathrm{SPC}} \sum_{j=1}^{\mathrm{FC}} \mathrm{IF}\left(t_{ij_{\mathrm{pub}}} - t_{i_{\mathrm{pub}}} \leq 3\right) \tag{7.2}$$

The *average publication year* for the patents in a domain includes patents that were published between 1 January 1976 and 1 July 2013. This measure is calculated using Eq. 7.3 where SPC is the simple patent count and $t_{i_{\mathrm{pub}}}$ is the publication year of patent $i$.

$$\frac{\sum_{i=1}^{\mathrm{SPC}} t_{i_{\mathrm{pub}}}}{\mathrm{SPC}} \tag{7.3}$$

After these two values are calculated for the domain of interest, they can be plugged into the regression model developed in Benson and Magee (2015b):

$$k = -31.1285 + 0.0155 * \mathrm{AvePubYear} + 0.1406 * \mathrm{Cite\,3} \tag{7.4}$$

The result is a simple number that represents the time-based technological improvement rate for the domain of interest.

## 7.3 Results, Discussion and Implications

The aim of this paper was to explore the results of the PTRI methodology applied to a number of potentially important technologies in the future. To act as a basis of what technologies would be important in the future, we used a premade list of the '10 breakthrough technologies of 2014' as noted by the MIT Technology Review (2014) as a basis for a list of potential transformational technologies that would be of interest to know an estimated technological improvement rate.

This section illustrates all elements of the PTRI Methodology described in the previous section using as cases the 10 technologies listed in the MIT Technology Review and additional five others. All 15 technologies will be translated into technological domains, representative patent sets will be selected using the COM methodology, and then, patent indicators will be calculated and technological improvement rates will be determined. The end result will be estimated technological improvement rates for 15 technological domains.

## 7.4 Defining the Domains

The first step is to illustrate the translation of the ten technologies into a list of technological domains. Table 7.1 shows the 10 technologies from the MIT Technology Review along with the 10 domains and a short description of the domain. The final 4 rows are additional technologies that the authors decided to include based upon their subjective potential importance in the upcoming near future and the academic and media interest paid to the domains.

The translation process to technological domains is illustrated by the technology 'Agricultural Drones' from the MIT Technology Review list that was determined to be slightly narrow in its scope as it was only focusing on one potential use for the automated air vehicles that they were intended to represent. Focusing first on the broad function, we arrive at remote flight control. Following this path further, while 'drones' themselves are a rather broad category, they do not represent a particularly specific technological domain in that the term drone could be interpreted in a number of ways (Wikipedia lists 11 possible interpretations for the term 'drone' not including the entertainment or music categories such as the movie *Star Wars: Attack of the Drones*). Thus, we added further clarity to the definition by referring to the technological domain as remote flight control technologies, with the specific generic purpose being remote flight control, and the underlying set of knowledge being a unique overlap of aeronautics, control theory, and signal transmission methods. Note that this new domain does not necessarily preclude manned aircraft,

**Table 7.1** Technical domains as inputs into PTRI

| Technology candidate (from MIT Technology Review) | Derived technological domain | Description of domain |
|---|---|---|
| Agricultural drones | Remote flight control technologies | Controlling flying vehicles from afar, including drones and advanced flight controls |
| Ultraprivate smartphones | Information security | Information security across all form factors |
| Brain mapping | Brain scanning | Determining brain features and structure using a number of tools (CT, MRI…) |
| Neuromorphic chips | Artificial neural network Computing | Computing architectures that resemble that of the human brain |
| Genome editing | Genome sequencing | Determining the genomes of specific strands of DNA |
| Microscale 3D printing | SLA 3D printing | Additive manufacturing using light to cure resins |
| Mobile collaboration | Online learning | Education in digital classrooms |
| Oculus rift | Digital representation | Digital modelling of reality (includes virtual reality as well as less immersive forms of digital representation of the physical world) |
| Agile robots | Robotics | Performance of physical functions by Automatic mechanical devices |
| Smart wind and solar power | Wind turbines | Energy generation from moving air. |
| | Solar PV | Energy generation using the photoelectric effect |
| – | Nuclear fusion | Energy generation relying directly on atomic fusion |
| – | Water purification | Removing salt from water using reverse osmosis |
| – | Food engineering | Chemical and genetic modifications for enhanced food production |
| – | Gaseous purification | More broad term for one enabling technology behind climate geo-engineering |

as there are plenty of reasons to control a vehicle remotely even when a pilot is sitting in the cockpit.

Other illustrations of translation are the transformation from ultraprivate smartphone to information security—as the smartphone form factor seems an unnecessary constraint for the analysis of the improvement rate of information security technologies. Admittedly, the most liberty was taken in translating mobile collaboration to online learning—this was done partially due to the lack of clarity over what exactly constitutes mobile collaboration and the recent intense emphasis on online learning and MOOCs; therefore. this 'translation' illustrates what could be termed a 'substitution' of near neighbour technologies.

## 7.5 Patent Sets Selected Using the COM

The next innovative aspect of the PTRI (Fig. 7.1) is to find relevant patent sets for each of the technological domains using the COM, as described at the top of page 4. Column 2 of Table 7.2 shows the patent classes that were used to define each domain, Column 3 the size of the overall patent set and column 4 the relevancy as determined by subjective reading of a sampling of 200 patents from each domain.

As noted earlier, each technological domain is represented by a set of patents that are defined by a combination of overlapping US and international patent codes. As an example, the 'remote flight control technologies' domain is defined by the overlap of either of the US codes 701/2 or 701/3 (data processing: vehicles, navigation, and relative location/2 remote control system/3 aeronautical vehicle) and the international patent code B64C (airplanes, helicopters). This overlap results in 328 patents that were qualitatively determined to be ~85 % relevant. This same process was repeated for all 15 technological domains, and the results are shown in Table 7.2.

## 7.6 Calculating the Patent Indicators and Using the PTRI Regression Model to Estimate Technological Improvement Rates

The next innovative aspect from Fig. 7.1 is digesting the patent information in order to calculate the patent indicators required by the PTRI regression model: average year of publication and number of citations received within 3 years of publication (Cite 3). These values for each of the domains are shown in columns 5 and 6 of Table 7.2. It is interesting to note the extremes of each patent indicator. In this study, the oldest average date of publication is 1992 (food engineering and nuclear fusion), while the newest average publication date is 2010 for information security. The large size of the information security patent set (13,607) and the very high relevance ratio (0.985) give credibility to this very recent average publication date and indicate that this is likely a very dynamic domain and that the recency is unlikely an artefact of the data. These numbers are in line with the oldest and newest average publication date of the 29 technological sets used to construct the PTRI regression model with 1992 and 2006, respectively (Benson and Magee 2015b).

The smallest Cite 3 technological domain had just 1.5 forward citations within the first 3 years of publication on average (food engineering), while the largest belongs to digital representation with 5.85 citations within 3 years, which is a higher value than any of the original 29 domains used to create the PTRI (4.62-MRI).

**Table 7.2** PTRI input and output

| Domain | Patent sets | Number of patents | Relevance ratio | Average date of publication of patents | Cite 3 | Predicted K |
|---|---|---|---|---|---|---|
| Artificial neural network computing | 706/15 AND G06F | 361 | 0.71 | 2007.3 | 3.49 | 0.407 |
| Brain scanning | (600 AND 382) AND A61B AND 'brain' | 284 | 0.93 | 2009.3 | 3.14 | 0.390 |
| Water purification | C02F1/44 AND 210 | 1033 | 0.63 | 2003.6 | 3.80 | 0.393 |
| Digital representation | 345/419 AND G06F3 | 486 | 0.655 | 2004.9 | 5.85 | 0.702 |
| Food engineering | 426 AND C12 N | 1865 | 0.96 | 1992.2 | 1.50 | −0.107 |
| Genome sequencing | (435/6.11 OR 435/6.12) AND C12Q | 3990 | 0.74 | 2006.7 | 2.15 | 0.209 |
| Gaseous purification | (95 AND 423) AND B01D | 1683 | 0.72 | 1993.1 | 2.40 | 0.034 |
| Information security | 726 AND H04L | 13,607 | 0.985 | 2010.1 | 3.52 | 0.454 |
| Nuclear fusion | (G21B OR H05H) AND 376 | 508 | 0.95 | 1992.4 | 1.52 | −0.102 |
| Online learning | G06Q50 AND 434 | 197 | 0.78 | 2001.8 | 6.62 | 0.76 |
| Remote flight control technologies | (701/2 OR 701/3) AND B64C | 328 | 0.855 | 2003.1 | 3.18 | 0.299 |
| Robotics | B25 J AND 901 | 4122 | 0.935 | 1994.6 | 3.74 | 0.245 |
| SLA 3D printing | 264/401 AND B29C35/08 | 251 | 0.93 | 2001.4 | 3.98 | 0.385 |
| Solar photovoltaic energy generation | 136 AND H01L | 5203 | 0.85 | 1998.6 | 2.73 | 0.165 |
| Wind turbine energy generation | (416 OR 290) AND F03D | 2498 | 0.94 | 2002.8 | 2.17 | 0.152 |

These patent indicators can now be plugged into Eq. 7.4 to calculate the estimated technological improvement rates for each of the 15 domains as shown in the final column of Table 7.2.

## 7.7 Discussion and Conclusions

The 15 cases presented here clearly illustrate all aspects of the newly created PTRI methodology. We briefly note here some areas for even further improvement while discussing the novel information provided by the PTRI. We first note that some of the k values are negative, which would seem to indicate that the particular technological domain is getting worse with time. Obviously, this explanation is logically inconsistent, and the more correct interpretation is that the PTRI model does a poor job of distinguishing among very slowly improving technologies and that any technology that is estimated as a negative improvement rate is simply a very slowly improving domain (<5 %). Additionally, the PTRI model as shown in Benson and Magee (2015b) tends to give estimates that fall within ±0.10 of the measured technological improvement rates. In the future, more accurate confidence intervals should be developed to accompany the estimated k. To demonstrate this further, some of the technologies that were predicted in this study have been measured before, and the comparison between the empirically measured values and the estimated values is shown in Table 7.3.

The agreement between the predicted values and the empirically measured values lends credibility to the predicted values shown above and is consistent with the relatively close correlation between the PTIR model and previously measured empirical values.

The highest technological improvement rate is digital representation with an estimated k of 0.7, which would indicate that its capabilities would more than double every year. An interesting finding of the 15 tests of the PTIR method illustrated here is that it is rather difficult to imagine a way to objectively measure the improvement rate for how well the digital world represents the real world; however, this high rate is not inconsistent with the subjective experiences of the

**Table 7.3** Estimated and measured ks

| Technical domain | Technical measure | Estimated k | Empirically measured k |
|---|---|---|---|
| Genome sequencing | (base pairs/$) | 0.21 | 0.29 |
| SLA 3D printing | (1/s*$(including build volume/machine size))) | 0.39 | 0.38 |
| Solar photovoltaic energy generation | ($W_p$/$) | 0.17 | 0.09 |
| Wind turbine energy generation | ($W_p$/$) | 0.15 | 0.09 |

rapidly changing digital world and the ever-increasing ways that people spend on a digital version of what used to be physical (i.e. social networking, talking, banking and watching entertainment). Thus, the cases show that the PTIR allows us to map technological improvement rates to technologies that may be improving but are hard to measure due to lack of metrics or data or other reasons.

These improvement rate estimates should be used, however, in conjunction with increased knowledge about the measures by which the technical domains improve. Table 7.3 shows a few examples of technical measures by which the domains improved, including more simple measures such as $W_p/\$$ for solar PV and wind turbines and more complex measures for 3D printing, which includes metrics for speed of printing (mm/s), resolution (1/mm) of the machine, cost ($, machine size) and flexibility (build volume), which when combined result in the 'highly complete' measure in Table 7.3 for 3D printing.

When evaluating technologies using the PTIR model, the measures can be estimated and can be somewhat more abstract, but must always include a benefit and a cost. For example, when considering water purification, the benefit of the process is clean water and the cost is energy or price. Thus, an appropriate measure for the improvement rate of water purification could be gallons of clean water per kWhr or per dollar.

The methodology described in this paper (PTRI) is novel in allowing comparison of improvement rates of a broad set of 15 technologies. The second highest $k$ values are grouped into a clump around 0.4 with information security, brain mapping, artificial neural networks, 3D printing and purification all within 0.06 of one another. These rather disparate technologies are predicted to improve at relatively rapid rates similar to those of Moore's law ($k = 0.36$). While some may not be surprised to see information security and neural networks improving at this rate due to their relation to information technology, the estimated rapid rate of growth for brain mapping, 3D printing and purification has less to do with the rapid rate of improvement in information technology yet is still estimated to be improving at a high rate.

Remote flight control, robotics, genome sequencing, solar PV and wind turbines make up the next grouping of technologies that have estimated improvement rates between 0.15 and 0.3, corresponding with a doubling of capability every 2.5–5 years. These technologies also seem to be rather disparate, yet all seem to have less of a pure reliance on information technology than does the top group.

The bottom dwellers, with estimated rates ranging from −0.1 to 0.03, include gaseous engineering, nuclear fusion and food engineering. As was mentioned previously, it is unlikely that these particular domains are decreasing in capability over time, and it is much more likely that all three of these domains have been improving at a very slow rate.

While the topic was touched upon briefly in the results section, the intent of this paper is not to look at commonalities between domains with high (or low) estimated technological improvement rates, as that topic is covered in depth in Benson and Magee (2015b); rather, the goal of this paper was to introduce the PTRI methodology into the FTA world as a tool that can be used to combine qualitative and quantitative data to provide numeric estimates of technological potential for the future. This tool can be especially useful for technical domains which are hard to measure or have scarce data such as may become more common as technology improves accelerates.

While this paper is mainly focused on demonstrating the potential of the PTRI for estimating quantitative technological growth rates, it will be important in practical use to include qualitative analysis to complement the quantitative estimates. For example, information security is estimated to be a fast-growing technological domain with a $k$ value of 0.45; likewise, purification is estimated to improve at a $k$ value of 0.39. These two values fall well within the rough confidence interval of +/−0.1, and therefore, it is reasonable to assume that they will improve at similar rates. Despite this fact, however, the results of the improvements could well be rather different.

Information security, while it may be improving quickly, is constantly having to compete with other people who are looking to break through that security, which relies on similar principles and may improve at a similar rate, leading to an arms race in information protection; therefore, while we would expect the capabilities of information security to increase drastically, we might not expect the number of information security breaches to decrease at the same rate due to the concurrent increasing capabilities of hackers and electronic thieves.

A different story can be told about water purification, and as was mentioned earlier, increases in purification capabilities should rapidly increase the capability to create drinking water using fewer resources. Thus, the high k for purification could indicate that the problem of water scarcity should not be a high risk if the purification technologies continue to improve at their estimated rates, which is a relatively safe bet considering the long-term stability of k for most technological domains.

The PTRI method, when combined with appropriate qualitative analysis, can be a powerful tool for policymakers, technological strategists and investors of many kinds. The development of more powerful patent analysis techniques to produce quantitative estimates of technological change can help decrease technological uncertainty for current and future technologies.

## References

Benson, C. L., & Magee, C. L. (2013). A hybrid keyword and patent class methodology for selecting relevant sets of patents for a technological field. *Scientometrics, 96*(1), 69–82. doi:10.1007/s11192-012-0930-3.

Benson, C., & Magee, C. (2014). On improvement rates for renewable energy technologies: Solar PV, wind turbines, capacitors, and batteries. *Renewable Energy*.

Benson, C. L., & Magee, C. L. (2015a). Technology structural implications from the extension of a patent search method. *Scientometrics, 103*(3), 1965–1985. doi:10.1007/s11192-014-1493-2.

Benson, C. L., & Magee, C. L. (2015b). Quantitative Determination of technological improvement from patent data. *PLoS One, 10*(4), e0121635. doi:10.1371/journal.pone.0121635.

Breakthrough Technologies. 2014. (2014, April). *MIT Technology Review*.
Gao, L., Porter, A. L., Wang, J., Fang, S., Zhang, X., Ma, T., et al. (2013). Technology life cycle analysis method based on patent documents. *Technological Forecasting and Social Change, 80*(3), 398–407. doi:10.1016/j.techfore.2012.10.003.
Haegeman, K., Marinelli, E., Scapolo, F., Ricci, A., & Sokolov, A. (2013). Quantitative and qualitative approaches in Future-oriented Technology Analysis (FTA): From combination to integration? *Technological Forecasting and Social Change, 80*(3), 386–397. doi:10.1016/j.techfore.2012.10.002.
Magee, C. L., Funk, J. L., Benson, C. L., & Basnet, S. (2014). *Quantitative empirical trends in technical performance* (No. ESD-WP-2014-22). Cambridge, MA.
Nagy, B., Farmer, J. D., Bui, Q. M., & Trancik, J. E. (2013). Statistical basis for predicting technological progress. *PLoS One, 8*(2), e52669. doi:10.1371/journal.pone.0052669.
Nemet, G. F. (2006). Beyond the learning curve: Factors influencing cost reductions in photovoltaics. *Energy Policy, 34*(17), 3218–3232. doi:10.1016/j.enpol.2005.06.020.
Patsnap. (2014). Patsnap patent search and analysis. Retrieved October 2014, from http://www.patsnap.com.
Rosenberg, N. (1982). *Inside the black box: Technology and economics*. Cambridge, MA: Cambridge University Press.
Sahal, D. (1979). A Theory of Progress Functions. *AIIE Transactions*, (November 2013), 37–41.
Schaper-Rinkel, P. (2013). The role of future-oriented technology analysis in the governance of emerging technologies: The example of nanotechnology. *Technological Forecasting and Social Change, 80*(3), 444–452. doi:10.1016/j.techfore.2012.10.007.

# Chapter 8
# Innovation and Design Process Ontology

**Cherie Courseault Trumbach, Christopher McKesson, Parisa Ghandehari, Lawrence DeCan and Owen Eslinger**

**Abstract** Many domain-specific ontologies exist. These ontologies are used in text mining processes to better understand text that is available within the specific domain. Example domains include specific business areas such as marketing or functional areas such as particular types of operations within the intelligence community. This paper makes a step toward developing a broad ontology for the innovation and design process as a domain. Such an ontology can be used to better understand the discussion that takes places in the design and development of new innovations and can be used to better understand the influences on that development. In many cases, the success, failure, or final path of a new innovation may not rest upon its technical merits but on the non-technical influences during the design and development process such as political influences. This paper uses examples within the shipbuilding domain in order to take steps toward building an Innovation and Design Process Ontology that can be applied to the Forecasting Innovation Pathways (FIP) framework as a means of capturing and understanding the influences on the technology delivery system.

**Keywords** Design process · Ontologies · Text data mining · Ship design

C.C. Trumbach (✉) · P. Ghandehari · L. DeCan
School of Naval Architecture and Marine Engineering, University of New Orleans, 2000 Lakeshore Drive, New Orleans, LA 70148, USA
e-mail: ctrumbac@uno.edu

C. McKesson
Department of Mechanical Engineering, University of British Columbia, 2050-6250 Applied Science Lane, Vancouver, BC V6T 1Z4, Canada

O. Eslinger
US Army Engineering Research and Development Center, 3909 Halls Ferry Road, Vicksburg, MS 39180, USA



T.U. Daim et al. (eds.), *Anticipating Future Innovation Pathways Through Large Data Analysis*, Innovation, Technology, and Knowledge Management, DOI 10.1007/978-3-319-39056-7_8

## 8.1 Ontology Background

In the last two decades, the term ontology has gained great significance in particular within the framework of knowledge management and informatics (Neches et al. 1991). A very early definition of the term ontology within the lexicon of informatics was published in 1991 by Neche et al. in the AI Magazine (Neches et al. 1991): "An ontology defines the basic terms and relations comprising the vocabulary of a topic area as well as the rules for combining terms and relations to define extensions to the vocabulary." This definition already refers to some of the most essential elements of an ontology. The vocabulary is being defined relative to a certain knowledge domain. In addition, it becomes apparent that this vocabulary is being composed of the essential terms of the domain as well as the relation between those terms. Also this definition describes the rules on how to combine the terms with each other in order to expand the vocabulary. However, this definition does not impose any immediate conclusions on the form in which this vocabulary should be illustrated (Spath and Stefanie 2011).

One of the best known and most cited definitions on ontologies was verbalized by Gruber in 1993 (Gruber 1993): "An ontology is an explicit specification of a conceptualization" and "A conceptualization is an abstract, simplified view of the world that we wish to represent for some purpose." Gruber's definition is not only one of the most cited but also one of the most discussed definitions in the relevant literature. Many authors generally agree with Gruber, but often they consider his interpretations as too universal. A great many further definitions therefore developed with the aim to expand the construction of Gruber. To this group, the following two definitions from Borst and Karp can be added as well (Borst 1997): "Ontologies are defined as a formal specification of a shared conceptualization." "An ontology is a specification of a conceptualization that is designed for reuse across multiple applications and implementations. Put another way, a specification of a conceptualization is a written, formal description of a set of concepts and relationships in a domain of interest."

The purpose of an ontology is to define an entity, attribute, and relationship among knowledge concepts within a specific domain using explicit descriptions and specifications that present an interoperable format that both humans and machines can understand, thereby realizing information sharing and reuse (Chen et al. in press). Ontologies have been widely applied in a variety of domains to represent information or knowledge models owing to the fact that their formal semantics can be unambiguously interpreted by humans and machines (Dong et al. 2008). As a matter of fact, some researchers have attempted to employ ontology-based techniques in manufacturing or in the production domain for integrating information. The early stages of product development need the combination of technological knowledge with identified product requirements, via a logical connection between those two domains. Borst (1997) addressed ontology-based assembly design to support collaborative product development so that design intent could be well understood by other designers, and the applications could reason about assembly information without any semantic ambiguity. Ahmed et al. (2007) developed an ontology for engineering design for information sharing among engineers to assist engineers in indexing, searching, and retrieving design information.

Another example is the field of mechanical engineering, where an ontology-based semantic network was successfully established as a homogeneous data basis for rapid prototyping (Diederich and Warschat 2007). Further, ontology-based applications can be found in academic knowledge portals, in data management and system integration, in electronic commerce, for the planning of production systems or for semantic web services in general (Fischer et al. 2005; Gao and Roller 1998; Schwartz 2014; Spath and Lentes 2005; Sure and Studer 2001).

In addition to illustrating technological knowledge, ontologies also offer advantages by simplifying the related identification of experts and connecting them with the technological knowledge. By using adequately modeled ontologies, new knowledge can be generated by inference. To be more specific, it is possible to search for technologies and applications in the ontology that comply with very specific criteria (Spath and Stefanie 2011).

Thus, the ontology model of an expert can be combined with a model of one or more knowledge fields and create a tool for the search for experts. This approach is not the combination of technology keywords, but it is rather problem oriented. This means that in the first step, the relevant technology is identified in the ontology by certain criteria and in a secondary search, the associated experts are identified immediately (see Fig. 8.1). With this approach, the search for experts becomes much more problem oriented and therefore more efficient (Spath and Stefanie 2011).

Product development demands the cooperation of experts from different parts of the organization with different expertise and varying levels of experience. As a result, effective communication is required (Effendi et al. 2002). Innovations are undisputedly based on the evolution of knowledge (Nelson and Winter 1982; Kogut and Zander 1992). Two current developments increasingly expand the importance of external sources of knowledge for the innovation activity of firms: First, the opening of innovation processes which is discussed under the term of "open innovation" in the field of innovation research (Chesbrough et al. 2006) and propagated in innovation management literature (Chesbrough 2003). Second, the

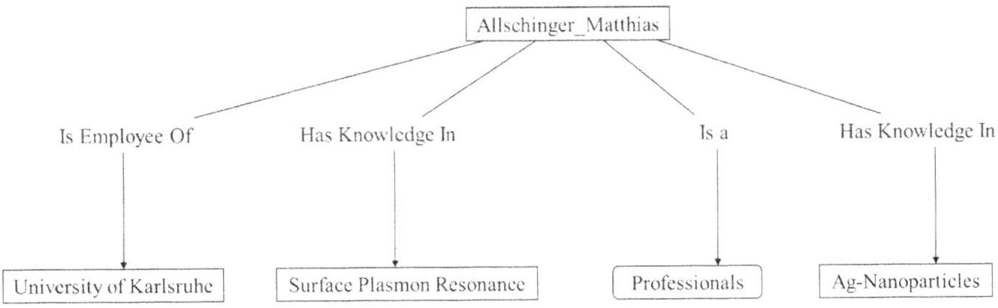


**Fig. 8.1** The connection between experts and the technological knowledge domain [a31]

acceleration of this opening through new information and communication technologies (Awazu et al. 2009). The trigger of these developments rests in the insight that the source of innovation may lie in the firm's external context (van Hippel 1988, 2005).

Data mining techniques assist in the understanding of the content of both the external contextual knowledge documents and internal data through document categorization and clustering (Porter 1991). For example, based on the set of keywords in a document, (Guo et al. 2012) used a genetic programming algorithm to calculate a fitness value to decide whether a document of interest can be linked to a specific user. Svingen (1998) constructed an unsupervised text classifier by using the Naive Bayes algorithm. They divided the documents into sentences and categories, each sentence using keyword lists and each category using a sentence similarity measure. The proposed method shows a similar degree of performance, compared with the traditional supervised learning methods. Lam and Han (2003) unified the strengths of k-nearest neighbor and linear classifiers to develop a generalized instance set (GIS) algorithm for automatic document categorization. To further enhance the performance of document classification, they proposed a meta-model framework, which uses the category feature characteristics derived from the training document set to capture inherent properties of a particular category. Consider the impact of these knowledge parsing activities to the design task. It has been established within the literature that 80 % of design engineers' activities are related to repetitive, routine, and mundane tasks, while the remaining 20 % are innovative tasks (Skarka 2007). Applying knowledge-based engineering (KBE) systems, including the use of ontologies to identify knowledge, could significantly reduce repetitive tasks (80 % automation) and allow product design engineers to gear their focus toward innovative design activities (Sanya and Shehab 2014).

## 8.2 Forecasting Innovation Pathways and the Technology Delivery System (TDS)

The Forecasting Innovation Pathways framework is a four-stage process that starts with understanding "New and Emerging Science and Technologies" or NESTs and its critical environment. The additional 3 stages include Tech Mining, Forecasting Likely Innovation Paths, and Synthesis and Reporting. Porter et al. (2010) breaks down those 4 stages into 10 steps. The first stage includes modeling the TDS in order to capture the organizational and contextual factors influencing the delivery of the technology and their dynamics. In Stage 1, where the nature of the technology is defined and TDS construction takes place is where a design ontology would be a useful tool. Placet and Fowler (2002) demonstrates how interest in an external goal such as sustainability can drive innovation in a technology domain such as the cement industry. While (Tait et al. 2014) concerns the drug industry, the principles and analysis of how regulatory demands and policy decisions affect the development decisions is pertinent to other areas and certainly to the naval ship design process. The purpose of the ontology is aligned with the goals of FIP in that the aim to reduce the uncertainty in technology development by better understanding the social, economic, and environmental changes. The question is how does this ontology accomplish this objective? An ontology can capture words that define the stages in the development process along with the sentiment and contextual terminology that may influence that development. The development of an innovation and design process ontology can aid in the understanding of the relevant contextual factors that would affect the innovation pathway, particularly where access to expert opinion or to key decision-makers is limited.

Wenk and Kuehn (1977) introduced the Technology Delivery System model in 1977 to represent the key elements, both institutional and contextual, that affect the delivery of a technology in a systematic manner. Figure 8.2 from [a2] shows the components of the system. There are three primary inputs: the technological inputs, inputs from the organizational landscape, and the environmental context. A fourth input is the feedback from the environment. In 2001, (Porter 1991) expressed the four elements in this manner: technological inputs, institutions, systems, and outcomes. (i) Technological inputs to the system include capital, natural resources, man power, tools, knowledge from the basic and applied research and human values, (ii) institutions refers to organizations and organizations, public and private, that play a role in the operation of the TDS or that modify and control its output, (iii) systems refers to processes by which institutions interact through information linkages, market, political, legal, and social means, and (iv) outcomes include both the intended and unintended effects on the social and physical environment.

Components (ii) and (iii) are primarily responsible for the delivery of the technology. These include the enterprises that deliver both the topic technology and the enabling technologies. What are the resources and commitments that the primary organization is able and willing to commit to the delivery of the technology product?

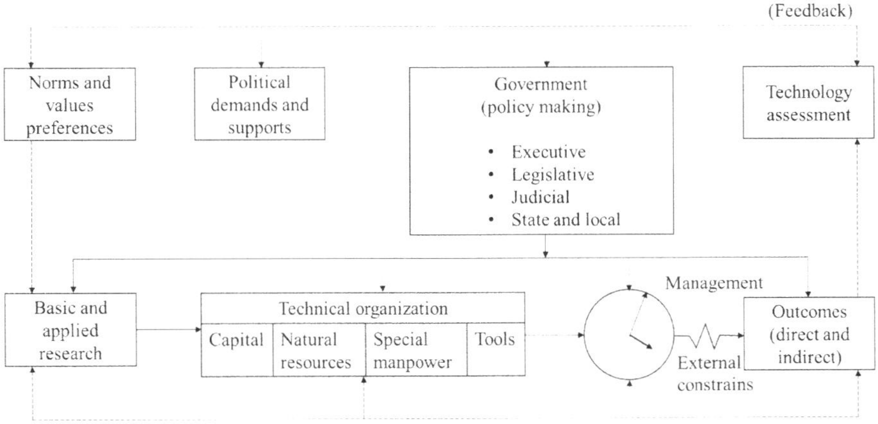


**Fig. 8.2** The technology delivery system

In delivering a technology, these elements are much easier to understand. Less understood are the influences outside of the delivery system, particularly in a turbulent environment. In an environment of high turbulence, user demands can have high turnover and their effects can be substantial. Such an environment is particularly difficult for large design products such as ships. Modeling the product development as a technology delivery system is a way to view the influences on the main delivery system, taking into account the pull of the user. In this ship design case, the users are Navy captains. The model also takes into consideration the contextual factors acting upon delivery including non-technical influences such as political elements, the interests of key decision-makers external to the direct process, regulations, and the impact of competitive forces. In this particular example, the impact of competitive forces could be the forces of within the shipbuilding industry or could be considered the forces of the enemy. Identifying ontological pattern related to innovation and the design process within the discussion of a technology can help decision-makers better understand the influence of these forces.

In order to demonstrate the efficacy of the TDS model, a citation within (Guo et al. 2012) presents a TDS explaining the difficult path of residential solar energy. Solar energy experienced significant government support for R&D efforts, oil prices were high, yet the technology did not take hold at the residential arena. By documenting the influences on delivery, five influences challenging the delivery of solar panels were uncovered, one of which was the lack of incentives for technology transfer within national labs, an issue solved by legislation. The others were the need for multiple entities to implement the panels such as manufacturers and homebuilders together, the risk averse nature of homebuilders, and the variety of building codes that exists.

The TDS model thus captures both the institutions directly involved in the development of a technology along with the assets held that contribute to the delivery of the technology as well as the institutions in society that are not directly involved in the development but influence it. In portraying technologies that are developed by government institutions, there is an interesting dilemma to consider. The government agency, in this case the Navy, is directly responsible for both the delivery of the technology and many of the influences affecting development as well as is the user. The context in which naval ship acquisition takes place includes identifying what capabilities it needs to meet the strategic direction and priorities set forth in high-level strategy and guidance documents such as the National Military Strategy, National Defense Strategy, and Quadrennial Defense Review. These documents take into consideration the long-term strategy of the Navy, current state of warfare, political and geopolitical influences, national financial influences, and changes in technology (Schwartz 2014). All these influences can help determine the mission a certain asset is designed to fulfill. However, changes in the political landscape internally and globally can alter that mission and ultimately affect the innovation pathway.

In our example, we are using the development and delivery of a ship asset within the DOD's acquisition structure. In developing this particular set of indicators, an ontology was developed that addresses the terminology of this domain. However, it is clear that the broad structure is relevant across multiple domains and relates to the early structure of the FIP framework. Within that framework, whether it be in industry or a government setting, the TDS provides a guideline to capture the institution and external factors affecting delivery, though the details may change from one industry or technology to another. So, for example, very different from findings in solar panels, the TDS provided the guideline described by Sure and Studer (2001) that identified how language barriers impacted the delivery of microcomputer technology in developing countries.

Technology mining is the second stage of the FIP framework but can also be utilized in modeling the TDS in Stage 1. It can play a role in populating the TDS as a means to uncover many of these relationships. For instance, Sterling and Theodor (1977) described the need for microcomputers in developing countries to have software written in native languages. Technology mining enables the rapid detection of patterns in research work that can be used to populate a technology road map. It can also be used to identify patterns in the discussion surrounding the development of the technology. Is the discussion positive or negative toward the development of the technology? Who has an interest in the technology and what are they saying about it? How strong is the underlying technological infrastructure that is developing the technology? Where are the enabling technologies in the development process? There are also a number of sources that can supply the feedback channels. Trade literature and search engines can provide information on the applications of the technology. Reviews and blogs provide insight into the response of the marketplace. In the case of naval ships, news media, GAO reports, and patent filings provide insight into the evaluation of the asset post-delivery.

In the context of our ship acquisition problem, an asset must go through a three-step process of identifying a required (needed) weapon system, establishing a budget, and acquiring the system. These three steps are organized as follows:

- The Joint Capabilities Integration and Development System (JCIDS)—for identifying requirements.
- The Planning, Programming, Budgeting, and Execution System (PPBE)—for allocating resources and budgeting.
- The Defense Acquisition System (DAS)—for developing and/or buying the item.

While each of these three steps is distinctly separate, there is a significant overlap in the responsibility. "The conventional wisdom notwithstanding, the process as spelled out in DOD's directives and instructions is fundamentally sound and could avoid its unending cost overruns, delays and performance failures, if it were implemented in a better informed and rigorously disciplined manner. The problem is not nearly as much in the laws and regulations as it is in the execution by the people who have been operating the system" (Christie 2014). The identification of the requirements and even the established mission of the program can often be elusive to the ship designers. In this effort, we address the need for decision-makers, particularly early stage concept designers to be "better informed." If there is a better understanding of the non-technical influences on the delivery process, then different decision may be made at the front end. The challenge is the wall that often divides various points in the delivery process. The integration of text mining in the TDS has been expanded and integrated into the discussion of technology mining by Porter as a means to provide a context to innovation indicators (Fig. 8.2). However, the link between the two is at a general level (Guo et al. 2012). With a long-term view in mind, the present study seeks to identify patterns and indicators that can be directly linked to components of the TDS.

## 8.3 The Ontology Development

We take this particular problem—the acquisition of naval ships—and describe the process as well as relate the developed ontology to a broader context. First, the ontology structure demonstrated in Fig. 8.3 was created. Some of the second-level categories have another level of sub-category that is not depicted here. There are six top-level categories: Influence Words, Design Words, Construction Words, Geography, Assets, and Person-Groups. This ontology is specific to this particular naval architecture example, but it can be generalized according to Fig. 8.4. The broader categories are the same, but in particular instances may have different terminology or a slightly different direction.

In our example, the Product Types is the asset built by the military. "Influence Words" are those words that reflect influence on the development of the technology. The sub-categories of Influence Words demonstrate the type of information that is relevant regarding the Technology Delivery System, for example "DC Political" words are those words that show the influence of politics in the nation's capital, including regulatory bodies, while "Strategy" includes terms that reflect the strategy of stakeholders which may include the primary enterprise or even standards organizations.

"Design Words" are those words that demonstrate the market pull of the technology. How will the market actually use the technology? How is that driving the discussion in design and how is are the specifications taking shape over time? In our specific shipbuilding example, the use of the technology can be construed as the "Mission Words." What mission will the asset cover? It is equivalent to the "User Application" sub-category in Fig. 8.4 and is the category that would require to most domain-specific language. However, each of these categories may have a broader framework with a domain-specific component added on.

In this example, there is also a category for "Construction," which reflects the terms and entities that are actually manufacturing the ship assets, corresponding to the manufacturing component of the delivery system.

The last three categories, Geography, Assets (Products), and Persons-Groups, have less of a stand-alone relationship with the delivery system but are the result of categorizing entity extractions that can be used in conjunction with the other categories. For example, matrices that cross Person-Groups with other categories

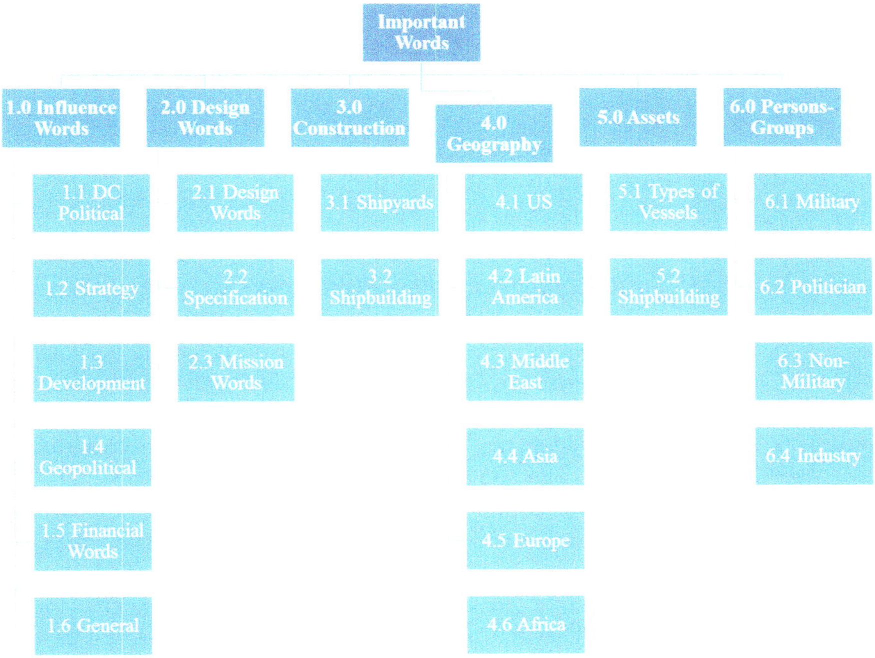


**Fig. 8.3** The important words ontology for naval vessel design

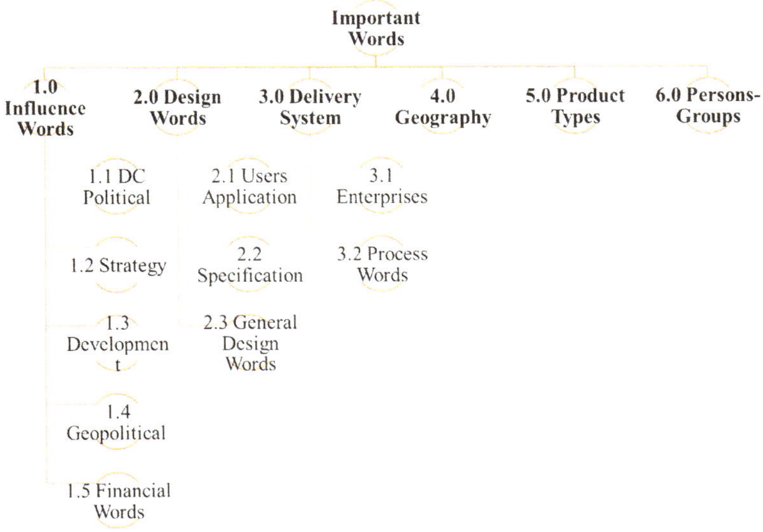


**Fig. 8.4** A generic important words ontology for new product development

easily enable the evaluation of the influences of different parties within the TDS structure. So, for example, the ontology allows an analyst to sort out the strategy terms for the primary enterprise and for external organizations.

### *8.3.1 The Shipbuilding Example*

For this project, we started with four examples of ship development projects, the Littoral Combat Ship (LCS), Mobile Landing Platform (MLP), Arsenal Ship, and Auxiliary Dry Cargo Ship (ADC). We first retrieved documents from government sources that told the narrative of these ships. The idea is to determine whether by using an ontology framework, we would be able to find indicators that corresponded to the actual narrative. This first proof-of-concept phase should be followed up to determine whether the indicators found would correspond to the narratives of many assets, in addition to these specific assets.

Next, searches were conducted in Compendex and Academic Search Complete. Compendex included technical records while Academic Search Complete included newspaper articles and articles from military magazines and conferences discussing the asset development. Figure 8.5 shows the search terms and number of documents downloaded from each source. To create the ontology, all of those files were combined into one ontology file in VantagePoint text mining software. The abstract phrases, full-text phrases, title phrases, and keywords were all combined and cleaned. The result was a list of 3500 terms each of which occur at least four times. The 3500 terms deemed "Important Words" became the basis for analysis. New term lists were created from these groups, and the subgroups were then created for each list. A set of VantagePoint "thesauri" were created at each level and combined into one large "ontology" thesaurus file. This ontology thesaurus file was run against each of the phrase lists of the four individual ship files (Fig. 8.6).

Figure 8.6 shows a snapshot of the software. In Fig. 8.7, we see indications of the driving political forces regarding the particular asset. Such analysis could lead us to determine who are the stakeholders and determine their positions on the technology. It may also aid in determining the level of profile the technology is garnering. In comparing technologies, the difference between one list and another is important. Figure 8.8 contains mission words, and at this level, these words are fairly broad and reflect the coverage of many assets in one document. However, some notable terminologies distinct in the high-frequency list of the LCS ship are words such as "mission module," "flexibility," and "mission package" which are terms that are not common to all the ships and reflect the unique missions of the LCS ship.

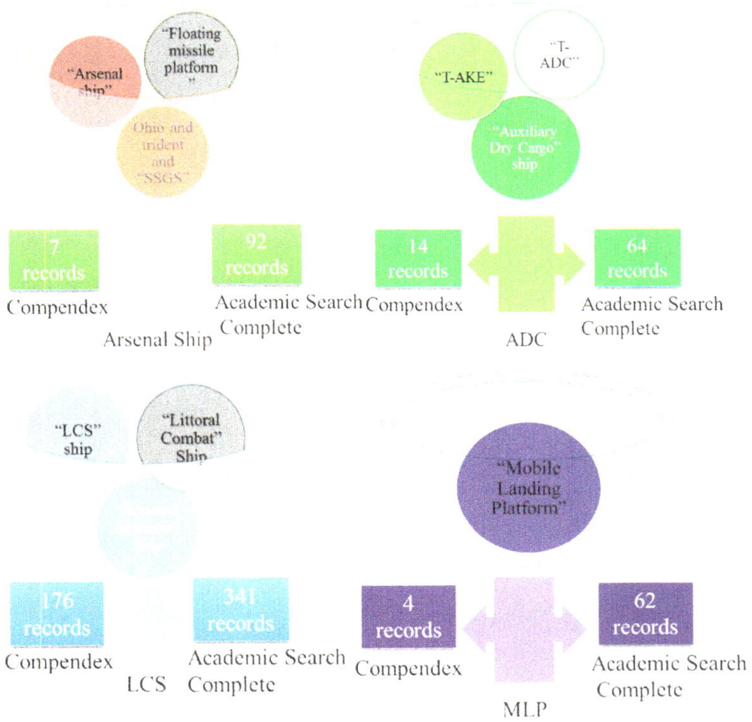


**Fig. 8.5** The search terms and number of documents downloaded from each source

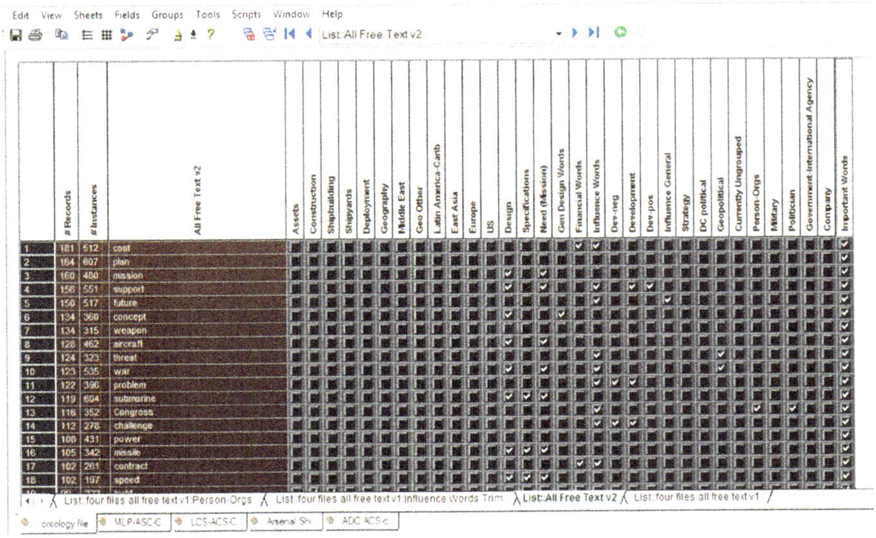

**Fig. 8.6** Snapshot of VantagePoint phrase lists and category tags

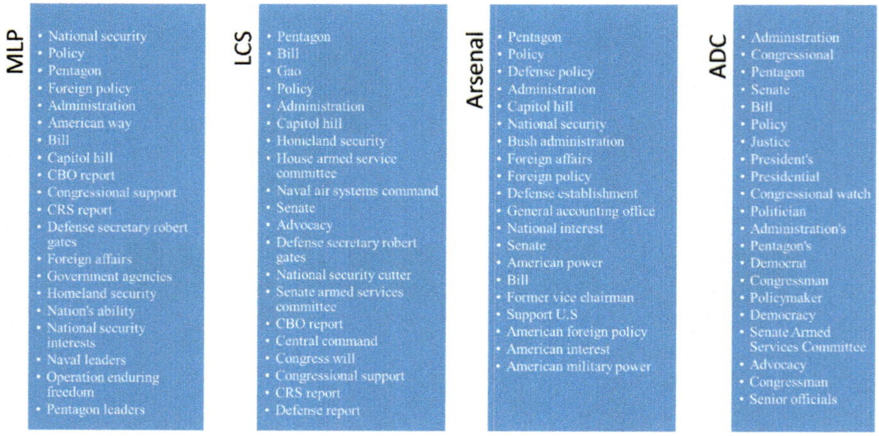


**Fig. 8.7** 1.1 DC Political: DC political words found most prominent in each of the ship cases

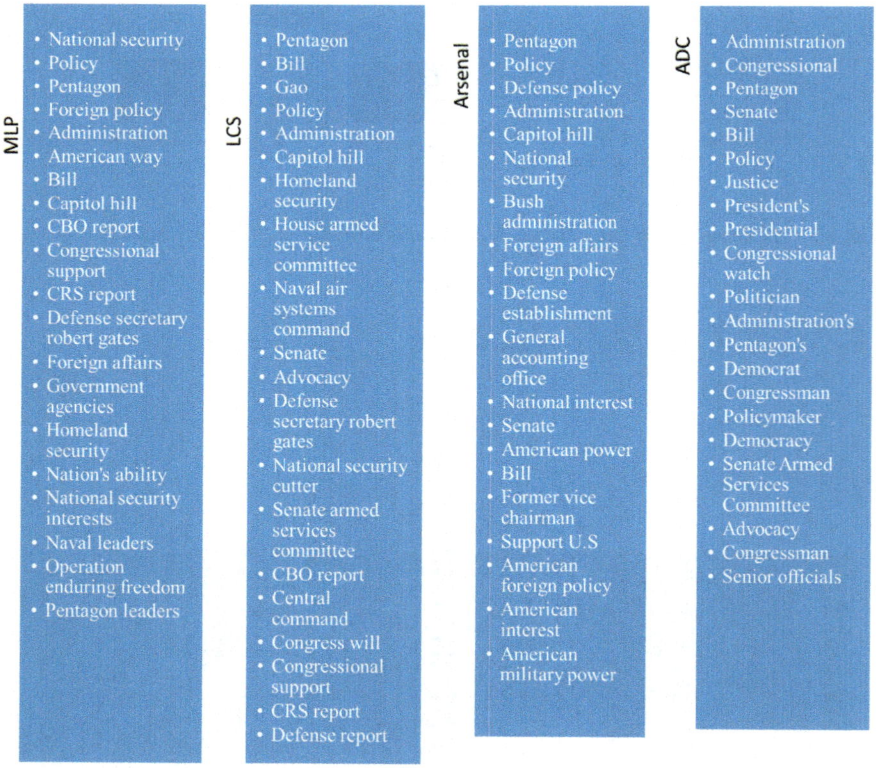


**Fig. 8.8** 2.3 Mission Words: Lists the mission words found most prominent in each of the ship cases

### 8.3.2 Example Technology Delivery System Indicators

Once the ontology is in place, it allows the analyst to develop indicators that can work within the TDS framework. Our first task was to determine whether indicators existed that mirrored the narrative stories. We present herein examples of results that we found in two contrasting narratives. The MLP had a significant amount of government discussion that took place outside of the public domain. A significant amount of testing was performed before the initial contracts were written. The articulation of the need for the ship and the gap it would fill was clear. There was also little change in the specification from initial design to production of the vessel. Changes in the contracts resulted in overall less costs, and the scheduling was on time. In contrast, the LCS had little research completed before the initial design contract was awarded. Much of the discussion about the ship took place in the public domain. The specifications were numerous at the onset and remained so. As a result, the costs continued to rise and the deadlines continued to move further into the future. The differences between the acquisition processes in the two ships are mirrored in derived indicators for the two ships. Tables 8.1 and 8.2 show the

**Table 8.1** Mobile landing platform comparison of narrative and indicators

| Year | Narrative | Indicator |
|---|---|---|
| 2005 | 15-year plan<br>Clear objectives communicated<br>Principle used post-Katrina | Minimal activity across all categories<br>Peak in new mission words |
| 2006 | Concept testing<br>The gap filled clearly communicated<br>Broad specs communicated | Minimal discussion |
| 2007 | Program position communicated<br>The gap filled clearly communicated<br>Broad specs communicated | Minimal discussion |
| 2008 | Contract out for detailed design<br>Additional testing | Drop in new mission words |
| 2009 | $3.5 million contract signed<br>Initial requirements set<br>Contract modification in line with acquisition reform | Minimal discussion |
| 2010 | Budget $120 million<br>Demonstration exercise completed<br>Contract modification to lower costs on 1st, 2nd, and 3rd ships | Peak in new mission words<br>Lots of discussion |
| 2011 | Production of MLP 1 initiated<br>Contract modification for lower costs on 3rd ship | Jump in spec words<br>Slight drop in discussion |
| 2012 | Detail design contract issued for MLP 3<br>4th ship proposed<br>Final design similar to initial design with some lessened capabilities that lowered costs | Drop in Discussion |
| 2013 | Delivery of MLP accepted for testing<br>Broad specs largely unchanged | Peak in spec words<br>Jump in discussion<br>Discussion takes a more positive turn |

**Table 8.2** Littoral combat ship comparative of narrative and indicators

| Year | Narrative | Indicator |
|---|---|---|
| 2001 | Broad concepts and specs announced | |
| 2002 | | Lots of new mission words |
| 2003 | Financial commitment made before analysis or clearly defining purpose<br>Experimental vessel launched | Peak in Mission Words with steady decline afterwards, though lots of words |
| 2004 | Production contract awarded without field testing concepts | |
| 2005 | | Peak in spec words |
| 2006 | LCS-1 expected launch<br>Expected cost of LCS-2 $220 million<br>Long list of specs | Lots of spec words |
| 2007 | Frequent design changes and 50 % cost overruns<br>Long list of specs | Drop in spec words |
| 2008 | LCS-1 launched 18 months late and double the price, 6 % overweight | Shift to more negative discussion |
| 2009 | Contracts for LCS-3 and LCS-4 renewed | Increase in already large number of mission words |
| 2010 | LCS-2 commissioned, 330 % over budget<br>Personnel assignment problems identified due to design | |
| 2011 | LCS-2 continued budget increase identified<br>Structural problems identified<br>Navy unsure how to use it<br>Questions of survivability | Dip in discussion<br>Jump in influence words and financial words |
| 2012 | LCS-3 and LCS-4 commissioned<br>Long list of differing specs<br>Many problems in trials | |
| 2013 | Launch and christening of LCS-5<br>Questions of survivability | A peak in spec words |

parallels between the narratives of each ship and the corresponding indicators from the analysis of the open source documentation.

The MLP initially had very little public discussion. During the period of little open source documentation, conceptual testing was taking place. The analysis of the open source documentation shows a consistent mission and few specification changes once the ship began discussion in the public sphere (see Fig. 8.9). 2009 marks the year that a design contract was first signed. As in the other examples and as seen in the difference between 2012 and 2013 here, there is typically a one-year lag between the time major events occur and a bump in the increase of discussion. One noticeable occurrence is that in the year of deployment, there is a jump in the number of documents expressing positive development terminology in comparison with negative development terminology. In the context of the TDS, this information provides insight into the influences affecting delivery.

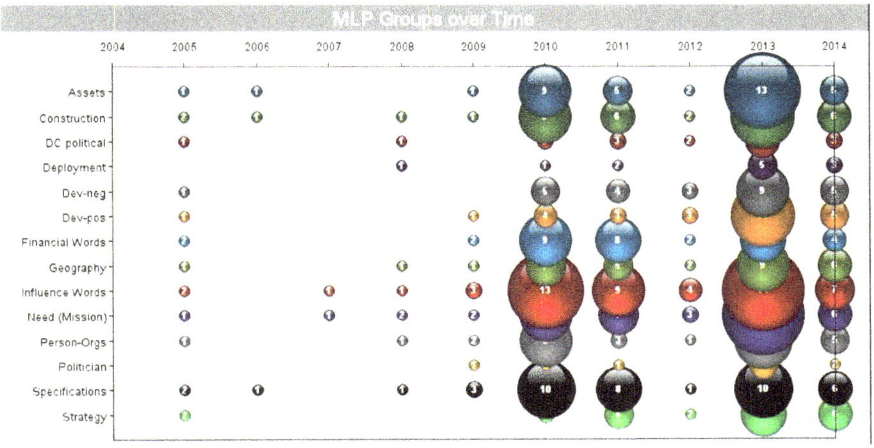


**Fig. 8.9** The narrative around the MLP ship, across the years 2004–2014

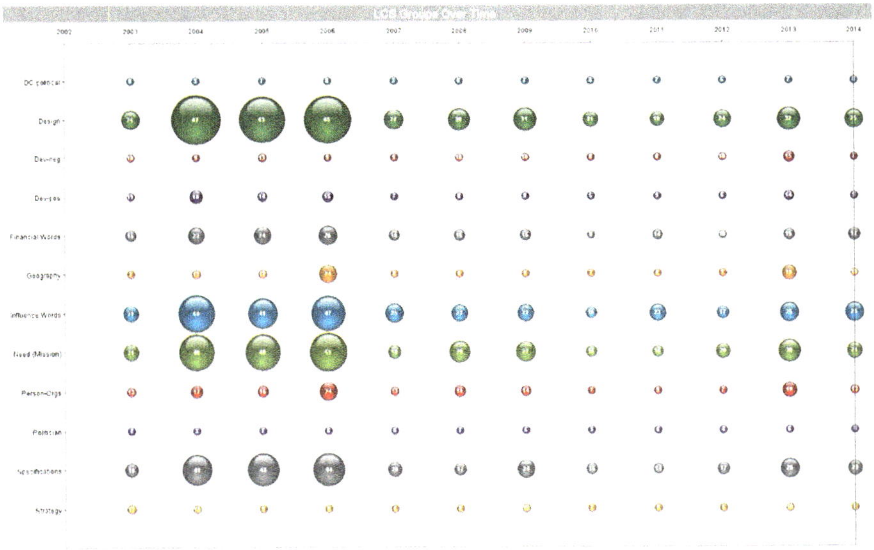

**Fig. 8.10** The narrative around the LCS ship, across the years 2003–2014

We can analyze who is speaking positively and negatively about development. The development and open source discussion of the MLP contrasts that of the LCS.

A review of the same chart for the LCS (Fig. 8.10) when compared to the narrative reveals that there was substantial open source discussion of the mission for this class of ship very early in the development process. Interestingly, as the narrative reveals a ship that had unrealistic expectations, numerous mission changes, and many problems, unlike the MLP that started with a very slightly negative

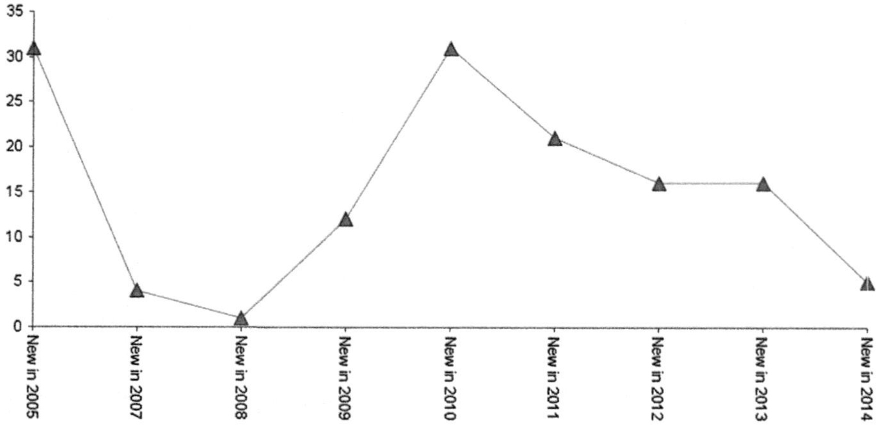


**Fig. 8.11** MLP new mission words per year

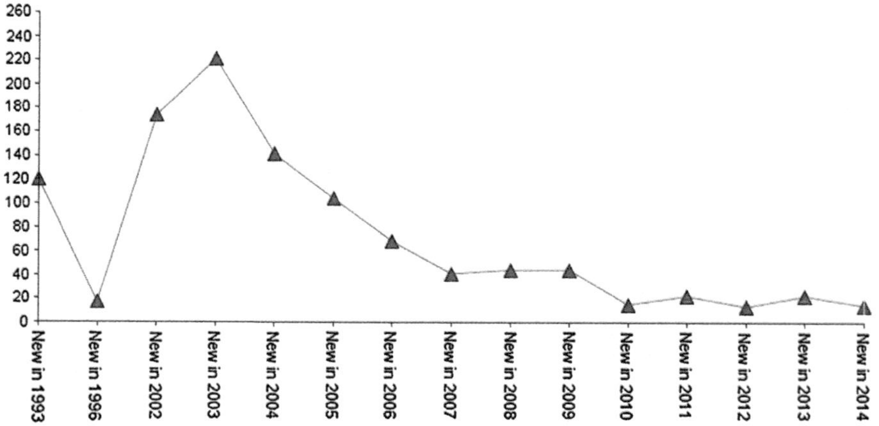


**Fig. 8.12** LCS new mission words per year

development discussion that shifted to a positive one, the LCS discussion started positively and fairly quickly, by 2006, the discussion terms became more negative. This is seen in the shrinking of the "dev pos" bubbles in Fig. 8.10. The number of documents and the number of mission-related phrases for the LCS were also are much higher throughout. Even accounting for the number of documents, the number of new mission words introduced each year for the LCS ship is significant. Compare Figs. 8.11 and 8.12 to see this clearly. In terms of a delivery system, this demonstrates a lack of strategic vision for the asset. Such flux would hinder the delivery and quality of the technology in question.

## 8.4 Conclusion

Within the Forecasting Innovation Framework, the Technology Delivery System model provides a mechanism by which the organizational, technological, and contextual influences on a technology can be visualized and better understood. Technology analysts have a sufficient challenge capturing the technical influences on the delivery of a technology. Even more challenging are the contextual influences. An ontology reflective of the language utilized in the steps in the design and development process is an important step through which these influences can be understood in order to identify potential innovation pathways. Identifying important entities and their potential influence on the technology can aid decision-makers in determining the appropriate strategy for technology delivery in turbulent environments. This research takes a step in that direction.

**Acknowledgments** The authors would like to acknowledge the support and help of Adrian Mackenna, NSWC Carderock Division—Code 8240, Steven Wynn B NAVSEA HQ, SEA 05, and Kelly Cooper, Office of Naval Research, Code 331.

## References

Ahmed, S., Kim, S., & Wallace, K. M. (2007). A methodology for creating ontologies for engineering design. *Journal of Computing and Information Science in Engineering, 7*, 132–140.

Awazu, Y., Baloh, P Desouza, KC, Kim, J. (2009). Information Communication technologies open up innovation. *Research Technology Management* 51–58.

Borst, W.N. (1997). Construction of Engineering Ontologies, Ph.D. thesis, University of Tweenty, Enschede, NL: Centre for Telematica and Information Technology.

Chen, Y.J., Chen, Y-M., Chu, H-C. Development of a mechanism for ontology-based product lifecycle knowledge integration. *Expert Systems with Applications*, in press.

Chesbrough, H. W. (2003). *Open innovation—the new imperative for creating and profiting from technology*. Boston: Harvard Business School Press.

Chesbrough, H. W., Vanhaverbeke, W., & West, J. (2006). *Open innovation: Researching a new paradigm*. Oxford: Oxford University Press.

Christie, T. (2014) Developing, buying and fielding superior weapon systems, refusing to misunderstand the defense acquisition problem, Center for Defense Information at POGO, 7 Oct 2014.

Diederich, M., & Warschat, J. (2007). Wissensrepräsentation und Kommunikation (RPD-IT-Infrastruktur). In *Entwicklung und Erprobung innovativer Produkte - Rapid Prototyping*, Berlin Heidelberg, Springer, 2007.

Dong, Y., Ming, D., & Rui, M. (2008). Development of a product configuration system with an ontology-based approach. *Computer Aided Design, 40*, 863–878.

Effendi, I., Henson, B., Agouridas, V., & de Pennington, A. (2002). Methods and tools for requirements engineering of made-to-order mechanical products ASME 2002 Design Engineering Technical Conferences and Computer and Information in Engineering Conference.

Fischer, T., Murphy, M., Tippmann, V., & Ayroumlou, M. (2005, July). Semantic web services enabling collaborative engineering, 11th International Conference on Human Computer Interaction, pp. 22–27. Las Vegas, Nevada, USA.

Gao, F., & Roller, D. (1998). Semantic based product model: Principles and representations. In D. Roller (Hrsg.): *Proceedings of 31th ISATA, volume automotive mechatronics design and engineering*, pp. 127–135. Croydon, England: Düsseldorf Trade Fair.

Gruber, T. R. (1993). A translation approach to portable ontologies. *Knowledge Acquisition, 5*(2), 199–220.

Guo, Y., Xu, C., Huang, L., & Porter, A. (2012). empirically informing a technology delivery system model for an emerging technology: illustrated for dye-sensitized solar cells. *R&D Management, 42*(2), 133–149.

Kogut, B., & Zander, U. (1992). Knowledge of the firm, combinative capabilities, and the replication of technology. *Org. Sci., 3*, 383–397.

Lam, W., & Han, Y. (2003). Automatic textual document categorization based on generalized instance sets and a metamodel. *IEEE Transactions on Pattern Analysis and Machine Intelligence, 25*(5), 628–633.

Neches, R., Fikes, R., Finin, T., Gruber, T., Patil, R., Senator, T., et al. (1991). Enabling technology for knowledge sharing. In AI Magazine, Fall.

Nelson, R. R., & Winter, S. G. (1982). *Evolutionary Theory of economic change*. Cambridge: Harvard University Press.

Placet, M., & Fowler, K. (2002, March) Toward a sustainable cement industry: How innovation can help the cement industry move toward more sustainable practices. An independent study commissioned by world business council for sustainable development.

Porter, A. L. (1991). *Forecasting and Management of Technology*. New York: Wiley, 1991. Print.

Porter, A., Guo, Y., Huang, L., & Robinson, D. (2010) *Forecasting innovation pathways: The case of the nano-enhanced solar cells*. Paper presentation. International Conference on Technological Innovation and Competitive Intelligence, Beijing.

Sanya, I. O., & Shehab, E. M. (2014). An ontology framework for developing platform-independent knowledge-based engineering systems in the aerospace industry. *International Journal of Production Research, 52*(20), 6192–6215.

Schwartz, M. (2014) Defense acquisitions: How DOD acquires weapon systems and recent efforts to reform the process, Specialist in Defense Acquisition May 23, 2014, Congressional Research Service. Federation Of American Scientists. Congressional Research Service, 23 May 2014. Web.

Skarka, W. (2007). Application of MOKA methodology in generative model creation using CATIA. *Engineering Applications of Artificial Intelligence, 20*(5), 677–690.

Spath, D., & Stefanie, B. (2011). Ontology-based technology model for the use in the early stage of product development. Technology Management in the Energy Smart World (PICMET), 2011 Proceedings of PICMET'11:. IEEE, 2011.

Spath, D., Lentes H.-P., & Lentes, J. (2005). Production Engineering: Research and development in Germany, annals of the German Academic Society for Production Engineering, Jg. 12, Nr. 2, S. pp. 117–120.

Sterling, Theodor D. "Acm Forum." Communications Of The ACM 20.12 (1977): 968-972. Business Source Complete. Web. 13 Feb 2015.

Sure, Y. & Studer, R. (2001) On-To-Knowledge Methodology. Employed and evaluated version, On-To-Knowledge EU IST-1999-10132 Project Deliverable D16 (WP5), http://www.ontoknowledge.org/downl/del16.pdf. Universität Karlsruhe 2001.

Svingen, B. Using genetic programming for document classification. *Proceedings of the Eleventh International Florida Artificial Intelligence Research Society Conference*, pp. 63–67. Florida 18–20 May 1998.

Tait, J., Bruce, A., Mittra, J., Purves, J., & Scannell, J. (2014, December) Independent review on anti-microbial resistance: Regulation-innovation interactions and the development of antimicrobial drugs and diagnostics for human and animal diseases.
van Hippel, E. (1988). *The sources of innovation*. New York: Oxford University Press.
van Hippel, E. (2005). *Democratizing innovation*. Oxford: MIT Press.
Wenk. & Kuehn. (1977). Technological delivery system.

# Chapter 9
# Generating Competitive Technical Intelligence Using Topical Analysis, Patent Citation Analysis, and Term Clumping Analysis

**Ying Huang, Yi Zhang, Jing Ma, Alan L. Porter, Xuefeng Wang and Ying Guo**

**Abstract** Because of the flexibility and complexity of Newly Emerging Science and Technologies (NESTs), traditional statistical analysis fails to capture technology evolution in detail. Tracking technology evolution pathways supports industrial, governmental, and academic decisions to guide future development trends. Patents are one of the most important NESTs data sources and are pertinent to developmental paths. This paper draws upon text analyses, augmented by expert knowledge, to identify key NESTs sub-domains and component technologies. We then complement those analyses with patent citation analysis to help track developmental progressions. We identify key sub-domain patents, associated with particular component technology trajectories, then extract pivotal patents via citation analysis. We compose evolutionary pathways by combining citation and topical intelligence obtained through term clumping. We demonstrate our approach with empirical analysis of dye-sensitized solar cells (DSSCs), as an example of a promising NESTs, contributing to the remarkable growth in the renewable energy industry. The systematic approach we proposed not only offers a macro-perspective

Y. Huang · Y. Zhang · J. Ma (✉) · X. Wang · Y. Guo
School of Management and Economics, Beijing Institute of Technology,
Beijing 100081, China
e-mail: wxf5122@gmail.com

Y. Zhang
Centre for Quantum Computation & Intelligent Systems,
University of Technology Sydney, Sydney 2007, NSW, Australia

A.L. Porter
School of Public Policy, Georgia Institute of Technology,
Atlanta 30332, GA, USA

A.L. Porter
Search Technology Inc., Atlanta 30092, GA, USA



T.U. Daim et al. (eds.), *Anticipating Future Innovation Pathways Through Large Data Analysis*, Innovation, Technology, and Knowledge Management,
DOI 10.1007/978-3-319-39056-7_9

covering technology development levels and future trends, but also makes R&D information accessible for micro-level probes as needed. We work to uncover developmental trends and to compile mentions of possible applications, and this study informs NESTs management by spotting prime opportunities for innovation.

**Keywords** Innovation pathways · Citation analysis · Text mining · Topic analysis · Dye-sensitized solar cells · Technology roadmapping

## 9.1 Introduction

Analyzing and tracking the historical and current stages of a technology are critical for gaining competitive advantage (Choi and Park 2009). Patents, as a fruitful data source, are main outputs of research and development that represent the characteristics of an emerging technology, and thus, they are important for technology innovation management (Huang et al. 2014b). A large portion of recent technical knowledge is available in patent documents, and the importance of exploiting this knowledge is constantly increasing. Additionally, the number of patents is rapidly increasing, and the pace of technological development is accelerating. Identifying core and emerging technologies is crucial for formulating technology strategies and policies that pursue promising technological opportunities (Cho and Shih 2011).

Compared with traditional technologies, Newly Emerging Science and Technologies (NESTs) have tremendous innovation potential (Robinson et al. 2013). Domain experts may become less reliable due to increasing data and the fragmentation of technology domains (Shibata et al. 2008). Data-based methods offer an appealing alternative to expert opinion. However, some patent-analytic methods only use simple bibliometric indicators—e.g., logistic growth curves (Milanez et al. 2014)—and compare the numbers of patents assigned to different entities—e.g., nations, companies, inventors, citations, and technological fields—over time (Watatani et al. 2013; Bengisu 2003; Harhoff et al. 2003). Such indicators are useful, but they cannot reflect micro-level technology changes, especially for NESTs that manifest an increasing technological complexity and a shortened technology life cycle (Han and Shin 2014). More researchers are making efforts to adopt advanced qualitative techniques, including morphological analysis (Yoon et al. 2008; Lee et al. 2007; Yoon and Park 2005), TRIZ (Zhang et al. 2014b; Yoon and Kim 2011a), conjoint analysis (Xin et al. 2010; Lee et al. 2008; Yoon and Park 2007), and technology roadmapping (Huang et al. 2014a; Zhang et al. 2013, 2014c; Choi et al. 2013; Lee et al. 2008b, 2009). Experts' involvement enhances the effective identification of competitive technical intelligence. However, such expert-based methods depend highly upon the experts' knowledge, experience, and motivation, while experts' biases and, sometimes, insufficient knowledge may create difficulties. If, however, domain experts can reflect upon knowledge derived from data mining, then more accurate conclusions can be obtained. Thus, our research combines expert-based methods with large-scale (text) data-based methods.

In this paper, we combine topical analysis, patent citation analysis, and term clumping analysis to generate competitive technical intelligence, which achieves a balance between data-driven and expert-influenced conclusions. "Tech mining" is a multi-step process to analyze Science, Technology and Innovation (ST&I) information resources by using text mining, visualization, and communication tools (Porter et al. 2011). Term clumping, a characteristic method of "Tech mining," cleans and consolidates topical content in the publication and patent records and then extracts topical content intelligence (Zhang et al. 2014a). Meanwhile, we explore patent development paths in a large-scale patent citation network by evaluating the weight of citations among patents to provide a more robust understanding of influential nodes. The combination of these methods can, to some extent, mitigate their respective drawbacks and make full use of their strengths in (1) obtaining technical core terms in domain areas, (2) identifying influential nodes of a directed citation network, and (3) discovering significant clues about technology hot spots now and technology development prospects in the future.

Our case study focuses on dye-sensitized solar cells (DSSCs), a third-generation photovoltaic technology. The outstanding features of this technology—e.g., low cost, ease of fabrication, environmental friendliness of raw materials, and relatively high efficiency—have attracted tremendous scientific and industrial attention.

The remainder of this paper consists of five sections. In the following section, we make a brief literature review to introduce some related works. In the section entitled "Data and Methodology," we describe the dataset and framework of our research. The next section, "Empirical Research: DSSCs," applies the proposed approach for developing technology evolution pathways, and this incorporates patent topical analysis, patent citation analysis, and term clumping analysis. Finally, in "Conclusions," we present remarks and directions for further study.

## 9.2 Related Works

### *9.2.1 Bibliometrics and Text Mining*

Bibliometrics, as the analysis of R&D literature (and other) information, has the ability to explore, organize, and analyze large volumes of historical data. It helps researchers find "hidden patterns" to support ST&I decision-making processes (Daim et al. 2006). Lacasa et al. (2003) applied patent statistics to trace technological changes in chemicals over long periods of time in Germany; Zhou et al. (2014) analyzed technology evolutionary pathways for sub-technologies within different time intervals according to the linkages among IPCs; Han and Shin (2014) applied bibliometric analysis and trend impact analysis to identify the gap between current and market-required performance by specifying key performance indicators, considering a reverse salient; Zhang et al. (2013) integrated bibliometrics analysis with qualitative methodologies, first to construct a hybrid model for composing technology roadmaps, and then to understand the macro-technology development status; Yoon et al. (2014) carried out bibliographic analyses to trace the development of printed electronics so as to better understand the technology's evolving characteristics and insights for future R&D pathways.

Text mining, which extracts knowledge from unstructured textual data, is useful for anticipating new technologies and new uses for existing technologies (Smalheiser 2001). This approach also enhances ST&I roadmapping through bibliometric analysis, such as co-word analysis (Kostoff and Schaller 2001). However, isolated text mining techniques may lack reliability, have technical limitations, and present challenges in understanding topical information and knowledge flows from patents.

### 9.2.2 Patent Citation Analysis

Patent citations represent the foundational knowledge for a specific inventive step and indicate potential technological significance, as they generally correlate with valuable innovations (Jaffe and Trajtenberg 2002). Kajikawa and Takeda (2008) identified the trends and current structure of biomass and biofuel research through citation network analysis; Boyack et al. (2009) mapped the structure and evolution of chemistry research using journal citation patterns; Choi and Park (2009) introduced a systematic algorithm to extract a patent-based technology development map from patent citation information and also to track the history of the technology development to help understand micro-processes of technological innovation; Chang et al. (2010) used patent network analysis to monitor the technological trends in the field of carbon nanotube field emission display (CNT-FED); Barirani et al. (2013) validated the use of patent citations as indicators for technological relatedness, as well as tools for evaluating an industry's development stage; Ho et al. (2014) applied a citation network analysis method to identify major research development trends, crucial technological issues, and possible resolutions to help improve R&D investments in fuel cell development.

Citation-based analysis has a fundamental limitation, in that it underestimates the importance of contemporary patents and may not work in rapidly evolving industries where technology life cycles (TLC) are increasingly short and too many new inventions are being patented throughout the world (Yoon and Kim 2011b).

### 9.2.3 Competitive Technical Intelligence

Technology opportunities analysis (TOA) was first proposed by Porter and Detampel (1995). It generates effective intelligence and insight into specific emerging technologies by using bibliometric methods that are augmented by expert opinion (Porter and Detampel 1995). Since the development of intelligence information provides a possibility for tapping potential technological innovations

(Ma et al. 2014), our research explores opportunities based on "Tech mining" (Porter and Cunningham 2005) to support innovation pathway foresight. Yoon and Kim (2011b) proposed a patent network based on semantic patent analysis, using subject–action–object (SAO) structures to identify the technical importance of patents, the characteristics of patent clusters, and the technical capabilities of competitors; Zhang et al. (2014a) blended bibliometrics and text mining techniques to explore key technological system components, current R&D emphases, and key players for a specified technology; Ma and Porter (2015) extracted topical intelligence from patent compilations to identify potential innovation pathways and technology opportunities in the Nano-Enabled Drug Delivery (NEDD) domain.

## 9.3 Data and Methodology

In this research, we offer a systematic approach to trace technology evolution. The framework, organized in five steps, is illustrated in Fig. 9.1. The five steps are as follows: (1) download related patents (data search and collection), (2) acquire sub-technology datasets, (3) obtain core technological keywords, (4) obtain complementary terms, and (5) trace technology evolution roadmap.

### *9.3.1 Data Search and Data Collection*

As the initial step of bibliometric analysis, devising an appropriate search strategy requires attention to assure quality data. However, it is notoriously difficult to define the boundary of a multidisciplinary and emerging field and to harvest the relevant publications and patents. In our opinion, the search strategy should balance between precision and recall. On one hand, we need to retrieve a dataset whose records are relevant to the targeted field. On the other hand, we cannot spend too much time refining the search strategy to eliminate noise: After a certain point, the retrieval results become asymptotically stable, despite additional efforts to refine the search. We also recognize the potential to clean data, as needed, after download, so recall takes some priority over precision in our approach.

In this paper, we choose DSSCs as our target technology field. All patents are collected from the Thomson Reuters Derwent Innovation Index (DII) database. The search terms used are shown in set "#1" in Table 9.1. Unlike previous research, we combine keywords with manual codes in order to achieve more effective results. Keywords are easy to understand, even for those who are not specialized in the field. Manual codes, applied by a subject expert, not only cover standard technology terms, but they are also able to represent a technology and reflect the group (e.g., a patent family, a group of related inventions filed in one or multiple patent authorities) that each record belongs to (Thomson Reuters 2015). Our search resulted in 5668 records, reflecting a group of related inventions filed in one or more patent authorities from 1991 to 2014 (retrieved on March 12, 2015).

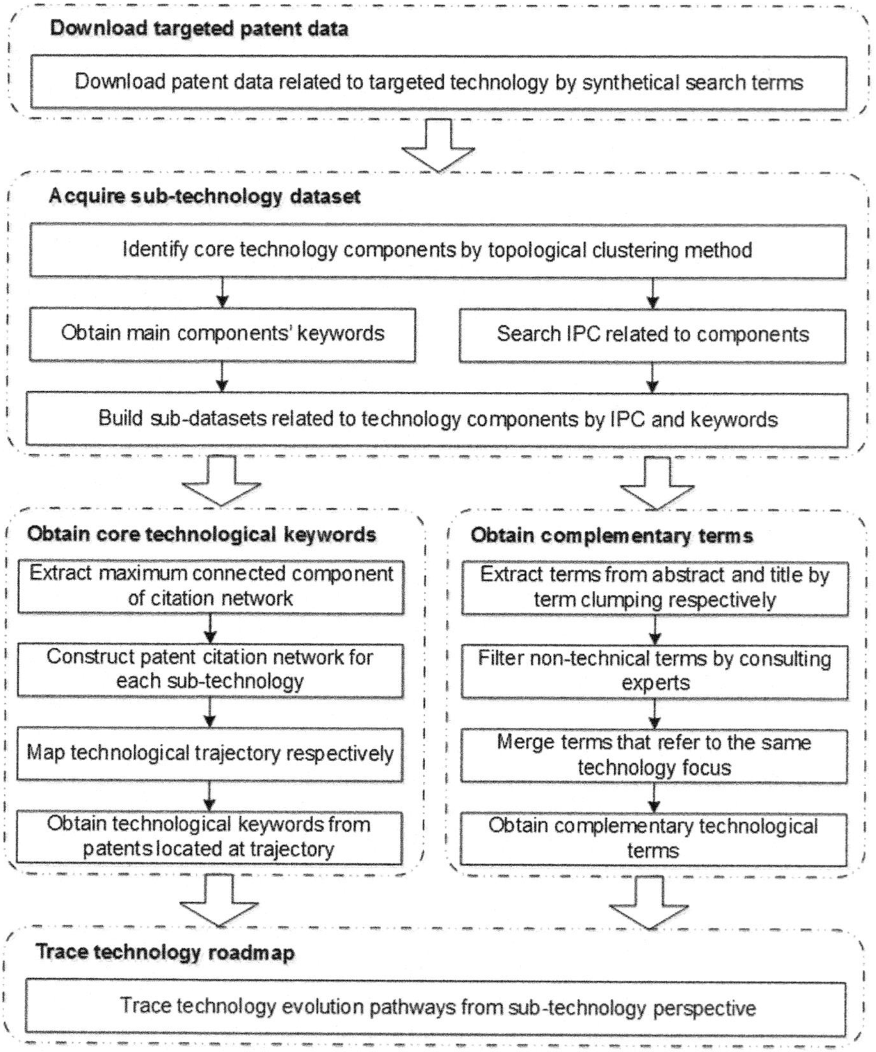


**Fig. 9.1** Framework for tracing technology evolution

## *9.3.2 Topical Analysis for Sub-technologies*

Unlike our previous work, in which we treated the domain technology as a whole, this paper divides the domain technology into sub-components for further exploration and more detailed technology intelligence. International Patent Classification (IPC) terms provide a hierarchical system of language-independent symbols for classifying patents according to the different areas of technology to which they pertain (http://www.wipo.int/classifications/ipc/). When seen as technology classes,

**Table 9.1** Search result of different sub-technologies of DSSCs

| Set | Number | Result |
|---|---|---|
| #1 | 5668 | TS = (((dye-sensiti*) or (dye* same sensiti*) or (pigment-sensiti*) or (pigment same sensiti*) or (dye* adj sense)) same ((solar or Photovoltaic or photoelectr* or (photo-electr*)) same (cell or cells or batter* or pool*))) AND MAN = (L03-E05B1 OR U12-A02A8 OR X15-A02D1 OR X16-A04) |
| #2 | >100,000 | IP = (H01L*) AND TS = (dioxide OR oxide) AND (anode)) |
| #3 | 2181 | #2 AND #1 |
| #4 | 31960 | IP = (C09B*) OR TS = (sensitiser OR sensitizer) |
| #5 | 615 | #4 AND #1 |
| #6 | 74849 | IP = (H01G* OR H01 M*) and TS = (electrolyte) |
| #7 | 1674 | #6 AND #1 |
| #8 | 97287 | IP = (H01L-031/0224) OR (IP = (H01 M*) AND TS = (electrode)) |
| #9 | 2348 | #8 AND #1 |
| #10 | 53 | #9 AND #7 AND #5 AND #3 |
| #11 | 4089 | #9 OR #7 OR #5 OR #3 |

*Note* Indexes = CDerwent, EDerwent, MDerwent Timespan = 1991–2014

IPCs are useful analytical units for exploiting the information in patent databases (Dibiaggio and Nesta 2005).

After downloading the patent data, we use VantagePoint, a professional text mining software (www.theVantagePoint.com), for topical clustering analysis. Natural language processing (NLP) helps extract a set of phrases and terms from specified textual fields (e.g., title and abstract). Next, we use ClusterSuite, a compilation of VantagePoint algorithms, to reduce noise, consolidate related items, and provide more refined topical information (O'Brien et al. 2013). We apply principal components analysis (PCA) to the top 200 informative phrases to identify the core technology components that are most often considered sub-technologies. One of the most important aspects of this research is record selection. In DSSCs, some high-frequency terms (e.g., semiconductor, solar cell) and common terms (e.g., photoelectric conversion, optoelectric transducer) should be removed. During the keyword selection process, we achieve better results with the help of domain experts. Thus, in this paper, related keywords and IPCs correspond to technology components that are identified with the help of domain experts, allowing us to obtain sub-technology datasets.

### 9.3.3 Patent Citation Analysis

Based on the sub-technology datasets, we construct the patent citation network consisting of both connected components and isolated nodes, respectively. We ignore the patents that never cite others or are never cited by others, and concentrate on the maximally connected component. We conduct this analysis in the following four steps:

1. Merge patents into record families. A patent family is the collection of patents in different countries referring to the same technical topic (Michel and Bettels 2001). Our first step is to merge the patent documents of a family into a single family record, which is identified by the priority patent number (the earliest patent in this family).
2. Construct the patent citation network. While conducting main path analysis for a given field of technology, only citations between patents within the technology field are considered. These effective citations are extracted from the merged family records. A patent citation network can be represented as a patent citation matrix. Nodes stand for the individual family records, and the arcs between two nodes are citations.
3. Identify the main paths of the patent citation network. In a citation network, the main trajectory is the path from a source vertex to a sink vertex that has the highest traversal weight on its arcs. Several methods have been proposed to extract main paths from the network of traversal weights. The method we use is "Search Path Count" (SPC) (Batagelj and Mrvar 2004), which extracts the main trajectory, meaning the path from a source vertex to a sink vertex with the highest traversal weight on its arcs.
4. Extract key technical intelligence from the patents located on the main path. The key patents located on the trajectory are obtained, and their technological keywords are extracted manually. This information is used to create an initial technology evolution road map.

### 9.3.4 *Term Clumping Analysis*

Although we have obtained the preliminary technology evolution pathways with patent citation analysis, the technology focus included in some less-cited patents might influence the whole technology development circle. One solution depends on the "terms" derived from NLP techniques. Phrases and terms retrieved in this way are large and "noisy," making them difficult to manually categorize. Using bibliometric and text mining techniques, this paper applies the semiautomatic "term clumping" steps, which generate better term lists for achieving competitive technical intelligence (Zhang et al. 2014a).

"Term clumping" includes four main phases: (1) common and basic term removal, e.g., instance, technology; (2) fuzzy word matching (combine terms with similar structures based on pattern commonality, such as stemming—e.g., sensitiser and sensitizer, and combine singular and plural forms of English words, e.g., dye and dyes); (3) extreme words removal [remove very common (top 5 %) and very rare (occurrence in single records)]; (4) combine term networks (combine select low-frequency phrases with the high-frequency phrases that appear in the same records, sharing terms).

After extracting the terms and phrases from titles and abstracts, some weakly correlated terms are removed after consulting with experts. At the same time, some keywords that indicate the same technology focus are merged to improve the integration level. The final keywords reflecting the technology focus are obtained to construct a technology evolution roadmap, building on former analysis experiences.

## 9.4 Empirical Research: DSSCs

### *9.4.1 DSSCs Patent Topical Analysis*

It is crucial to identify the main sub-systems and the evolution road map for key topics of DSSCs. In the first step, we use VantagePoint's NLP to extract nouns and phrases to obtain a keyword list from the titles and abstracts of 5668 records. A total of 1562 terms are obtained by the ClusterSuite process of term clumping analysis. In this context, we select the top 200 terms as the high-level terms and use Factor Maps (PCA) to reduce the number of items for further topical analyses (Newman et al. 2014). Seventeen clusters are generated, and most of them are related to each other. Based on experts' review, our preliminary cluster result effectively reflects the major characteristics of DSSCs. The four major sub-technologies are as follows: photoanode, electrolyte, counter-electrode, and sensitizer (Fig. 9.2).

The photoanode components generally cover various nano-structured materials, such as titanium dioxide and zinc oxide, and these keywords often include "dioxide" or "oxide." Obviously, "anode" is another keyword. IPC H01L relates to semiconductor devices and semiconductors technologies, which are the key parts for photoanodes. In the sensitizers group, "dyes" is practically synonymous with "sensitizers," and C09B specializes in "organic dyes or closely related compounds for producing dyes." For the electrolyte classification, H01G stands for electrolyte light-sensitive or temperature-sensitive devices, so most of the patents related to electrolyte are included. Additionally, because of the close relationship between electrolyte and counter-electrode, some patents are also located in the field of H01M, which mainly focuses on electrodes and voltage generators. Under the counter-electrode component, aside from H01M, there is a specialized classification named H01L-031/0224, which is related to counter-electrodes. The detailed search terms and their results are shown in Table 9.1.

We use keywords and IPCs to generate the sub-technology datasets from the original database. The advantage of this method is that it can get a more accurate dataset than the approaches that only use IPCs (Zhou et al. 2014) or keywords (Ma et al. 2014). The disadvantage of this method is the inherent risk that some patents are excluded. Overall, this approach helps us to obtain meaningful and useful, albeit incomplete, data in a reasonable way. In the end, 72.14 % (4089 records) of the patents were divided into the four sub-technologies.

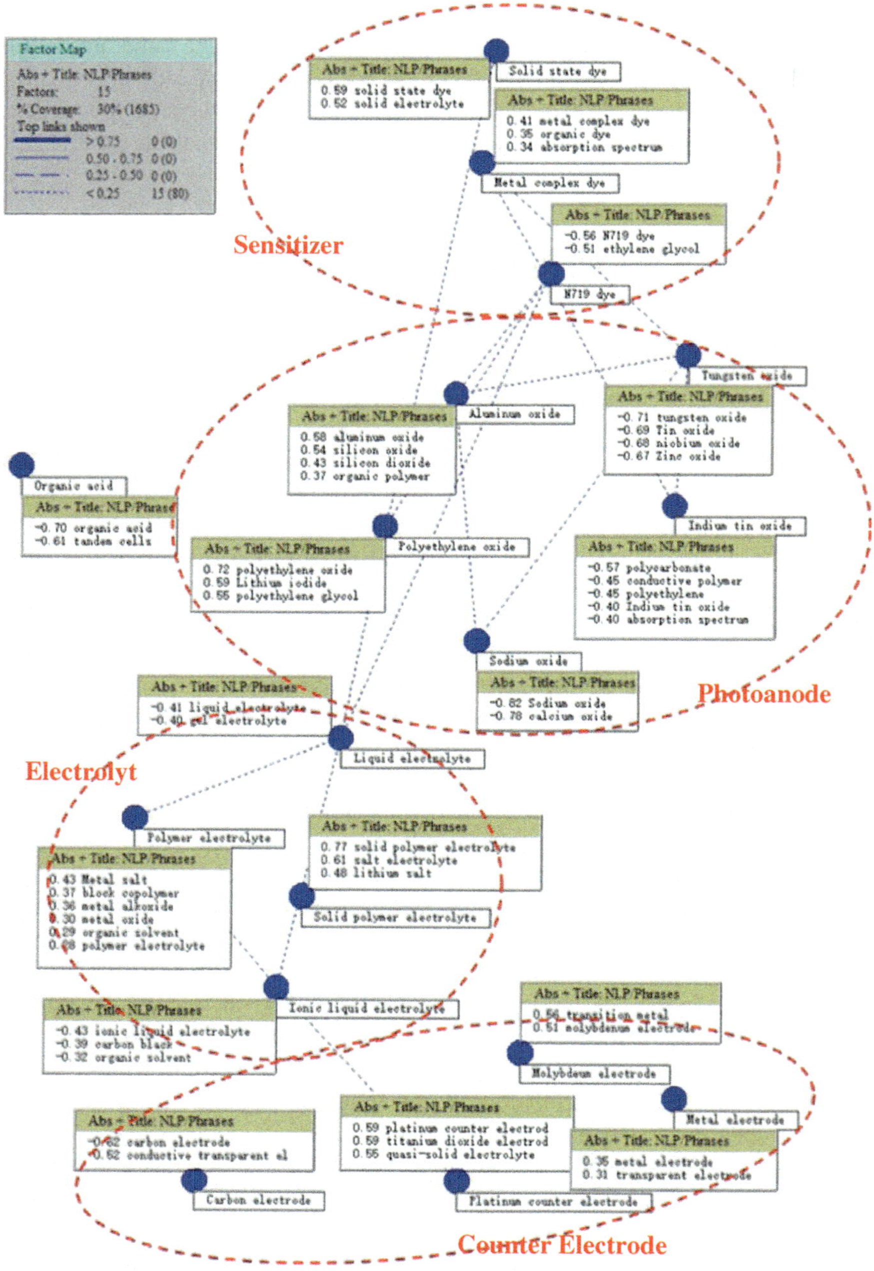


**Fig. 9.2** Factor map of DSSCs (based on the top 200 topical terms)

**Table 9.2** Characteristics of the sub-technology in DSSCs

| # | Topics | Records | Keywords | Avg. year | Growth curve |
|---|---|---|---|---|---|
| 1. | Photoanode | 2181 | Aluminum oxide; Tungsten oxide; Indium tin oxide; Polyethylene oxide | 2009.21 | 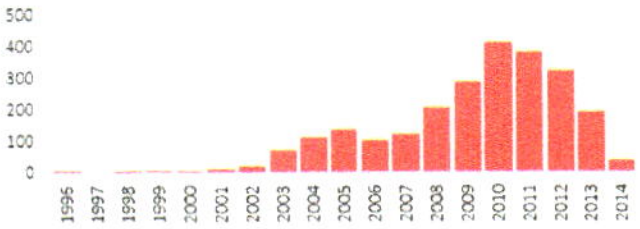  |
| 2. | Sensitizer | 615 | Solid state dye; Metal complex dye; N719 dye | 2009.83 | 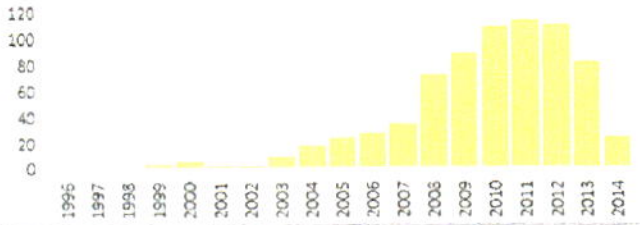  |
| 3. | Electrolyte | 1674 | Liquid electrolyte; Polymer electrolyte; Ionic liquid electrolyte; Solid polymer electrolyte | 2008.66 | 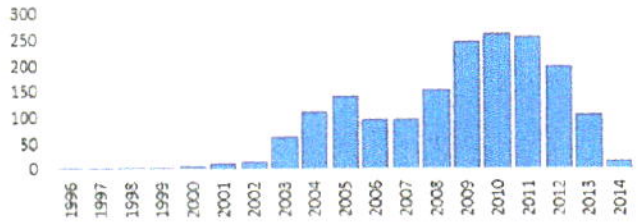  |
| 3. | Counter-electrode | 2348 | Carbon electrode; Molybdenum electrode; Platinum counter electrode; Metal electrode | 2008.43 | 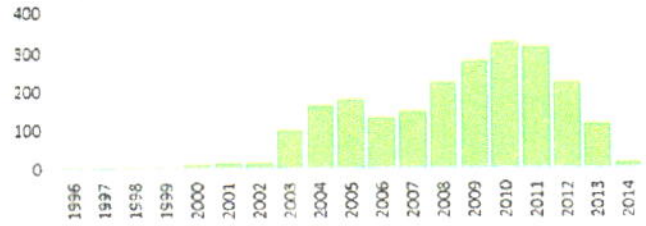  |

Table 9.2 represents the characteristics of the sub-technologies in DSSCs. Photoanode and counter-electrode are the two largest clusters among them, reaching 2181 and 2348, respectively, and both show similar growth curves. Sensitizer shows remarkably different growth curves when compared to the other sub-technologies; the patents drastically increase after 2008. In this paper, we calculate the average priority year of patents in each cluster to track emerging fields in DSSCs research. Among these four sub-technologies, photoanode and sensitizer are relatively new and emerging, and we speculate that more and more patents related to these two fields will appear.

## 9.4.2 *DSSCs Patent Citation Analysis*

The citation network of patents provides a representation of the innovation process (Érdi et al. 2013). In the development process, technology is present on different development tracks. Therefore, this context highlights the dynamic nature of technology in the development process in order to improve the accuracy of the analysis of sub-technologies in their technology evolution roadmaps. As previous research has shown, the whole of DSSCs development can be divided into three stages: emerging stage (1991–2001), growth stage (2002–2009), and maturity stage (2010–2014) [The previous research refines a maturity stage from 2010 to 2012 for choosing the time span from 1991 to 2012, but we choose the time span from 1991 to 2014 in this paper to obtain more newly data.] (Huang et al. 2014b). In Table 9.3, it is not hard to see the evolution of DSSCs citation behaviors from the 1990s to

**Table 9.3** Maximum patent citation networks of sub-technologies in DSSCs at different stages

| | 1991–2001 | 1991–2009 | 1991–2014 |
|---|---|---|---|
| Photoanode |  |  | 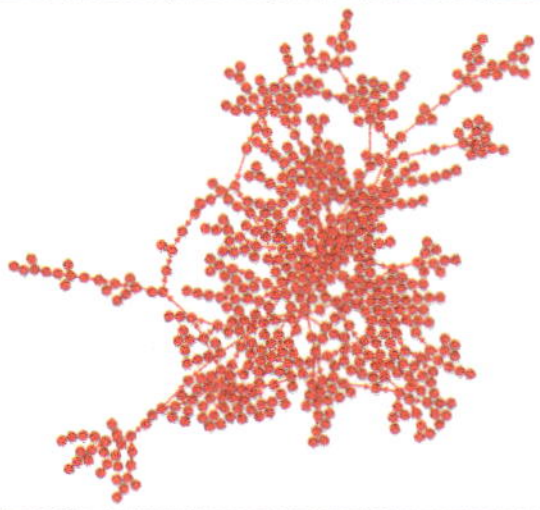 |
| Sensitizer | 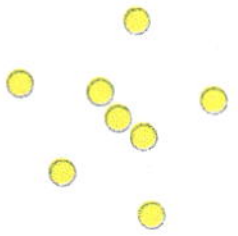 | 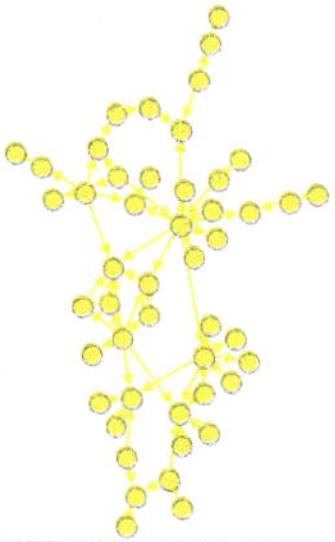 | 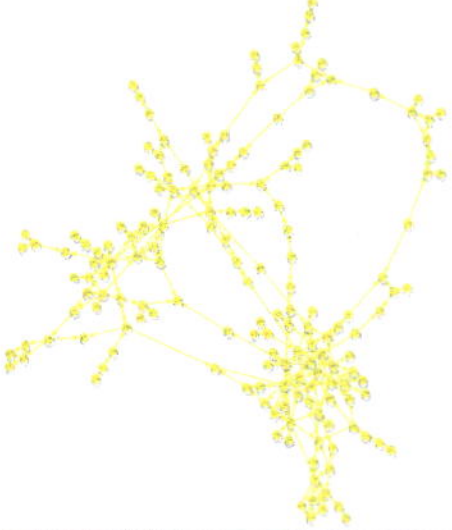 |
| Electrolyte |  | 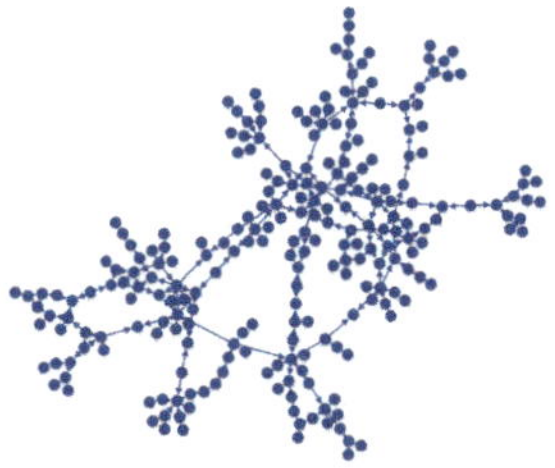 | 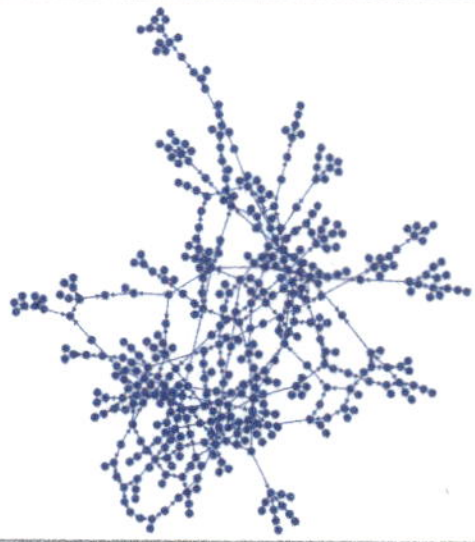 |
| Counter-electrode |  | 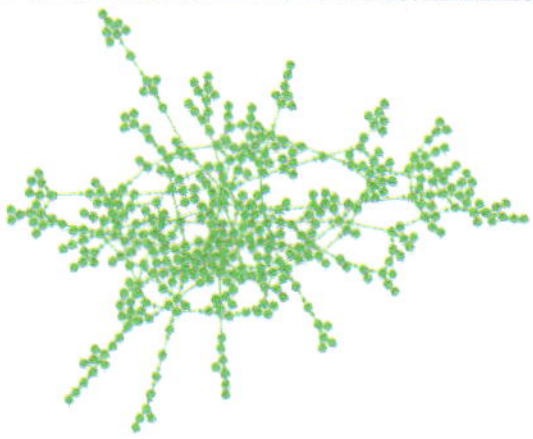 |  |

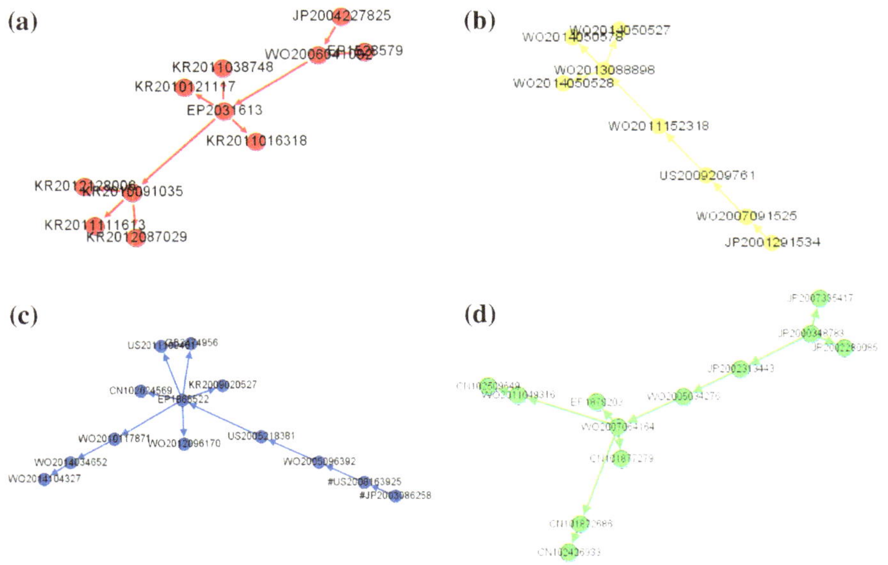


**Fig. 9.3** Main technological trajectories of sub-technologies in DSSCs from 1991 to 2014: **a** photoanode, **b** sensitizer, **c** electrolyte, and **d** counter-electrode

2014. We can see that the technology will continue to produce a more complex network of references as time goes on. This will include early inventions, which will be cited more frequently by later inventions. Likewise, recent patents will be cited by more patents in the future. In addition, analyzing the technology evolution will be a big challenge if only property analysis is used. Therefore, in this paper, we will extract the technology trajectory by means of the main path analysis.

The patents located on the main technological trajectory for sub-technologies in DSSCs from 1991 to 2014 are shown in Fig. 9.3. For photoanode, there are 11 patents selected as the characteristic nodes to represent the evolution pathway, and its trajectory has three clusters. The sensitizer trajectory includes 8 patents and has an arrowhead-shaped citation path. For electrolyte, 13 patents are located on the technology trajectory where they produce a sudden change, driven by EP1865522, which takes advantage of the electrolyte characteristics therein. For counter-electrode, the technological trajectory includes 12 patents, among which is a Korean patent (Patent No. WO2007064164) that uses carbon nanotube electrode, which produces a much better result than the previous platinum electrode. From that point on, carbon counter-electrodes play an important role in improving photo-electric conversion efficiency.

After identifying the patents located on the technological trajectories for the four sub-technologies, we read titles and abstracts and manually extract the technology focus. Some patents may reflect the same technology focus for their citation relationships. In this case, we prefer to seek the different foci to help researchers and

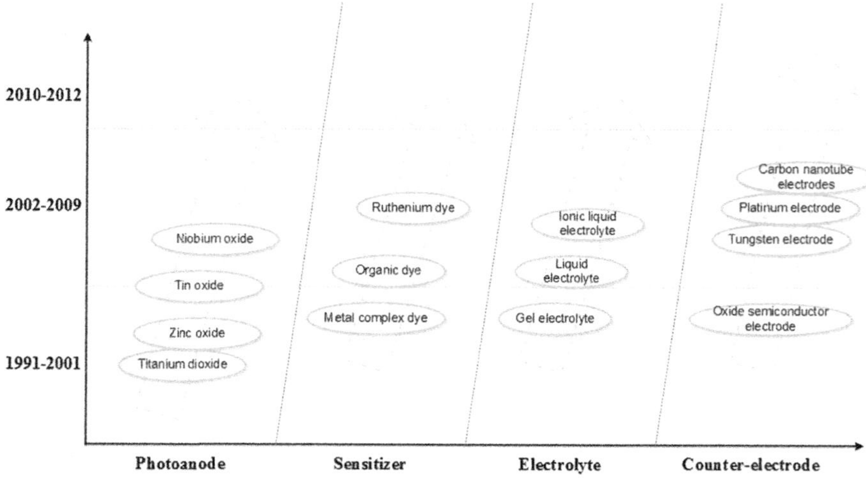


**Fig. 9.4** Preliminary technology evolution pathways of DSSCs (based on the citation analysis)

decision-makers identify future research trends and technology opportunities. These main technological focuses are displayed by time period in Fig. 9.4.

To further ensure accurate results, we use the technology focus to retrieve the dataset of sub-technologies. Moreover, we ask experts to help evaluate the importance of the technology focus using an "Importance index." Table 9.4 indicates that most of the technology foci are influential in their actual field, and this evaluation, to some extent, is supported by experts' understanding. Based on such analyses, we have grounds to believe that extracting technology foci from the main citation paths of the patent citations network is both feasible and meaningful.

### 9.4.3 DSSCs Term Clumping Analysis

Using the important yet limited information extracted from the trajectory, we use term clumping for each sub-technology to enhance the technology evolution roadmap and to better identify opportunities and possibilities for future trends. The results of the term clumping for DSSCs are shown in Table 9.5.

After the term clumping process, we are left with four lists of keywords, from which we choose those that are most related to respective technology foci. Then, we extract the first year that these terms appeared in the patent documents. The results are shown in Table 9.6.

Based on the preliminary technology evolution (shown in Fig. 9.4), we complement DSSCs sub-technologies keywords to construct the complete technology evolution roadmap (shown in Fig. 9.5). Among these four sub-technologies, the smallest focus is counter-electrode. Platinum counter-electrodes, as the most widely used form for counter-electrodes, have the best performance, but its high cost

**Table 9.4** Evaluation of the technology focus included in patents located on main technological trajectories

| Sub-technology | Records | Technology focus | Coverage (%) | Importance index |
|---|---|---|---|---|
| Photoanode | 616 | Titanium dioxide | 28.24 | ☆☆☆☆☆ |
| | 184 | Zinc oxide | 8.44 | ☆☆☆☆ |
| | 116 | Tin oxide | 5.32 | ☆☆☆☆ |
| | 22 | Niobium oxide | 1.01 | ☆☆☆ |
| Sensitizer | 69 | Organic dye | 11.22 | ☆☆☆☆☆ |
| | 34 | Metal complex dye | 5.53 | ☆☆☆☆ |
| | 28 | Ruthenium complex dye | 4.55 | ☆☆☆ |
| Electrolyte | 108 | Liquid electrolyte | 6.45 | ☆☆☆☆ |
| | 104 | Ionic liquid electrolyte | 6.21 | ☆☆☆ |
| | 21 | Gel electrolyte | 1.25 | ☆☆☆☆ |
| Counter-electrode | 270 | Oxide semiconductor electrode | 11.40 | ☆☆☆ |
| | 146 | Platinum electrode | 1.39 | ☆☆☆☆☆ |
| | 77 | Carbon electrode | 3.25 | ☆☆☆☆☆ |
| | 68 | Tungsten oxide | 0.46 | ☆☆☆ |

*Note* The number of stars indicates importance. The more stars it earns, the more importance the technology focus presents. Five stars is the highest in our evaluation

**Table 9.5** Term clumping results for DSSCs

| Process | All | Photoanode | Sensitizer | Electrolyte | Counter-electrode |
|---|---|---|---|---|---|
| Original phases | 75563 | 33361 | 11585 | 24175 | 28570 |
| Common and basic removal | 63958 | 27775 | 8816 | 20136 | 24131 |
| Fuzzy words matching | 41924 | 21129 | 7137 | 20135 | 18294 |
| Extreme words removal | 9238 | 4864 | 1683 | 4696 | 4534 |
| Combine term networks | 1562 | 748 | 424 | 411 | 613 |

*Note* The number shown in the table represents the number of remaining terms after the process

**Table 9.6** Keywords of sub-technologies in DSSCs by term clumping method

| Component | Keywords |
|---|---|
| Photoanode | Indium tin oxide (2002); Silicon oxide (2004); Magnesium oxide (2004); Tungsten oxide (2007); Zirconium oxide (2008); Calcium oxide (2008); Lanthanum oxide (2008); Yttrium oxide (2008); Cerium oxide (2008); Molybdenum oxide (2008); Lithium oxide (2009); Aluminum zinc oxide (2010) |
| Sensitizer | Polymethine dye (2003); Porphyrin-based dye (2005); Solid state dye (2007); Fluorescent dye (2007); N719 dye (2008); Merocyanine dye (2008); Natural dye (2009); Cyanine dye (2009); Phenothiazine dye (2009) |
| Electrolyte | Non-aqueous electrolyte (1999); Polymer electrolyte (2001); Gel-like electrolyte (2001); Solid electrolyte (2002); Redox electrolyte (2003); Quasi-solid electrolyte (2004); Salt electrolyte (2004) |
| Counter-electrode | Conductive polymer electrode (2003); Metal electrode (2004) |

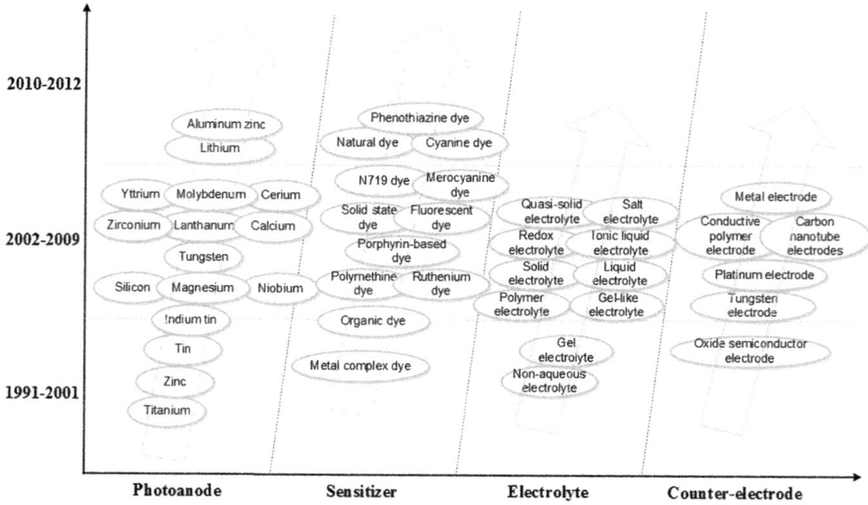


**Fig. 9.5** Final technology evolution pathway of DSSCs

restricts industrialization. Novel materials, such as different nano-structured carbon materials, conductive polymers, and their composite counter-electrodes, are drawing attention because of their low cost and high activity. The electrolyte relates closely to the cell's stability, and both counter-electrode and electrolyte have not achieved a special breakthrough in recent years. In the beginning, the high energy conversion efficiency of DSSCs was achieved with conventional liquid electrolytes, which involve a serious problem of stability. With the development of quasi-solid state and solid-state electrolytes, the stability of DSSCs may improve remarkably. Dye sensitizer, which greatly affects the photoelectronic efficiency of solar cells, is an important research focus in the field of cell materials. According to research of the past twenty years, the sensitizers used in DSSCs are mainly divided into two types: metal complex dye and organic dye. Regarding these two types of sensitizers utilized in DSSCs, some different structures and improved dye have been developed. The photoanode plays a role in a cell's performance. Its function is to load sensitizers and collect and transport electrons. Currently, a series of semiconductor materials, including TiO2, ZnO, $Nb_2O_5$, $SnO_2$ , are being pursued. We believe that, in the near future, gains in both sensitizers and photoanodes will further improve the efficiency of DSSCs and promote industrialization of these third-generation photovoltaic cells.

## 9.5 Conclusions

Efficiently generating intelligence on NESTs is an essential topic both for academia and for industry. Patents, as a major public resource, offer a wealth of information for conducting technology opportunity analysis. In this paper, we trace technology evolution roadmaps and identify potential low-cost, high efficiency opportunities in the DSSCs field by combing patent citation network analysis with "Tech mining" from the view of non-domain experts.

For a comprehensive technology, it is hard to capture the whole technology evolution at a micro-level, so the perspective of sub-technologies becomes necessary. In this paper, with the help of text analytic software, we apply PCA to the top 200 phrases consolidated from selected titles and abstracts. After obtaining the 17 factors and their keywords, we use these keywords and IPCs to extract the sub-technology datasets from the original database. Finally, with the guidance of an expert, we identify four sub-technologies: photoanode, sensitizer, electrolyte, and counter-electrode. Overall, this approach helps us to obtain meaningful and useful, albeit incomplete, data in a reasonable way.

Technological hot topics always play vital roles in the technology development process. Discovering these hot topics leads to a better understanding of the technology attributes. In this context, we first gain the maximum connected components from the patent citation network of all four datasets, respectively. Then, we get the key patents located on the main citation trajectory and manually extract the technological keywords. The results show that, by using this method, we can obtain essential technology information in the respective fields. This opinion, to some extent, is supported by the judgment of experts. Therefore, extracting the main citation path to obtain the technology focus from the patent citations network is both feasible and meaningful.

Tech mining is a multi-step process for analysis of ST&I information resources by using bibliometrics, text analytics, and visualization tools. Term clumping as an effective method of text analysis helps us to extract terms from titles and abstracts, which provides a abundant basis to further identify highly relevant keywords reflecting the technology focus. We can then go on to improve the accuracy of constructing technology evolution pathways. In this paper, we generated lists of 12, 9, 7, and 2 keywords for photoanode, sensitizer, electrolyte, and counter-electrode, respectively, which makes a great contribution to drawing technology evolution pathways for DSSCs.

This paper combines patent citation network analysis with "Tech mining" to trace technology evolution roadmaps, generate a framework of DSSCs development pathways, and discover potential opportunities. The methodology proposed in this paper has some limitations. On the one hand, we combine related keywords and IPCs to obtain a dataset for each sub-technology, but the results are influenced by the keywords we choose and the IPC range we set. Also, different experts and researchers vary in their opinions. On the other hand, though this paper identifies the core hot topics and important technology focuses, the connection between different technology focuses is hard to grasp without domain specialists.

Some possible future research topics related to this work are as follows: First, publications, as another important source of technological information, can be merged to better trace the technology evolution pathways; Second, more attention should be given to finding reasonable and reliable ways to identify the sub-technologies of a technology field, especially for the NESTs; Third, seeking the relationship and evolution between different technology focuses will be an interesting direction to explore because it is important for policymakers, technology managers, and entrepreneurs to better predict the future trends based on recent topics.

**Acknowledgments** We acknowledge support from the US National Science Foundation (NSF) (Award No. 1064146), MOE (Ministry of Education in China) Project of Humanities and Social Sciences (Award No. 13YJC630042) and the National High Technology Research and Development Program of China (Grant No. 2014AA015105). Besides, we are grateful for the scholarship provided by the China Scholarship Council (CSC Student ID 201406030005). The findings and observations contained in this paper are those of the authors and do not necessarily reflect the views of the supporters.

## References

Batagelj, V., & Mrvar, A. (2004). *Pajek-analysis and visualization of large networks*. Berlin Heidelberg: Springer.

Barirani, A., Agard, B., & Beaudry, C. (2013). Discovering and assessing fields of expertise in nanomedicine: A patent co-citation network perspective. *Scientometrics, 94*(3), 1111–1136.

Bengisu, M. (2003). Critical and emerging technologies in materials, manufacturing, and industrial engineering: A study for priority setting. *Scientometrics, 58*(3), 473–487.

Boyack, K. W., Börner, K., & Klavans, R. (2009). Mapping the structure and evolution of chemistry research. *Scientometrics, 79*(1), 45–60.

Chang, P. L., Wu, C. C., & Leu, H. J. (2010). Using patent analyses to monitor the technological trends in an emerging field of technology: A case of carbon nanotube field emission display. *Scientometrics, 82*(1), 5–19.

Cho, T. S., & Shih, H. Y. (2011). Patent citation network analysis of core and emerging technologies in Taiwan: 1997–2008. *Scientometrics, 89*(3), 795–811.

Choi, C., & Park, Y. (2009). Monitoring the organic structure of technology based on the patent development paths. *Technological Forecasting and Social Change, 76*(6), 754–768.

Choi, S., Kim, H., Yoon, J., Kim, K., & Lee, J. Y. (2013). An SAO-based text-mining approach for technology roadmapping using patent information. *R&D Management, 43*(1), 52–74.

Daim, T. U., Rueda, G., Martin, H., & Gerdsri, P. (2006). Forecasting emerging technologies: Use of bibliometrics and patent analysis. *Technological Forecasting and Social Change, 73*(8), 981–1012.

Dibiaggio, L., & Nesta, L. (2005). Patents statistics, knowledge specialisation and the organisation of competencies. *Revue d'économie industrielle, 110*(1), 103–126.

Érdi, P., Makovi, K., Somogyvári, Z., Strandburg, K., Tobochnik, J., Volf, P., & Zalányi, L. (2013). Prediction of emerging technologies based on analysis of the US patent citation network. *Scientometrics, 95*(1), 225–242.

Harhoff, D., Scherer, F. M., & Vopel, K. (2003). Citations, family size, opposition and the value of patent rights. *Research Policy, 32*(8), 1343–1363.

Han, K., & Shin, J. (2014). A systematic way of identifying and forecasting technological reverse salients using QFD, bibliometrics, and trend impact analysis: A carbon nanotube biosensor case. *Technovation, 34*(9), 559–570.

Ho, J. C., Saw, E. C., Lu, L. Y., & Liu, J. S. (2014). Technological barriers and research trends in fuel cell technologies: A citation network analysis. *Technological Forecasting and Social Change, 82*, 66–79.

Huang, L., Zhang, Y., Guo, Y., Zhu, D., & Porter, A. L. (2014a). Four dimensional Science and Technology planning: A new approach based on bibliometrics and technology roadmapping. *Technological Forecasting and Social Change, 81*, 39–48.

Huang Y., Zhu F. Guo Y., Porter, A.L., & Zhu, D.(2014b). Identifying technology evolution pathways based on tech mining and patent citation network- illustrated for dye-sensitized solar cells. *Proceedings-the 5th International Conference on Future-Oriented Technology Analysis (FTA)*. Brussels, Belgium.

Jaffe, A. B., & Trajtenberg, M. (2002). *Patents, citations, and innovations: A window on the knowledge economy*. MIT press.

Kajikawa, Y., & Takeda, Y. (2008). Structure of research on biomass and bio-fuels: A citation-based approach. *Technological Forecasting and Social Change, 75*(9), 1349–1359.

Kostoff, R. N., & Schaller, R. R. (2001). Science and technology roadmaps. *Engineering Management, IEEE Transactions on, 48*(2), 132–143.

Lacasa, I. D., Grupp, H., & Schmoch, U. (2003). Tracing technological change over long periods in Germany in chemicals using patent statistics. *Scientometrics, 57*(2), 175–195.

Lee, C., Seol, H., & Park, Y. (2007). Identifying new IT-based service concepts based on the technological strength: A text mining and morphology analysis approach. *The 4th International Conference on Fuzzy Systems and Knowledge Discovery*, (Vol. 4, pp. 36–40).

Lee, C. Y., Lee, J. D., & Kim, Y. (2008a). Demand forecasting for new technology with a short history in a competitive environment: The case of the home networking market in South Korea. *Technological Forecasting and Social Change, 75*(1), 91–106.

Lee, S., Lee, S., Seol, H., & Park, Y. (2008b). Using patent information for designing new product and technology: Keyword based technology roadmapping. *R&D Management, 38*(2), 169–188.

Lee, S., Yoon, B., Lee, C., & Park, J. (2009). Business planning based on technological capabilities: Patent analysis for technology-driven roadmapping. *Technological Forecasting and Social Change, 76*(6), 769–786.

Ma, J., & Porter, A. L. (2015). Analyzing patent topical information to identify technology pathways and potential opportunities. *Scientometrics, 102*(1), 811–827.

Ma, T., Porter, A. L., Guo, Y., Ready, J., Xu, C., & Gao, L. (2014). A technology opportunities analysis model: Applied to dye-sensitised solar cells for China. *Technology Analysis & Strategic Management, 26*(1), 87–104.

Milanez, D. H., de Faria, L. I. L., do Amaral, R. M., Leiva, D. R., & Gregolin, J. A. R. (2014). Patents in nanotechnology: An analysis using macro-indicators and forecasting curves. *Scientometrics, 101*(2), 1097–1112.

Michel, J., & Bettels, B. (2001). Patent citation analysis. A closer look at the basic input data from patent search reports. *Scientometrics, 51*(1), 185–201.

Newman, N. C., Porter, A. L., Newman, D., Trumbach, C. C., & Bolan, S. D. (2014). Comparing methods to extract technical content for technological intelligence. *Journal of Engineering and Technology Management, 32*, 97–109.

O'Brien, J. J., Carley, S., & Porter, A. L. (2013). Keyword field cleaning through ClusterSuite: A termclumping tool for VantagePoint software. *Poster presented at 3rd Global Tech Mining Conference*. Atlanta, USA.

Porter, A. L., & Cunningham, S. W. (2005). *Tech mining: Exploiting new technologies for competitive advantage*. New York: Wiley.

Porter, A. L., & Detampel, M. J. (1995). Technology opportunities analysis. *Technological Forecasting and Social Change, 49*(3), 237–255.

Porter, A. L., Guo, Y., & Chiavatta, D. (2011). Tech mining: Text mining and visualization tools, as applied to nanoenhanced solar cells. *Wiley Interdisciplinary Reviews: Data Mining and Knowledge Discovery, 1*(2), 172–181.

Robinson, D. K., Huang, L., Guo, Y., & Porter, A. L. (2013). Forecasting Innovation Pathways (FIP) for new and emerging science and technologies. *Technological Forecasting and Social Change, 80*(2), 267–285.

Shibata, N., Kajikawa, Y., Takeda, Y., & Matsushima, K. (2008). Detecting emerging research fronts based on topological measures in citation networks of scientific publications. *Technovation, 28*(11), 758–775.

Smalheiser, N. R. (2001). Predicting emerging technologies with the aid of text-based data mining: The micro approach. *Technovation, 21*(10), 689–693.

Thomson Reuters. DWPI Manual Code Revision. (2015). Retrieved January 21, 2015, from http://ip-science.thomsonreuters.com/m/pdfs/DWPI_mcr_Jan2015.pdf

Watatani, K., Xie, Z., Nakatsuji, N., & Sengoku, S. (2013). Global competencies of regional stem cell research: Bibliometrics for investigating and forecasting research trends. *Regenerative Medicine, 8*(5), 659–668.

Xin, L., Jiwu, W., Lucheng, H., Jiang, L., & Jian, L. (2010). Empirical research on the technology opportunities analysis based on morphology analysis and conjoint analysis. *Foresight, 12*(2), 66–76.

Yoon, B., & Park, Y. (2005). A systematic approach for identifying technology opportunities: Keyword-based morphology analysis. *Technological Forecasting and Social Change, 72*(2), 145–160.

Yoon, B., & Park, Y. (2007). Development of new technology forecasting algorithm: Hybrid approach for morphology analysis and conjoint analysis of patent information. *IEEE Transactions on Engineering Management, 54*(3), 588–599.

Yoon, B., Phaal, R., & Probert, D. (2008). Morphology analysis for technology roadmapping: Application of text mining. *R&D Management, 38*(1), 51–68.

Yoon, J., & Kim, K. (2011a). An automated method for identifying TRIZ evolution trends from patents. *Expert Systems with Applications, 38*(12), 15540–15548.

Yoon, J., & Kim, K. (2011b). Identifying rapidly evolving technological trends for R&D planning using SAO-based semantic patent networks. *Scientometrics, 88*(1), 213–228.

Yoon, J., Park, Y., Kim, M., Lee, J., & Lee, D. (2014). Tracing evolving trends in printed electronics using patent information. *Journal of nanoparticle research, 16*(7), 1–15.

Zhou, X., Zhang, Y., Porter, A. L., Guo, Y., & Zhu, D. (2014). A patent analysis method to trace technology evolutionary pathways. *Scientometrics, 100*(3), 705–721.

Zhang, Y., Guo, Y., Wang, X., Zhu, D., & Porter, A. L. (2013). A hybrid visualisation model for technology roadmapping: Bibliometrics, qualitative methodology and empirical study. *Technology Analysis & Strategic Management, 25*(6), 707–724.

Zhang, Y., Porter, A. L., Hu, Z., Guo, Y., & Newman, N. C. (2014a). "Term clumping" for technical intelligence: A case study on dye-sensitized solar cells. *Technological Forecasting and Social Change, 85*, 26–39.

Zhang, Y., Zhou, X., Porter, A. L., & Gomila, J. M. V. (2014b). How to combine term clumping and technology roadmapping for newly emerging science & technology competitive intelligence: "problem & solution" pattern based semantic TRIZ tool and case study. *Scientometrics, 101*(2), 1375–1389.

Zhang, Y., Zhou, X., Porter, A. L., Gomila, J. M. V., & Yan, A. (2014c). Triple Helix innovation in China's dye-sensitized solar cell industry: Hybrid methods with semantic TRIZ and technology roadmapping. *Scientometrics, 99*(1), 55–75.

# Chapter 10
# Identifying Targets for Technology Mergers and Acquisitions Using Patent Information and Semantic Analysis

**Lu Huang, Lining Shang, Kangrui Wang, Alan L. Porter and Yi Zhang**

**Abstract** Technology plays an increasingly important role in today's enterprise competition. Technology mergers and acquisitions (Tech M&A), as an effective way to acquire the external technology resources rapidly, have attracted attention from researchers for their potential realization of value through synergy. A big challenge is how to identify appropriate targets to support the effective technology integration. In this study, we developed a model of target selection of Tech M&A from the perspective of technology relatedness and R&D capability. We present results for the Tech M&A case in China's cloud computing industry.

**Keywords** Technology merger and acquisition · Patent analysis · Semantic analysis · Cloud computing

## 10.1 Introduction

Rapid technological change and diverse customer needs make firms face increasing pressure of innovation. When enhancing the innovative capabilities, even the largest and most technologically self-sufficient firms do not always have the time to build their own new technologies from scratch. Getting external technology resources to enhance existing technology portfolios has been a preferred choice for firms.

L. Huang (✉) · L. Shang · K. Wang · Y. Zhang
School of Management and Economics, Beijing Institute of Technology,
Beijing, People's Republic of China
e-mail: huagnlu628@163.com

A.L. Porter
School of Public Policy, Georgia Institute of Technology,
Technology Policy & Assessment Ctr., Atlanta, GA, USA

A.L. Porter
Search Technology Inc., Norcross, GA, USA

Y. Zhang
Centre for Quantum Computation & Intelligent Systems,
University of Technology Sydney, Sydney, Australia



T.U. Daim et al. (eds.), *Anticipating Future Innovation Pathways Through Large Data Analysis*, Innovation, Technology, and Knowledge Management,
DOI 10.1007/978-3-319-39056-7_10

Technology mergers and acquisitions (Tech M&A), as an effective way to get external technology resources, have been a hot topic for innovation management (Sears and Hoetker 2014; Lodh and Battaggion 2014). Tech M&A enables firms to get quick access to the research frontier in the field of competence (Yoon et al. 2013) and facilitates firms to enter new technology areas with lower time cost and reduced R&D failure risk (Hussinger 2010). The main effect of Tech M&A was to achieve technological synergy to enhance acquirer's innovative capability (Di Guardo et al. 2015). The research on Tech M&A can be divided into three stages:

In the first stage, scholars found that in some cases, firms could develop fast after acquiring some small technology-based firms. Granstrand et al. (1982) strove to conclude what were key factors to Tech M&A success based on 13 M&A events in high-tech industries.

In the second stage, researchers began to explore motivations of Tech M&A and evaluate performance, especially from the perspective of finance. Scholars used multi-dimensional indexes and chose various time frames to evaluate acquisition performance (Loughran and Vijh 1997; Kohers and Kohers 2000).

Now, in the third stage, research on Tech M&A tends to be diversified, including Tech M&A integration, Tech M&A mode, and Tech M&A target selection. Paruchuri et al. (2006) analyzed the relationship between the research personnel and innovation output during Tech M&A integration. Wei and Tian (2011) identified attributes of target companies and proposed a theory to support the decision making of acquiring companies through four in-depth case studies conducted across three primary sectors in the medical technology industry. Lin (2012) tested an acquisition–learning–innovation framework and found that unrelated acquisitions also enhance exploration in an era of technology fermentation. Research at this stage mainly focuses on performance evaluation after Tech M&A. Few studies have been conducted on target selection pre-acquisition.

The volume of Tech M&A events has been steadily increasing in the recent years. However, it is not easy to realize Tech M&A successfully. The failure rate of Tech M&A is pretty high—between 70 and 90 % (Christensen et al. 2011). Taking account of the $2 trillion transactions of M&A every year, the failures are extremely costly. Tech M&A success or failure can be determined and influenced by many factors—e.g., strategic formulation, technology relatedness, and financial status. But the most fundamental step to increase the success of M&A is to select the right target companies, which are well matched to the strategic purpose of a given M&A action (Kengelbach and Roos 2011).

Existing studies on identifying M&A targets concentrate primarily on the development or application of financial and managerial variables (i.e., firm size, cash flow, and debt-to-equity ratio), neglecting consideration of the technological perspective (Ragothaman et al. 2003; Ali-Yrkkö et al. 2005).

Patents, as an important source for the management of technology in both industry and science, are useful sources for technology analysis. Traditional methods are mainly based on International Patent Classification (IPC) and citation analyses, without considering the text of patents, which constrain the analysis depth. Recently, the proliferation of patents worldwide has increased the demand for the more advanced quantitative patents analyses to support expert evaluation processes for decision making (Yoon and Kim 2012; Yoon et al. 2013). In this paper, we introduce semantic analysis to devise a new framework to analyze technology relatedness, including technology similarity and technology complementarity of Tech M&A. We apply our method to Huawei Technologies Co. Ltd (Huawei), China's leading firm in the field of cloud computing, for Tech M&A needs.

## 10.2 Challenges and Methods

Tech M&A, by its very nature, is a method to get external technology resources. The primary factor in target selection of Tech M&A is technical relatedness. However, few methods have been proposed to analyze it. A big challenge that corporate managers and government policy makers are facing is how to confirm a methodological architecture to help them identify the appropriate targets to support effective technology integration.

Our research is based on the following driving questions:

1. How do we use a quantitative method to measure technology relatedness?
2. What factors should be considered for effective technology integration based on the analysis of technology relatedness?
3. How to devise a comprehensive method from the perspective of technology relatedness and technology integration on post-acquisition stage?

In this study, we try to provide a detailed guidance for identifying potential Tech M&A targets from a technological perspective based on patent information. Patents have long been considered to be up-to-date and valuable information sources in technology, and careful analysis of patents could provide information of not only technological competitiveness, but also overall technological opportunity in the specific technology areas. Therefore, the technological capabilities of a corporation can be represented by its set of patents.

In this study, we divide our method of Tech M&A target selection into three steps based on patent analysis. Figure 10.1 shows the process.

Step 1: Technology Similarity Analysis—The purpose of this step was to calculate the technology similarity between the acquirer and targets, and reduce the selection scope for Step 2. First, IPCs of each patent will be extracted, and the degree of overlap will be regarded as the preliminary evaluation of consistency of technology area. Second, we measure technology similarity through Subject—Action—Objective (SAO) analysis of the USE field in the abstract of patents after choosing the potential candidates that show commonality with the acquirer.

Step 2: Technology Complementarity Analysis—Technology morphology analysis is introduced to help the complementarity analysis. First, we extract keywords from patent texts and arrange them according to the related technology.

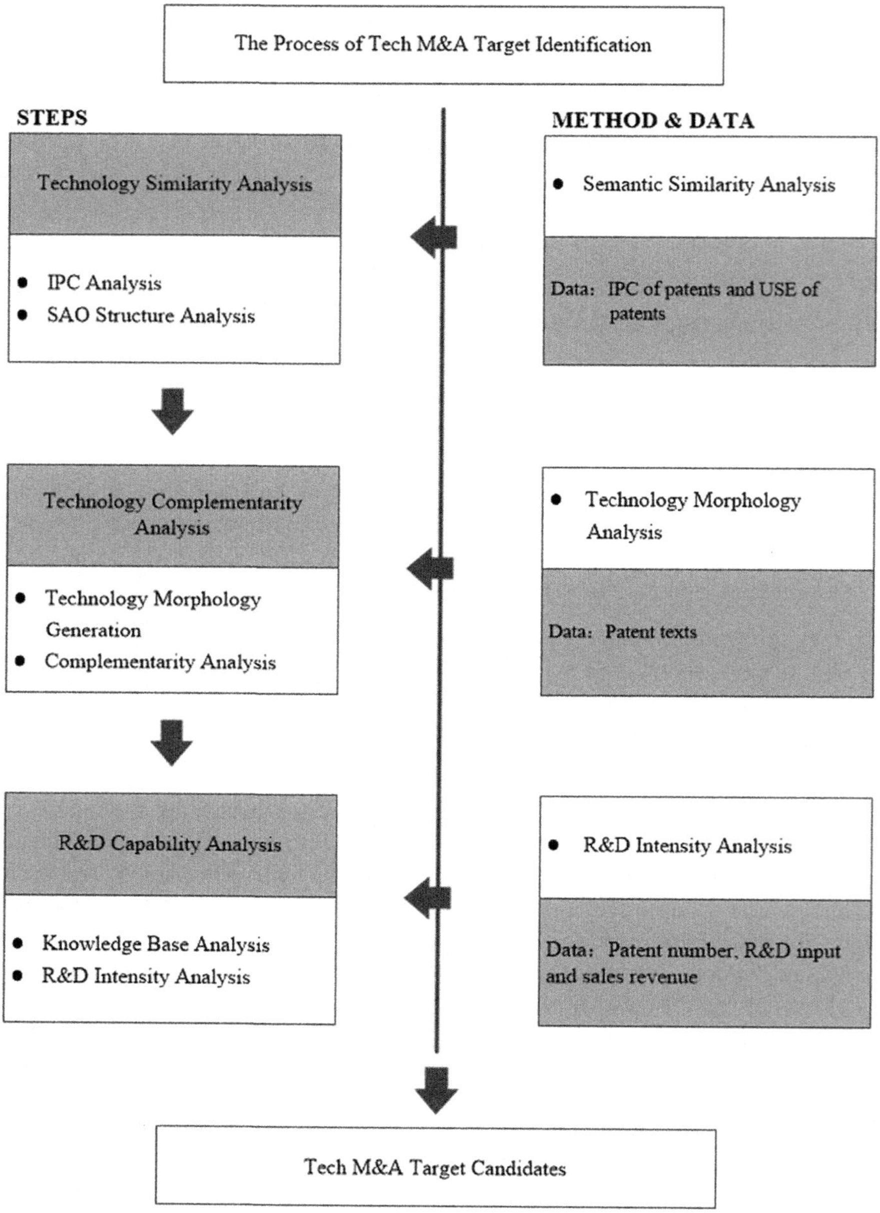


**Fig. 10.1** Tech M&A target identification process

Second, we calculate the complementarity of each technology combination with the help of expert experience.

Step 3: R&D capability analysis—We use knowledge base and R&D intensity as indicators to make further selection of potential targets after the first two steps.

For technology similarity analysis, we extract the IPCs of every patent and measure the consistency of technology area with the method of Makri (2010). After that, NLP tools are used to extract the SAO structures from the patents collected with the help of an open API. In order to identify the semantic similarity of SAO structures, a semantic knowledge base, WordNet, is introduced to calculate the similarity between two words or phrases in the SAO structures.

For technology complementarity analysis, first, we extract the keywords from patents and map them into their associated morphology. Second, we evaluate the technology complementarity level with the help of experts and then calculate technology complementarity.

For R&D capability analysis, we emphasize the view of knowledge base and R&D intensity of potential targets. Knowledge base of a firm is measured by the related number of patents, and R&D intensity is measured by a firm's ratio of expenditures on R&D to the firm's sales.

## 10.3 Case Study

### *10.3.1 Patent Collection*

Rapid technology renewal keeps emerging technology a hotspot in Tech M&A for years, and we chose a representative one—cloud computing technology, focusing on opportunities within China. Patents for cloud computing were collected from Derwent Innovations Index (DII), employing the search strategy devised by Owens and Logue (2012), from 2000 to 2012.

We got 621 patents on cloud computing in China after data cleaning.

Table 10.1 lists the top 10 assignees. According to technology development strategy and the layout of the cloud computing technology area, we chose Huawei Corp (Huawei) as an acquirer to search for Tech M&A targets. Huawei is a leading global provider of Information Communication Technology solutions and is also the most professional one-stop cloud service provider in China. Huawei has been expanding the layout of cloud computing since 2010, and in the year 2011, Huawei acquired Huasy Firm for 5300 million dollars to enhance the security of cloud platforms. Tech M&A is regarded as an important way to achieve the rapid development of related technologies on cloud computing for Huawei.

**Table 10.1** Top 10 assignees in cloud computing

| Assignee names | Patent number | Percentage (%) |
|---|---|---|
| ZTE Corp | 52 | 8.4 |
| Microsoft Corp | 41 | 6.6 |
| Inspur Electronic Information Co. Ltd. | 33 | 5.3 |
| LI Z (Individual) | 24 | 3.9 |
| **Huawei Crop** | **23** | **3.7** |
| IBM Corp | 21 | 2.7 |
| Univ Qinghua | 19 | 2.1 |
| Hon Hai Precision Ind Co. Ltd. | 19 | 1.9 |
| Guangdong Electronics Ind Inst Co. Ltd. | 17 | 1.4 |
| Univ Beijing Aeronautics & Astronautics | 15 | 1.4 |

### *10.3.2 Technology Similarity Analysis*

Literature on Tech M&A suggests that the maximum benefits from an acquisition can be realized when technology portfolios of both firms are related (Hussinger 2010; Gupta 2013; Ahn et al. 2014). We applied Makri's theory (Makri et al. 2010) to make a preliminary assessment of technology similarity between the Acquirer and the Target (A&T) with the help of IPC analysis. IPC of patents shows the distribution of technology areas. Technology similarity of firms with high consistency of technology distribution will probably be higher than the others'. The measure of technology similarity is described below. The Total Patent A&T in the formula means the total number of patents of both the A&T.

$$\text{Technology Similarity} = \frac{\text{Overlap All Patent Class}}{\text{Total Patent A\&T}} \times \frac{\text{Total Acquirer Patent In Common Classes}}{\text{Total Acquirer Patent}} \tag{10.1}$$

We extracted IPC information from the patents collected to illustrate technology similarity. We refined the data first. Individual assignees and firms whose total number of cloud patents was fewer than 6 were excluded. Second, we defined the degree of "common classes." For example, there are three patents $P_1$, $P_2$, $P_3$ with IPCs: H04L29/08, H04L29/06, H04H60/72. $P_1$ and $P_2$ represent similarity because they are under the same subcategory H04L29, and the combination with $P_3$ does not. In this way, we calculated the technology similarity between the acquirer and the potential target for each of the leading Chinese firms in the domain. The results are shown in Table 10.2. The left part of the table is the potential target list, and the right part is the evaluation of technology area. We found that GCI SCI & Technology Co. Ltd. (GCI), Shanghai Hechen Information Technology (Hechen), ZTE Corp (ZTE), and Shuguang Cloud Computing Technology Co. (Shuguang)

**Table 10.2** Preliminary technology similarity analysis

| Potential targets | Preliminary technology similarity analysis |
|---|---|
| GCI SCI & Technology Co. Ltd. | 0.064 |
| Shanghai Hechen Information Technology | 0.061 |
| ZTE Corp | 0.058 |
| Shuguang Cloud Computing Technology Co. | 0.053 |
| Inspur Electronic Information Co. Ltd. | 0.048 |
| Hon Hai Precision Ind Co. Ltd. | 0.036 |
| Microsoft Corp | 0.024 |
| IBM Corp | 0.014 |
| Beijing Z & W Technology Consulting Co. Ltd. | 0.011 |
| Yulong Computer Telecom Technology | 0.009 |
| Shenzhen Zidong Technology Co. Ltd. | 0.008 |

offer high consistency in the technology area. The IPC distribution is densely located in H40L29 and G06F09.

We further analyzed technology similarity from the perspective of the SAO structure of patent text. The SAO structure can express the precise meaning and can thus represent technological key concepts and key findings in the patent. Moehrle et al. (2005) proposed a method of using patent-based inventor profiles to guide human resource decisions. Park et al. (2013) used semantic patent maps to identify technological competition trends for R&D planning. We extracted the USE field from the patent abstracts and then transformed the content to SAO structures (Table 10.3). After filtering out some duplicated SAO structures using a set of Stop Words (2015), we got the data ready for semantic analysis.

WordNet-based semantic similarity between two SAO structures is computed by using the C# library (Simpson and Dao 2015). We set a threshold value as 0.7 to determine whether the two SAO structures are the same according to semantic similarity calculation results and the advice of experts. If the result is more than $s$, the two structures can be considered the same. For any two SAO structures ($SAO_i$ and $SAO_j$), we determined the Similarity (SIM) between them as follows:

$$\mathrm{SIM}(\mathrm{SAO}_i, \mathrm{SAO}_j) = \begin{cases} 1, & \mathrm{if}(\mathrm{Measure}(\mathrm{SAO}_i, \mathrm{SAO}_j)) \geq s \\ 0 & \mathrm{otherwise} \end{cases} \tag{10.2}$$

The USE of a patent includes more than one SAO structure. We defined the semantic similarity between the patents as the basis of how many SAO structures the two patents share. Suppose that there are two patents $P_1$ and $P_2$, and we denote that $Num_{SAO}(P_1)$ is the number of SAO structures in patent $P_1$, $Num_{SAO}(P_2)$ is the number of SAO structures in patent $P_2$, and $Num_{SAO}(P_1, P_2)$ is the number of the semantically identical SAO structures shared by patents $P_1$ and $P_2$. The Patent Similarity (PSIM) can be described as follows:

**Table 10.3** Sample of extracted SAO structure from patents

| S (Subject) | A (Action) | O (Object) |
|---|---|---|
| Method | Execute | Software application, e.g., batch application and user-interactive application, on a computer system, according to a SLA |
| System for creating a composite public cloud | Delivery | Hosted services |
| Method | Schedule | Cloud computing open platform |
| Virtualized desktop application display platform | Used | Cooperative computing of an electric power system |
| System | Control | Quantum microscopy instrument |
| Issuing network invoice | Based | Cloud computing and data asynchronous transmission technology |
| Method | Protect | Data and privacy of user in cloud environment |
| Multi-tenant service providers | Request | Dynamic platform reconfiguration |
| Distributed systems on a set of computer processors | Perform | Coordinated upgrades |
| Experiment cloud platform system | Manage | Computer calculation and software resources |

$$\mathrm{PISM}(\mathrm{P}_1, \mathrm{P}_2) = \frac{2 \times \mathrm{Num}_{\mathrm{SAO}}(\mathrm{P}_1, \mathrm{P}_2)}{\mathrm{Num}_{\mathrm{SAO}}(\mathrm{P}_1) + \mathrm{Num}_{\mathrm{SAO}}(\mathrm{P}_2)} \tag{10.3}$$

After measuring the similarity between patents, we took the pairwise average similarity of patents owned by two different firms as the technology similarity. For any two firms ($F_1$, $F_2$), Firm Technology Similarity (FSIM) could be measured as follows:

$$\mathrm{FSIM}(\mathrm{F}_1, \mathrm{F}_2) = \frac{\sum_{i=1}^{i=\mathrm{PN}(\mathrm{F}_1)} \left( \sum_{j=1}^{j=\mathrm{PN}(\mathrm{F}_2)} \mathrm{PSIM}(\mathrm{P}_i, \mathrm{P}_j) \right)}{\mathrm{PN}(\mathrm{F}_1) \times \mathrm{PN}(\mathrm{F}_2)} \tag{10.4}$$

Here, PN($F_1$) and PN($F_2$) are the patents of the two firms ($F_1$ and $F_2$), respectively, and the PSIM($P_i$, $P_j$) means the PSIM of the two firms. Figure 10.2 illustrates the degree of technology similarity between each other, and the first column is the technology similarity with Huawei.

We found that the three firms with highest technology similarity with Huawei were GCI, ZTE, and Hechen. The result matched with the IPC analysis that technology similarity of firms with high consistency of technology distribution is higher. We chose the firms whose technology similarity with Huawei was not less than 0.7 for technology complementarity (Table 10.4).

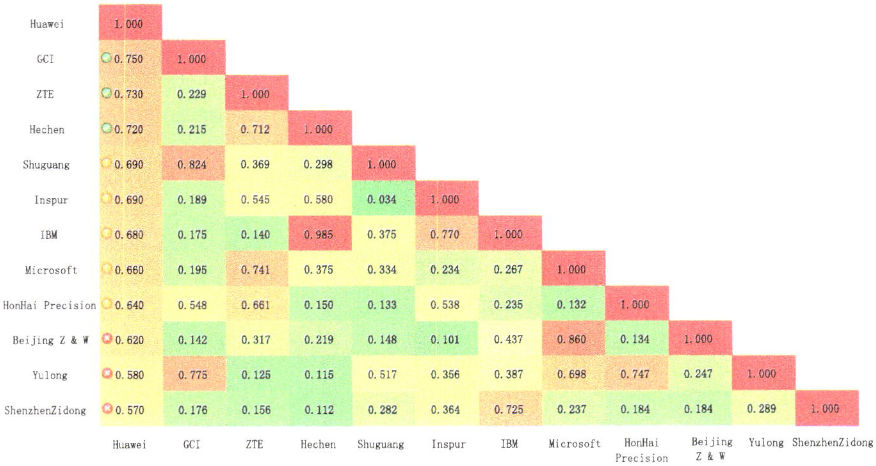


**Fig. 10.2** Degree of technology similarity between each other of the selected firms

**Table 10.4** Technology similarity evaluation of potential targets

| Potential targets | Technology similarity |
|---|---|
| **GCI SCI & Technology Co. Ltd.** | **0.75** |
| **ZTE Corp** | **0.73** |
| **Shanghai Hechen Information Technology** | **0.72** |
| Shuguang Cloud Computing Technology Co. | 0.69 |
| Inspur Electronic Information Co. Ltd. | 0.69 |
| IBM Corp | 0.68 |
| Microsoft Corp | 0.66 |
| Hon Hai Precision Ind Co. Ltd. | 0.64 |
| Beijing Z & W Technology Consulting Co. Ltd. | 0.62 |
| Yulong Computer Telecom Technology | 0.58 |
| Shenzhen Zidong Technology Co. Ltd. | 0.57 |

### *10.3.3 Technology Complementarity Analysis*

Analysis of technology complementarity is based on the results of the technology similarity analysis. Technology complementarity is considered as an important driver of invention (Golombek and Hoel 2004). Acquiring complementary technologies can promote exploratory learning within the organization, which may accelerate the process of innovation (Cassiman and Veugelers 2006). Different from technology similarity, complementary technology contributes to post-merger invention performance by stimulating higher quality and more novel inventions (Miozzo et al. 2011). Sparse research has been conducted on the measurement of technology complementarity at the pre-acquisition stage. In this paper, we apply morphology analysis for this. Technology Morphology Analysis was introduced to patent assessment by Yoon and Park (2005) and now is widely used for technology opportunities analysis. Technology complementarity before Tech M&A can be evaluated by analyzing different technology morphological combinations with the help of expert experience.

First, we converted the patents into structured data using keyword vectors according to their frequency of occurrence and with reference to technology dictionaries. Second, we set words associated with a specific technology and appearing frequently as keywords. Then, we mapped the patent keywords into their associated morphology and got Table 10.5, which shows the main technologies, subdivision technologies, and the corresponding keywords of cloud computing from the patents.

After that, we generated a table of subdivision technology complementarity according to the experts' assessment for complementarity between Huawei and the potential targets. If the firms had related patents, we added a group of lines to indicate that they had related technologies. The groups of vertical lines and horizontal lines reflected the technology distribution of Huawei and the target, respectively. The depth of background color of each cell showed the complementarity level, which was divided into 3 layers, and the white background meant the two technologies had no complementarity. The crossing lines with red background meant the two firms had complementary technologies. The table was symmetrical, so we took the lower triangular region for analysis. We set the degree

**Table 10.5** Technology morphology of cloud computing

| Main technology composition | | Sample keywords |
|---|---|---|
| Display technology | Based on plug-in | Flash, Silverlight, JavaFX |
| | Based on browser | HTML5, Ajax, CSS3 |
| Multi-tenancy | Shared nothing | Separate database, separate schema |
| | Shared hardware | Shared database, separated data storage, additional storage subsystems |
| | Shared everything | Shared schema, network monitoring, shared schema |
| Virtualization | Platform virtualization | Virtual machine monitor, hypervisor, host OS |
| | Resource virtualization | Load balancing, monitoring resources |
| | Application virtualization | Virtual terminal, remote access, application access |
| Security | Application security | Anti-virus services, network security monitoring, DDoS attack warning |
| | Platform security | Access control management, security API, network security |
| | Infrastructure security | Secure hypervisor, full disk encryption, secure virtual machine |
| Distributed Storage | | Horizontal scalability, area network storage, secret sharing |

of complementarity to three levels marked by the depth of color, and the three levels were set as 1, 2, and 3 for calculation. To measure the technology complementarity, we just needed to take the cells with crossing lines and background into consideration. Again, we supposed there were two firms $F_1$ and $F_2$. We denoted the related patent number of $F_1$ and $F_2$ in the $i$th cell with crossing lines and red background as $F_1PN(i)$ and $F_2PN(i)$. The complementarity formula between the two firms is below:

$$\text{Complementarity} = \sum_{i \in D} (F_1PN(i) \times F_2PN(i) \times \text{complementarity level}) \quad (10.5)$$

where $D$ means the technology areas in which the technologies of the two firms are complementary.

After calculating the complementarity of all the potential targets with the acquirer, we normalize the result by calculating the percentage of each complementarity result in the sum of all the complementarity results.

We took GCI (GCI SCI & Technology Co. Ltd.) as an example. The technology complementarity of the two firms is shown in Table 10.6. The two firms had complementary technologies in the area of display technology area and security technology area, of which the cells appear with crossing lines and red background. For GCI, 1 patent was on display technology based on plug-in; 2 patents were based on browsers, and 2 patents on application virtualization. For Huawei, 3 patents were on platform security. Using (5), we calculated the technology complementarity as 39. We could compute the technology complementarity for the remaining firms in the same way. We then carried out a normalization process. Technology complementarity of other potential targets is listed in Table 10.7. ZTE had the highest complementarity with Huawei, the second being GCI, and the third being Hechen.

**Table 10.6** Technology complementarity of Huawei and GCI

| Huawei \ GCI | Based on plug-in | Based on browser | Shared nothing | Shared hardware | Shared everything | Platform virtualization | Resource virtualization | Application virtualization | Application security | Platform security | Infrastructure security | Distributed Storage |
|---|---|---|---|---|---|---|---|---|---|---|---|---|
| Based on plug-in | | | | | | | | | | | | |
| Based on browser | | | | | | | | | | | | |
| Shared nothing | | | | | | | | | | | | |
| Shared hardware | | | | | | | | | | | | |
| Shared everything | | | | | | | | | | | | |
| Platform virtualization | | | | | | | | | | | | |
| Resource virtualization | | | | | | | | | | | | |
| Application virtualization | | | | | | | | | | | | |
| Application security | | | | | | | | | | | | |
| Platform security | | | | | | | | | | | | |
| Infrastructure security | | | | | | | | | | | | |
| Distributed Storage | | | | | | | | | | | | |

Note: complementarity level 1 shown by ■ complementarity level 2 shown by ■ and complementarity level 3 shown by

**Table 10.7** Technology complementarity evaluation of potential targets

| Potential Targets | Technology complementarity |
|---|---|
| ZTE Corp | 0.56 |
| GCI SCI & Technology Co. Ltd. | 0.23 |
| Shanghai Hechen Information Technology | 0.21 |

**Table 10.8** Appropriate target candidates of Tech M&A for Huawei

| Potential targets | Knowledge base | R&D intensity |
|---|---|---|
| ZTE Corp | 52 | 0.12 |
| Shanghai Hechen Information Technology | 13 | 0.06 |
| GCI SCI & Technology Co. Ltd. | 10 | 0.05 |

### *10.3.4 R&D Capability Analysis*

Wu and Reuer (2014) indicated that R&D capability is an important factor of technology integration and innovation after Tech M&A. Acquiring firms with high R&D capability will promote technology integration and technology synergy creation after Tech M&A (Benitez and Ray 2012). In this study, we used the absolute size of the knowledge base measured by the number of related patents and R&D intensity to evaluate the R&D capability of candidate acquisitions. R&D intensity was defined as the ratio of expenditures by a firm on R&D to the firm's sales. We used the average of three years' ratio. All of the three firms could be target candidates for Huawei from the perspective of Tech M&A, and ZTE was the most appropriate target. Considering the scales of the three firms, if Huawei hoped to become the leading firm through Tech M&A, ZTE Corp could be the preferred target; if Huawei hoped to enhance subdivision technologies in cloud computing, Shanghai Hechen and GCI would be the better choices (Table 10.8).

## 10.4 Conclusions

This paper presented a framework to identify and evaluate companies from the technological perspective to support M&A target selection decision making. The paper took technology similarity, technology complementarity, and R&D intensity as main indicators to evaluate potential targets. We introduced patent text analysis to generate a more comprehensive method for technology relatedness evaluation.

First, technology similarity was preliminarily evaluated according to patent IPCs. Further analysis was conducted using SAO-based semantic similarity analysis based on patent text. The approach enables one to extract the technological key concepts and key findings in patents and can complement the IPC-based analysis. Firms with high technology similarity with the acquirer can be selected.

Second, technology morphology analysis was introduced to analyze the technology complementarity between the targets and the acquirer. Keywords from patent text were mapped into their associated morphology. Technology complementarity level was set by the experts to all the possible technology combinations. The technology complementarity could be computed according to the patent distribution and the corresponding technology complementarity level. Thus, firms were further selected.

Third, this paper used R&D capability, including the absolute size of knowledge base and R&D intensity, to help choose the targets for an acquirer. We verified the usability and practicality of the method by applying it to patents related to cloud computing technologies and selected Huawei Technologies Co. Ltd. as an example to assess Tech M&A in the cloud computing technology area.

During the whole analysis process, we kept in contact with department of cloud computing of Huawei. Huawei showed interest in our research results, especially the measurement of the technology similarity and technology complementarity. Huawei extended requests for our research, including further technology similarity and complementarity analyses and the analysis of technology development trends of cloud computing, to support the firm's present work.

However, there are a few limitations in the study. Some doubts remain regarding the reliability of patent data. Sometimes, patent data cannot reflect the core technology of a firm because an emerging technology is not patented. The analysis based on patents does not take tacit knowledge into account per se. Another limitation is to what extent the framework can be applicable to other industries. Firms in some kinds of industries may not have many patents, though they have complex knowledge. Other indicators, such as the stage of technology development and the range of multiple sector interests of the players, should be considered in further studies.

**Acknowledgments** This research was undertaken at Georgia Tech and Beijing Institute of Technology drawing on support from the National Science Foundation (NSF) through the Science of Science Policy Program—"Measuring and Tracking Research Knowledge Integration" (Georgia Tech; Award No. 0830207), and the National Science Foundation of China Young Funds (Beijing Institute of Technology; Award No. 71103015). The findings and observations contained in this paper are those of the authors and do not reflect the views of the NSF and NSFC.

## References

Ahn, J. M., Minshall, T., & Mortara, L. (2014). Open innovation: An approach for enhancing performance in innovative SMEs. Available at SSRN 2431205.

Ali-Yrkkö, J., Hyytinen, A., & Pajarinen, M. (2005). Does patenting increase the probability of being acquired? Evidence from cross-border and domestic acquisitions. *Applied Financial Economics, 15*(14), 1007–1017.

Benitez-Amado, J., & Ray, G. (2012). Introducing IT-enabled business flexibility and IT integration in the acquirer's M&A performance equation.

Cassiman, B., & Veugelers, R. (2006). In search of complementarity in innovation strategy: Internal R&D and external knowledge acquisition. *Management Science, 52*, 68–82.

Christensen, C. M., Alton, R., Rising, C., & Waldeck, A. (2011). The big idea: The new M&A playbook. *Harvard Business Review, 89*(3), 48–57.
Di Guardo, C., Harrigan, K. R., & Marku, E. (2015). Quantity at expense of quality? Measuring the effects of technological M&A on innovation and firm performance.
Golombek, R., & Hoel, M. (2004) Unilateral emission reductions and cross-country technology spillovers. *Advances in Economic Analysis & Policy, 3.*
Gupta, P. K. (2013). Mergers and acquisitions (M&A): The strategic concepts for the nuptials of corporate sector. *Innovative Journal of Business and Management, 1*(4).
Hussinger, K. (2010). On the importance of technology relatedness: SMEs versus large acquisition targets. *Technovation, 30*(1), 57–64.
Kengelbach, J., & Roos, A. W. (2011). *Riding the next wave in M&A: Where are the opportunities to create value?* Boston Consulting Group, Incorporated.
Kohers, N., & Kohers, T. (2000). The value creation potential of high-tech mergers. *Financial Analysts Journal, 56*(3), 40–51.
Lin, L. H. (2012). Innovation performance of Taiwanese information firms: An acquisition–learning–innovation framework. *Total Quality Management & Business Excellence, 23*(9–10), 1135–1151.
Lodh, S., & Battaggion, M. R. (2014). Technological breadth and depth of knowledge in innovation: The role of mergers and acquisitions in biotech. *Industrial and Corporate Change*, dtu013.
Loughran, T., & Vijh, A. M. (1997). Do long-term shareholders benefit from corporate acquisitions? *Journal of Finance*, 1765–1790.
Makri, M., Hitt, M. A., & Lane, P. J. (2010). Complementary technologies, knowledge relatedness, and invention outcomes in high technology mergers and acquisitions. *Strategic Management Journal, 31*(6), 602–628.
Miozzo, M., DiVito, L., & Desyllas, P. (2011). Cross-border acquisitions of science-based firms: Their effect on innovation in the acquired firm and the local science and technology system. In *DRUID Conference* (Vol. 18), June 16.
Moehrle, M., & Geritz, A. (2004). Developing acquisition strategies based on patent maps. In *13th Iamot*, Washington, DC.
Owens, T., & Logue, F. (2012). Cloud computing the Irish perspective.
Park, H., Yoon, J., & Kim, K. (2013). Identification and evaluation of corporations for merger and acquisition strategies using patent information and text mining. *Scientometrics, 97*(3), 883–909.
Paruchuri, S., Nerkar, A., & Hambrick, D. C. (2006). Acquisition integration and productivity losses in the technical core: Disruption of inventors in acquired companies. *Organization Science, 17*(5), 545–562.
Ragothaman, S., Naik, B., & Ramakrishnan, K. (2003). Predicting corporate acquisitions: An application of uncertain reasoning using rule induction. *Information Systems Frontiers, 5*(4), 401–412.
Sears, J., & Hoetker, G. (2014). Technological overlap, technological capabilities, and resource recombination in technological acquisitions. *Strategic Management Journal, 35*(1), 48–67.
Simpson, T., & Dao, T. (2015). WordNet-based semantic similarity measurement. Retrieved January 12, 2015. Word Wide Web.
Stop Words Project. (2015). Stop-words. Retrieved January 21, 2015. Word Wide Web: https://code.google.com/p/stop-words/
Wei, T., & Tian, X. (2011). How to select target firms in M&As? Evidence from the Medical Technology Industry. IJEI: *International Journal of Engineering and Industries, 2*(2), 8–26.
Wu, C. W., & Reuer, J. J. (2014). Effects of R&D investments and signals on international acquisitions: Evidence from IPO firms. *Academy of Management, 2014*(1), 14479 (Academy of Management Proceedings).
Yoon, B., & Park, Y. (2005). A systematic approach for identifying technology opportunities: Keyword-based morphology analysis. *Technological Forecasting and Social Change, 72*(2), 145–160.

# Chapter 11
# Identifying Technological Topic Changes in Patent Claims Using Topic Modeling

**Hongshu Chen, Yi Zhang and Donghua Zhu**

**Abstract** Patent claims usually embody the core technological scope and the most essential terms to define the protection of an invention, which makes them the ideal resource for patent topic identification and theme changes analysis. However, conducting content analysis manually on massive technical terms is very time-consuming and laborious. Even with the help of traditional text mining techniques, it is still difficult to model topic changes over time, because single keywords alone are usually too general or ambiguous to represent a concept. Moreover, term frequency that used to rank keywords cannot separate polysemous words that are actually describing a different concept. To address this issue, this research proposes a topic change identification approach based on latent dirichlet allocation, to model and analyze topic changes and topic-based trend with minimal human intervention. After textual data cleaning, underlying semantic topics hidden in large archives of patent claims are revealed automatically. Topics are defined by probability distributions over words instead of terms and their frequency, so that polysemy is allowed. A case study using patents published in the United States Patent and Trademark Office (USPTO) from 2009 to 2013 with Australia as their assignee country is presented, to demonstrate the validity of the proposed topic change identification approach. The experimental result shows that the proposed approach can be used as an automatic tool to provide machine-identified topic changes for more efficient and effective R&D management assistance.

**Keywords** Tech mining · Topic modeling · Patent analysis

H. Chen (✉) · Y. Zhang
Decision Systems & e-Service Intelligence Research Lab,
Centre for Quantum Computation & Intelligent Systems,
Faculty of Engineering and Information Technology, University of Technology Sydney,
Sydney, Australia
e-mail: Hongshu.Chen@uts.edu.au

H. Chen · Y. Zhang · D. Zhu
School of Management and Economics, Beijing Institute of Technology,
Beijing, People's Republic of China



T.U. Daim et al. (eds.), *Anticipating Future Innovation Pathways Through Large Data Analysis*, Innovation, Technology, and Knowledge Management,
DOI 10.1007/978-3-319-39056-7_11

## 11.1 Introduction

Patent claims are often argued as a valuable source for the detection of technological changes and to gain technological insight (Campbell 1983; Ernst 1997; WIPO 2004). As an important part of unstructured segments of a patent document, claims hold explicit information and implicit knowledge revealing technological concepts, topics, and related R&D activities with concise, but precise language (Xie and Miyazaki 2013; WIPO 2002). Since manually conducting content analysis on massive patent documents is very time-consuming and laborious, in recent years, one of the fundamental changes to research in R&D management is the access to extremely powerful information techniques and a vast amount of digital and textual data (Daim et al. 2011). In particular, for efficient patent analysis, automatic approaches to assist domain experts and decision makers to discover and understand large volumes of patent documents have drawn increasing attention and still are in great demand (Abbas et al. 2014).

Much effort has been devoted to reveal latent knowledge from the textual data of patent documents. Watts and Porter (1997) suggested an approach to investigate terminological trends by tracking the historical change of keywords. Yoon and Park (2005) presented a keyword-based morphology study to identify the detailed configurations of promising technology. Zhang et al. (2014) introduced a term clumping approach based on principal components analysis to explore keywords and main phrases in abstract from scientific literature. In addition, text analytics have already been applied to technology intelligence application *TrendPerceptor* (Yoon and Kim 2012), *Techpioneer* (Yoon 2008), *VantagePoint* (Zhu and Porter 2002), and *Aureka* (Trippe 2003) to determine hidden concepts and relationships, where clustering, classification and mapping techniques were used to support further content analysis of technological documents. However, before most of these applications are applied, usually several sets of keywords need to be defined in advance, which still derive from the opinion and knowledge of domain experts. Moreover, the outcomes of majority traditional text mining techniques are based on single keywords with ranking, yet these words alone are usually too general or misleading for indicating a concept, especially when there are polysemous words actually describing different themes (Tseng et al. 2007).

To overcome the above-mentioned limitations, this research proposes a topic change identification approach using a well-known topic modeling approach, latent dirichlet allocation. Unsupervised topic modeling is applied to vast amounts of target patent claims, providing a corpus structure with minimal human intervention. There is no preset classification or keywords list for this approach, and the results are discovered in a completely unsupervised way. In addition, instead of using single terms, topics are represented by probability distributions over words. The actual semantic meaning of a topic is able to be delivered in this way, and at the same time, the polysemous words, which are actually depicting different concepts, can also be separated. After revealing topics from patent sub-collections of different years, a topic change model is presented to identify topic changes over time.

Finally, to demonstrate the performance of our proposed approach, patents published during years 2009 to year 2013 in the United States Patent and Trademark Office (USPTO) with Australia as their assignee country are selected to present a case study. The experimental result demonstrates that the proposed approach is able to provide machine-identified topic changes automatically without any presetting of keywords. The outcomes of our approach will be used to serve R&D management assistance.

This paper is organized as follows: the first section reviews related research developments by introducing patent data in technological research and latent dirichlet allocation. Methodology Section describes the proposed topic change identification approach step by step. Case Study Section carries out experiments using USPTO patents to demonstrate the proposed approach in a real patent analysis context. The conclusions and future study are addressed in the last section.

## 11.2 Literature Review

### *11.2.1 Patent Data in Tech Mining*

Patent documents are composed of structured information and unstructured descriptions of inventions. Analytical approaches based on structured data of patents, such as issue date, inventor, assignees, or International Patent Classification, have played the major role in both theoretical and practical research to gain insight of technology development in certain area (Lai and Wu 2005; Sheikh et al. 2011; Nishijima et al. 2013). However, the unstructured data in patent documents, such as abstracts, claims, and descriptions, usually contain much more abundant information than the structured sections, since they contain significant characteristics, detailed functionalities, or major contributions of technologies. Therefore, there has been a lot of interest in applying text mining techniques to conduct tech mining and set domain analysts free from studying and understanding massive amounts of technological content since the last decade (Tseng et al. 2007; Camus and Brancaleon 2003; Porter 2005).

Among all the unstructured segments of a patent file, patent claims play a role of embodying all the important technical features of an invention with the most essential technological terms to define the protection (Tong and Frame 1994). On one hand, they reveal the core inventive topics and the major technological scope of a patent; on the other hand, claims are written in concise, but precise language, which make them the best resource for identifying technological topics and facilitating patent document analysis (Xie and Miyazaki 2013; WIPO 2002; Yang and Soo 2012; Novelli 2014).

A patent claim usually consists of three parts: a preamble that serves as an introductory section to recite the primary purpose, function, or properties; a transition phrase, such as comprising, having including, consisting of, etc.; a “body” that contains the elements or steps that together describe the invention (Yang and Soo 2012; USPTO 2012; Sheldon 1995). This research utilizes patent claims as the main source of topic change analysis. Among patent databases from different countries, the United States Patent and Trademark Office (USPTO) database is mostly used because patents submitted in other countries are often also simultaneously submitted in the United States (USPTO 2015).

### *11.2.2 Latent Dirichlet Allocation*

Latent dirichlet allocation (LDA) (Blei et al. 2003) is a probabilistic model that aims to estimate the properties of multinomial observations by unsupervised learning. It gives an estimation of the latent semantic topics hidden in large archives of documents and calculates the probabilities of how various documents belong to different topics. LDA has been used as an efficient tool to assist topic discovery and analysis, in practice. For example, Griffiths and Steyvers (2004) applied LDA-based topic modeling to discover the hot topics covered by papers in Proceedings of the National Academy of Sciences of the United States of America (PNAS); Yang et al. (2013) proposed a topic expertise model (TEM) based on LDA to jointly model topics and expertise for community question answering (CQA) with stack overflow data; Kim and Oh (2011) proposed a framework based on LDA to identify important topics and their meaningful structure within the news archives on the Web.

The graphical model of LDA is presented in Fig. 11.1, showing three rectangular plates where: $D$ denotes the overall documents in a corpus; $K$ indicates the topic numbers for $D$; and $N_d$ stands for the term number of $d$th document in document collection $D$. Each node in the figure stands for a random variable in the generative process of LDA, while the plates indicate replication. In the left part of the figure, $\vec{\vartheta}_d$ stands for the topic proportions for the $d$th document. For document

**Fig. 11.1** The graphical model of latent dirichlet allocation

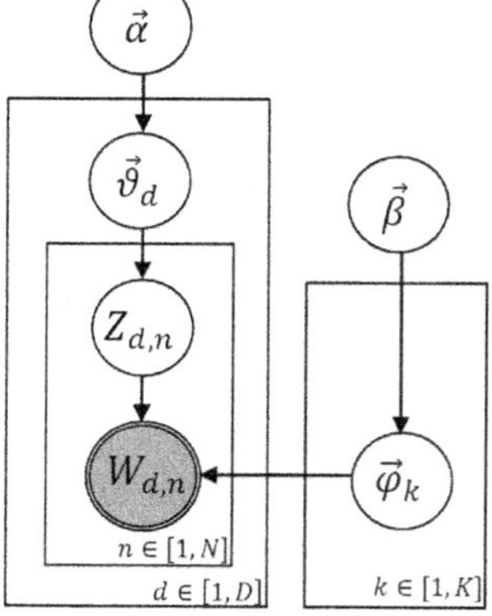

$d$, the topic assignments are $Z_d$, where $Z_{d,n}$ indicates the topic assignment of the $n$th word in the $d$th document. On the right of the figure, the topics themselves are illustrated by $\vec{\varphi}_{1:K}$, where each $\vec{\varphi}_k$ is a distribution over vocabularies. All of the unshaded circles indicate hidden nodes. The shaded circles, on the contrary, are observable nodes, where $W_{d,n}$ stands for the $n$th word in document $d$. Finally, $\alpha$ and $\beta$ are two hyperparameters that determine the amount of smoothing applied to the topic distributions for each document and the word distributions for each topic (Blei et al. 2003; Steyvers and Griffiths 2007; Blei 2012; Heinrich 2005).

The parameters of LDA need to be estimated by an iterative approach. Among existing approaches, Gibbs sampling is one of the most commonly used methods. It is an approximate inference algorithm based on the Markov chain Monte Carlo (MCMC) and has been widely used to estimate the assignment of words to topics by observed data (Griffiths and Steyvers 2004; Noel and Peterson 2014; Lukins et al. 2010). Gibbs sampling produces different results each time in executing LDA, so that the topic estimations are slightly different even with exactly the same setting of input and parameters; yet on the whole, the results of different experiments will not change much.

## 11.3 Methodology

This section explains the details of our proposed topic change identification approach. The framework is given first; each detailed step is illustrated subsequently.

### *11.3.1 Framework*

The overall framework of our proposed topic change identification approach is shown in Fig. 11.2. First of all, users need to initiate a search statement to declare their domain analytic requirements and address a group of target patents in USPTO database. Patent ID, title, claims, issue time, assignees, United States Patent Classification (USPC), and other information of target patents are then crawled into a database waiting for further analysis. To identify topic changes over time, the whole patent collection is divided into several sub-collections first and labeled with their corresponding issue year. Subsequently, for each sub-collection, patent claims and titles, embodying essential technical terms, and USPC, providing a general understanding of the domain classification, are extracted from the target patents database separately. The two plates in the figure indicate replication.

Textual data composed by claims and titles, after data segmentation and cleaning, are then placed into a series of words exclusion modules to filter out the most common function words, high-frequency words that commonly appeared in

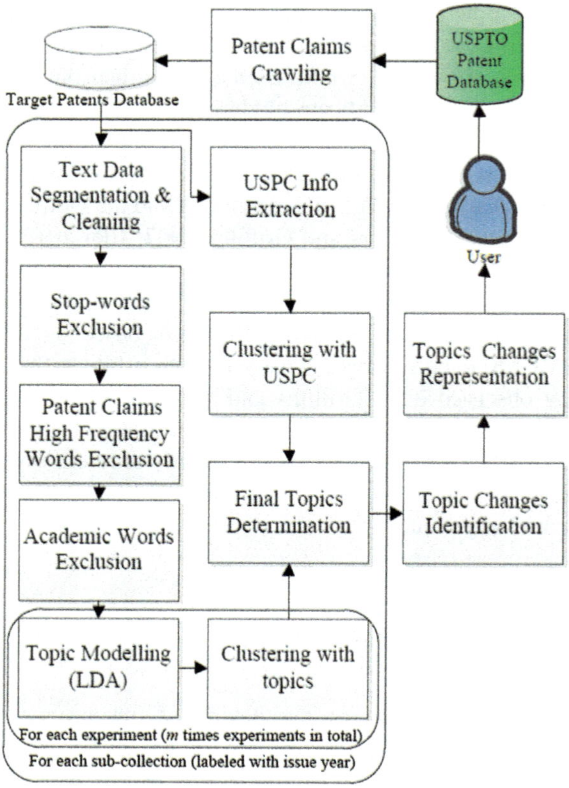


**Fig. 11.2** The framework of the proposed topic change identification approach

patent claims, and academic words with vague and general meanings. Then, the prepared text will be passed to the topic modeling module. Meanwhile, the USPC information of the corresponding patents is extracted to assist final topic determination. As mentioned, the randomness introduced by the initiation of the sampling will affect the final result of LDA. To acquire the most reliable topics of the corpus, we utilize USPC as a measurement to evaluate results from *m* times experiments. Patents are clustered with both their USPC and topic proportions. The final topic modeling result is the one trial that provides the most similar clusters to the USPC clustering outcome. Finally, with all the topics estimated from patent sub-collections of different years, topic changes over time can be identified and presented to users.

## *11.3.2 Patent Corpus Text Cleaning*

Patent claims are a special kind of textual data that contain plenty of technical terms, specific words serving as transition phrases, and numerous academic words

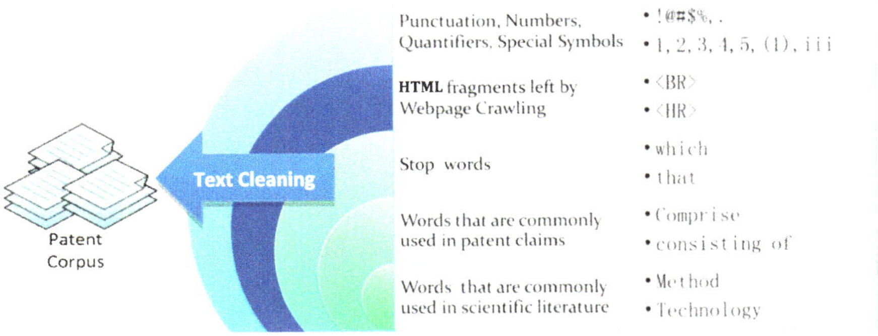


**Fig. 11.3** Relationships between sub-collections and topics

that describe invention outcomes. Among all the terms that one claim may contain, only technical terms provide the most meaningful information that reflects technological topics and innovations. Therefore, for our patent corpus, each sub-collection, as shown in Fig. 11.3, before modeling topics with LDA, except all the punctuations, numbers, and HTML fragments left by webpage crawling, we also utilize three modules to remove general words from the corpus of patents as follows:

- Stop words such as *the, that, and these*;
- High-frequency words in patent claims such as *claimed, comprising, and invention*;
- General academic words such as *research, approach, and data.*

The stop words list we applied is from an information retrieval Resources link from Stanford University (David et al. 2004); the patent claim commonly used phrases are summarized from a Transitional Phrase page on Wikipedia (2014); the general academic words list is provided by the University of Nottingham, we select the top 100 most frequent academic words and remove them from our final corpus (Haywood 2003; Zhang et al. 2014).

## *11.3.3 Topic Modeling*

LDA utilizes a probability distribution over words, instead of a single term, to define a concept, delivering better semantic meaning of the topic and, at the same time, allowing polysemy. Thus, it is very suitable for "understanding" the content of large corpuses such as emails, news, scientific papers, and our main data source here, patent claims. After removing all commonly used words from the corpus, we utilize LDA to generate several groups of topics for each patent sub-collection in the corpus, which is labeled by its corresponding issue year. In a sub-collection, the claims and title of each patent constitute one document, and the number of

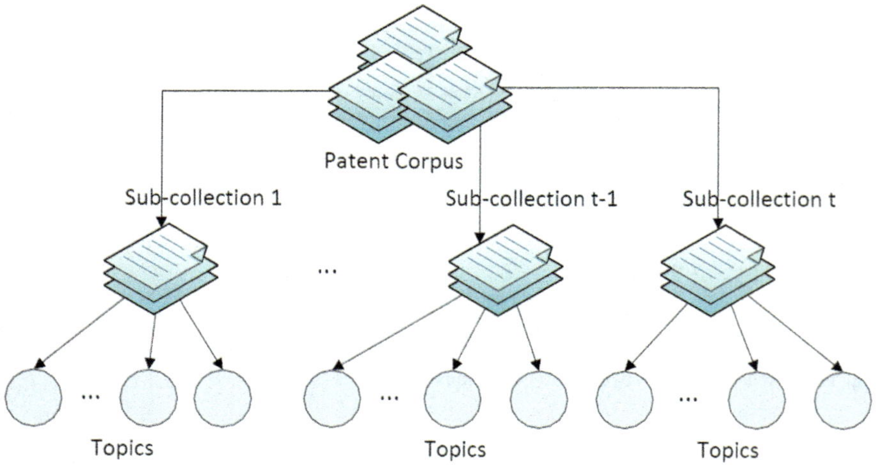


**Fig. 11.4** Relationships between sub-collections and topics

documents equals the number of patents; the USPC and other structural information are stored alone in a single file to assist further topic determination. All the textual documents in the corpus are seen as mixtures of a number of topics; each topic is seen as a distribution over various vocabularies. Here, we present the global topics as $\vec{P}_{1:t} = (\vec{P}_1, \vec{P}_2, \ldots, \vec{P}_i, \ldots, \vec{P}_t)$, where $\vec{P}_i$ stand for the topics of the $i$th sub-collection of the corpus. The relationship between sub-collections and topics is illustrated in Fig. 11.4.

Since we know nothing about the word distributions composing the topics and the topic distributions composing the documents, before topic modeling, assumptions need to be first drawn to determine the parameters $k, \alpha, \beta$ of LDA. According to previous research, hyperparameters $\alpha, \beta$ of the dirichlet distribution in LDA have a smoothing effect on multinomial parameters; that is, the lower the values of $\alpha$ and $\beta$ are, the more decisive topic associations there will be (Heinrich 2005). This research sets $\alpha = 0.5$ and $\beta = 0.1$, which are commonly used in LDA applications (Koltcov et al. 2014). For the setting of $K$, higher $K$ will reduce the topical granularity but increase the processing time significantly. Therefore, during the implementation, $K$ needs to be decided case by case, balancing user requirement and time consumption. Different parameter settings may improve modeling performance, yet optimizing these parameters is beyond the scope of this paper.

### 11.3.4 Final Topics Determination

We then apply Gibbs sampling to infer the needed distributions in LDA. Since the initial values of variables are determined randomly in Gibbs sampling, the outputs of LDA in multiple experiments with a same corpus are slightly different. To ensure the final topic modeling estimation as reliable as possible, evaluation criteria will be needed for the topics finalization. In this research, we select USPC as the criteria. As a predefined classification hierarchy built on domain expert judgments, USPC provides a general understanding of the technical domain of concern to one patent. Because patents covering similar topics are usually assigned to a same main USPC, thus here we use the main USPC to judge which estimation is closer to the actual topic structure.

For a sub-collection of corpus, multiple LDA experiments will produce a number of topic distribution matrixes, each indicating the topic distribution proportions of patent documents in the corresponding trial. As shown in the approach framework, Fig. 11.2, there will be $m$ times experiments for every sub-collection; and after performing each time run, patents in the sub-collection are clustered with their calculated topic distributions using the hierarchical clustering approach (Steinbach et al. 2000). Meanwhile, the same group of patents will be also clustered with USPC information. The closer the two clustering results are, the more reliable the topic modeling result is.

Specifically, the values of indexes Jaccard et al. and F1 of $m$ times experiments are used to measure the similarity of the two clustering results, one by topics and the other by USPC. The three indices are listed as follows (Halkidi et al. 2001):

$$J = a/(a + b + c), \tag{11.1}$$

$$\mathrm{FM} = a/\sqrt{r_1 \cdot r_2}, \tag{11.2}$$

$$F_\beta = \frac{(\beta^2 + 1) \cdot r_1 \cdot r_2}{\beta^2 \cdot r_1 + r_2}, \tag{11.3}$$

where $J$ stands for Jaccard coefficient, FM indicates Folkes & Mallows index, $F_\beta$ presents the $F1$ indice. In addition, $r_1 = a/(a+b)$, $r_2 = a/(a+c)$, where $a$ represents the number of patents that belong to the same cluster of topics and to the same USPC in our case, $b$ is the number of patents that belong to the same cluster of topics but to different USPC, and $c$ is the number of patents that belong to different clusters of topics but to the same USPC. The topic modeling result that provides the highest index values is the optimal one.

### *11.3.5 Topic Change Identification*

After locating the final topics and words underlying the sub-collections of our corpus, we are able to identify the topic change over time. As show in Fig. 11.5, we compare two groups of topics deriving from different corpus sub-collections, calculating the similarity of words between each topic in $\vec{P}_i$ and all the topics in $\vec{P}_{i-1}$, in a traversal way. If two topics under different sub-collections contain

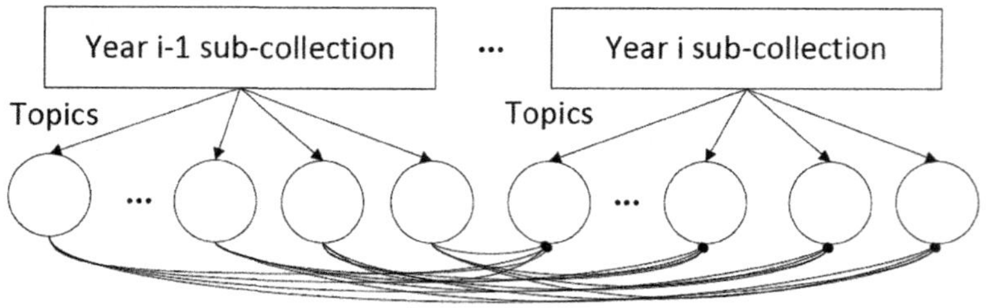


**Fig. 11.5** Topic change identification model

approximately the same group of words, then we believe that these two topics are actually one topic evolving from year to year. However, if the majority of words comprising two topics are very different, then we believe these are two different topics. Finally, for documents sub-collection of year $i$, if there is no similar topic can be matched in the previous year, year $i-1$, then the unmatched topic in the later year can be seen as a newly important one, which means it became more hot in the year $i$.

## *11.3.6 Topic-Based Trend Estimation*

If we already identified a topic evolving from year to year, besides discovering how the detailed content of the topic evolves from year to year with the above model, we can also use the topic distribution matrix to generate historical topic-based trend and forecast future trend. As an important part of LDA outcomes, the topic distribution matrix $\vec{\vartheta}$ provides the estimated result that how all the topics distribute over the document collection. The summation of each row of the matrix equals 1. The sum values of each column, however, are different. The larger the sum of a column, the more important the corresponding topic is. Since the patents are issued along a time line, if we add up a group of elements in a column that associates with patents published in a same time interval (month or year), the summation can be used to present the weight of the topic in that time frame. Thus we can then get a temporal-weight matrix to reveal the importance of selected topics in different month or years.

After the temporal-weight matrix is achieved, we calculate the weight changes in a least-squares sense to estimate the general trend of the target topics. The temporal-weight values of each topic are fitted to a univariate quadratic polynomial, $y = ax^2 + bx + c$, where $y$ stands for the topic weight, and $x$ represents the time. We utilize the coefficients $a$ and $b$ to measure developing trends of topics, since $a$ controls the speed of increase (or decrease) of the quadratic function, $-b/2a$ control the axis of symmetry. For instance, if coefficient $a$ is positive and the symmetry is on the left of $y$-axis, we consider the corresponding topic has a growing trend where the greater $a$ is, the faster the growth will be.

## 11.4 Case Study

### 11.4.1 Data Collection

To demonstrate the performance of our proposed approach, patents published during years 2009 to year 2013 in USPTO (http://www.uspto.gov/) with Australia as their assignee country are selected to present a case study. There are 7071 target patents covering 343 different main USPC[1,2]. Their patent ID, titles, issue time, inventors, Assignees, United States Patent Classification (USPC), International Patent Classification (IPC), and most importantly, their claims are clawed from USPTO and placed in a patents tool for further processing. The claims and title for each patent constitute one document in our corpus, which totals 7071 documents on the whole. Then, the whole document collection was divided into five sub-collections to present technological feature and essential terms of inventions by Australia assignees in the past five years. The detailed documents number was published every year from 2009 to 2010; the term number and USPC number in each corresponding sub-collection are shown in Table 11.1. Although the number of documents declined from year 2011, the term number kept rising, which implies that the average complexity of patent claims description is increasing in the resent three years. We also observe that the number of USPC in 2010 had a visible growth, suggesting that there may be a group of new topics appearing in year 2010 comparing with year 2009.

### 11.4.2 Topic Set Determination

Before topic modeling, as mentioned, a number of parameters need to be set first, including the number of topics $K$ and $\alpha, \beta$ of dirichlet distribution. In the case study, we applied $K = 10$ with model hyperparameters $\alpha = 0.5, \beta = 0.1$ to our target documents, to balance the topical granularity, convenience of understanding, and

**Table 11.1** The number of documents, terms, and USPC of patents published each year

| Year | Doc No. | Term No. | USPC No. |
|---|---|---|---|
| 2009 | 1174 | 19,796 | 199 |
| 2010 | 1613 | 24,726 | 233 |
| 2011 | 1746 | 23,757 | 228 |
| 2012 | 1256 | 25,102 | 233 |
| 2013 | 1282 | 29,714 | 227 |

[1]Data accessed in March 2014.

[2]All plant patents are seen as having one same USPC for calculation convenience.

**Table 11.2** Indexes information for the final chosen experiment result

| Year | Index | E 1 | E 2 | E 3 | E 4 | E 5 |
|---|---|---|---|---|---|---|
| 2009 | FM | 0.2376 | 0.2803 | **0.2845** | 0.2739 | 0.1948 |
| 2009 | DJC | 0.1217 | 0.1500 | **0.1505** | 0.1436 | 0.0962 |
| 2009 | F1 | 0.2169 | 0.2608 | **0.2616** | 0.2511 | 0.1755 |
| 2010 | FM | 0.2668 | 0.2152 | 0.2253 | 0.3125 | **0.3688** |
| 2010 | DJC | 0.1357 | 0.1037 | 0.1077 | 0.1634 | **0.2017** |
| 2010 | F1 | 0.2389 | 0.1880 | 0.1944 | 0.2809 | **0.3356** |
| 2011 | FM | 0.2521 | 0.2484 | 0.2334 | **0.2604** | 0.2541 |
| 2011 | DJC | 0.1334 | 0.1300 | 0.1166 | **0.1342** | 0.1294 |
| 2011 | F1 | 0.2354 | 0.2301 | 0.2089 | **0.2366** | 0.2292 |
| 2012 | FM | 0.3060 | **0.3202** | 0.2773 | 0.2820 | 0.2686 |
| 2012 | DJC | 0.1756 | **0.1853** | 0.1539 | 0.1632 | 0.1521 |
| 2012 | F1 | 0.2987 | **0.3127** | 0.2667 | 0.2806 | 0.2640 |
| 2013 | FM | 0.2984 | 0.2989 | **0.3356** | 0.3177 | 0.3086 |
| 2013 | DJC | 0.1753 | 0.1749 | **0.1986** | 0.1876 | 0.1794 |
| 2013 | F1 | 0.2983 | 0.2977 | **0.3313** | 0.3159 | 0.3042 |

the speed of processing. There are 10 topics describing the essential technological content and feature for each year; and every topic is presented with 10 words given highest probability by this topic.

Indices Folkes & Mallows (FM), Jaccard (DJC), and F1 are calculated after we clustered the patents using both topic assignment and main USPC information. Observation for each year was performed 5 ($m = 5$) runs with 2000 iterations of Gibbs sampling. The detailed index values of five times experiments are listed in Table 11.2, where we can observe directly that the 3rd experiment (E3) of documents sub-collection in 2009, the 5th experiment of documents sub-collection in 2010 (E5), the 4th experiment of documents sub-collection in 2011 (E4), the 2nd experiment (E2) of documents sub-collection in 2012, and the 3rd experiment (E3) of documents sub-collection in 2013 have the largest value of all three indexes among all experimental trials. We believe that these models can fit the observation better and the topics and parameters provided by the five trials are our final topic modeling result.

Since there is no preset classification or domain knowledge assistance needed, the topic modeling results are discovered in an unsupervised way. In the past five years, patents owned by Australia assignees cover several important technological topics, such as print head and nozzle, alkyl compound, pressure apparatus, and antibody sequence. The more the topic words are taken into consideration to describe a topic, the more clear and specific the topical semantic meaning will be. Specifically, the topics for each year are presented as follows. The order of the topics is random, and the numbers behind words are the probability values of corresponding topic words. Details of all the topics, the top 10 ranked words and their corresponding probabilities, are shown in Table 11.3 in the Appendix.

**Table 11.3** The top 10 ranked words of all the topics from years 2009 to 2013 and their corresponding probabilities

| Word | Probability | Word | Probability | Word | Probability | Word | Probability | Word | Probability |
|---|---|---|---|---|---|---|---|---|---|
| **Year 2009** | | | | | | | | | |
| *Topic 1* | | *Topic 2* | | *Topic 3* | | *Topic 4* | | *Topic 5* | |
| Printhead | 0.0418 | Device | 0.0244 | Ink | 0.0442 | Step | 0.0116 | Portion | 0.0246 |
| Ink | 0.0353 | Image | 0.0217 | Ejection | 0.0336 | Composition | 0.0095 | Body | 0.0150 |
| Print | 0.0333 | Coded | 0.0209 | Nozzle | 0.0334 | Gas | 0.0088 | Assembly | 0.0132 |
| Printer | 0.0252 | System | 0.0195 | Inkjet | 0.0307 | Leach | 0.0081 | Surface | 0.0110 |
| Media | 0.0229 | Sensing | 0.0181 | Printhead | 0.0245 | Material | 0.0065 | Extending | 0.0092 |
| Cartridge | 0.0138 | Digital | 0.0132 | Drop | 0.0229 | Acid | 0.0064 | Wall | 0.0091 |
| Module | 0.0137 | Computer | 0.0105 | Apparatus | 0.0224 | Fuel | 0.0063 | Mask | 0.0081 |
| Printing | 0.0135 | Camera | 0.0101 | Actuator | 0.0220 | Water | 0.0059 | Adapted | 0.0076 |
| Assembly | 0.0132 | Identity | 0.0092 | Element | 0.0191 | Polymer | 0.0058 | Substantially | 0.0072 |
| Configured | 0.0124 | Position | 0.0086 | Chamber | 0.0189 | Ph | 0.0055 | Support | 0.0069 |
| *Topic 6* | | *Topic 7* | | *Topic 8* | | *Topic 9* | | *Topic 10* | |
| Support | 0.0152 | Compound | 0.0183 | System | 0.0116 | Signal | 0.0278 | Antibody | 0.0379 |
| Roller | 0.0142 | Formula | 0.0111 | Material | 0.0090 | Sensor | 0.0108 | Fragment | 0.0246 |
| Device | 0.0122 | Alkyl | 0.0109 | Game | 0.0088 | Signals | 0.0107 | Sequence | 0.0220 |
| Drive | 0.0109 | Independently | 0.0102 | Plurality | 0.0087 | Frequency | 0.0089 | Human | 0.0219 |
| Assembly | 0.0101 | Layer | 0.0098 | Computer | 0.0079 | Device | 0.0087 | Acid | 0.0177 |
| Mechanism | 0.0082 | Optionally | 0.0095 | Gaming | 0.0073 | Input | 0.0084 | Peptide | 0.0175 |
| Surface | 0.0080 | Base | 0.0088 | Entry | 0.0072 | Output | 0.0081 | Mature | 0.0164 |
| Frame | 0.0075 | Detector | 0.0087 | Torque | 0.0063 | Apparatus | 0.0081 | Cell | 0.0157 |
| Position | 0.0071 | Substituted | 0.0087 | Object | 0.0058 | Processing | 0.0071 | Binding | 0.0138 |
| Mounted | 0.0067 | Reflector | 0.0087 | Service | 0.0054 | Power | 0.0067 | Amino | 0.0133 |

(continued)

**Table 11.3** (continued)

| Word | Probability | Word | Probability | Word | Probability | Word | Probability | Word | Probability |
|---|---|---|---|---|---|---|---|---|---|
| **Year 2010** | | | | | | | | | |
| *Topic 1* | | *Topic 2* | | *Topic 3* | | *Topic 4* | | *Topic 5* | |
| Portion | 0.0217 | Signal | 0.0240 | Ink | 0.0518 | **Material** | **0.0144** | **Memory** | **0.0253** |
| Surface | 0.0126 | Light | 0.0131 | Printhead | 0.0476 | **Step** | **0.0136** | **Computer** | **0.0191** |
| Outer | 0.0095 | System | 0.0121 | Nozzle | 0.0214 | **Water** | **0.0101** | **Plurality** | **0.0161** |
| Assembly | 0.0090 | Optical | 0.0104 | Inkjet | 0.0183 | **Layer** | **0.0101** | **Network** | **0.0155** |
| Body | 0.0088 | Device | 0.0104 | Print | 0.0176 | **Metal** | **0.0088** | **Single** | **0.0143** |
| Extending | 0.0086 | Image | 0.0083 | Assembly | 0.0172 | **Polymer** | **0.0081** | **Application** | **0.0141** |
| Wall | 0.0080 | Power | 0.0076 | Printer | 0.0156 | **Form** | **0.0070** | **Program** | **0.0133** |
| Support | 0.0076 | Frequency | 0.0076 | Media | 0.0127 | **Defined** | **0.0067** | **System** | **0.0117** |
| Upper | 0.0073 | Output | 0.0069 | Ejection | 0.0126 | **Composition** | **0.0066** | **Local** | **0.0103** |
| Frame | 0.0071 | Sensor | 0.0067 | Configured | 0.0110 | **Concentration** | **0.0063** | **Computers** | **0.0097** |
| *Topic 6* | | *Topic 7* | | *Topic 8* | | *Topic 9* | | *Topic 10* | |
| Device | 0.0269 | Acid | 0.0199 | Apparatus | 0.037 | **Compound** | **0.0184** | **System** | **0.0175** |
| Coded | 0.0252 | Sequence | 0.0172 | Air | 0.0214 | **Substituted** | **0.0183** | **Device** | **0.0154** |
| System | 0.0245 | Plant | 0.0159 | Pressure | 0.0164 | **Independently** | **0.0140** | **Electrode** | **0.0146** |
| Print | 0.0190 | Nucleic | 0.0152 | Fluid | 0.0148 | **Alkyl** | **0.0096** | **Apparatus** | **0.0107** |
| Computer | 0.0168 | Seq | 0.0146 | Valve | 0.0144 | **Formula** | **0.0094** | **Signal** | **0.0105** |
| Sensing | 0.0161 | Cell | 0.0136 | Flow | 0.0140 | **Optionally** | **0.0092** | **Configured** | **0.0095** |
| User | 0.0149 | Antibody | 0.0117 | Chamber | 0.0131 | **Aryl** | **0.0065** | **Euphorbia** | **0.0095** |
| Media | 0.0115 | Fragment | 0.0088 | System | 0.0129 | **Moiety** | **0.0051** | **Array** | **0.0079** |
| Mobile | 0.0109 | Binding | 0.0086 | Inlet | 0.0083 | **Composition** | **0.0049** | **Patient** | **0.0074** |
| Indicative | 0.0101 | Polypeptide | 0.0086 | Outlet | 0.0071 | **Hydrogen** | **0.0046** | **Processing** | **0.0071** |

(continued)

**Table 11.3** (continued)

| Word | Probability | Word | Probability | Word | Probability | Word | Probability | Word | Probability |
|---|---|---|---|---|---|---|---|---|---|
| **Year 2011** | | | | | | | | | |
| *Topic 1* | | *Topic 2* | | *Topic 3* | | *Topic 4* | | *Topic 5* | |
| Material | 0.0188 | Portion | 0.0260 | Ink | 0.0579 | Sequence | 0.0234 | Optionally | 0.0228 |
| Layer | 0.0166 | Assembly | 0.0202 | Printhead | 0.0457 | Acid | 0.0201 | Substituted | 0.0224 |
| Step | 0.0130 | Mask | 0.0113 | Nozzle | 0.0282 | Seq | 0.0179 | Compound | 0.0159 |
| Composition | 0.0083 | Support | 0.0110 | Inkjet | 0.0170 | Amino | 0.0138 | Alkyl | 0.0142 |
| Range | 0.0070 | Frame | 0.0105 | Assembly | 0.0163 | Cell | 0.0130 | Lens | 0.0102 |
| Polymer | 0.0064 | Surface | 0.0095 | Chamber | 0.0118 | Plant | 0.0120 | Independently | 0.0089 |
| Coating | 0.0060 | Outer | 0.0087 | Integrated | 0.0116 | Gene | 0.0113 | Optical | 0.0079 |
| Metal | 0.0058 | Wall | 0.0084 | Printer | 0.0113 | Fragment | 0.0096 | Aryl | 0.0074 |
| Solution | 0.0057 | Extending | 0.0071 | Fluid | 0.0107 | Cells | 0.0085 | Zone | 0.0070 |
| Forming | 0.0056 | Body | 0.0069 | Plurality | 0.0103 | Isolated | 0.0084 | Lower | 0.0067 |
| *Topic 6* | | *Topic 7* | | *Topic 8* | | *Topic 9* | | *Topic 10* | |
| Apparatus | 0.0226 | Signal | 0.0203 | Print | 0.0449 | System | 0.0289 | **System** | **0.0108** |
| Flow | 0.0191 | Light | 0.0133 | Media | 0.0296 | Coded | 0.0211 | **Step** | **0.0099** |
| Air | 0.0180 | Power | 0.0120 | Printer | 0.0177 | Device | 0.0207 | **Apparatus** | **0.0096** |
| Gas | 0.0180 | Device | 0.0114 | Image | 0.0170 | Computer | 0.0186 | **Plurality** | **0.0084** |
| Water | 0.0178 | Wireless | 0.0103 | Controller | 0.0148 | Memory | 0.0140 | **Pressure** | **0.0078** |
| Pressure | 0.0161 | Apparatus | 0.0090 | Module | 0.0141 | Sensing | 0.0130 | **Determining** | **0.0076** |
| Valve | 0.0158 | Source | 0.0090 | Game | 0.0131 | Plurality | 0.0114 | **Processing** | **0.0066** |
| Device | 0.0129 | Plurality | 0.0078 | Gaming | 0.0129 | Identity | 0.0109 | **Monitoring** | **0.0058** |
| Fluid | 0.0124 | Electrical | 0.0078 | Configured | 0.0127 | Indicative | 0.0101 | **Time** | **0.0057** |
| Humidifier | 0.0110 | Optical | 0.0074 | Printing | 0.0120 | Position | 0.0086 | **Determined** | **0.0055** |

(continued)

**Table 11.3** (continued)

| Word | Probability | Word | Probability | Word | Probability | Word | Probability | Word | Probability |
|---|---|---|---|---|---|---|---|---|---|
| **Year 2012** | | | | | | | | | |
| *Topic 1* | | *Topic 2* | | *Topic 3* | | *Topic 4* | | *Topic 5* | |
| Signal | 0.0325 | Fluid | 0.0209 | Portion | 0.024 | **Gaming** | **0.0513** | **Light** | **0.0145** |
| Configured | 0.0165 | Gas | 0.0172 | Assembly | 0.0213 | **Game** | **0.0504** | **Plurality** | **0.0114** |
| Frequency | 0.0132 | Flow | 0.0151 | Support | 0.0126 | **System** | **0.0205** | **System** | **0.0107** |
| Optical | 0.0116 | Chamber | 0.0145 | Mask | 0.0106 | **Symbols** | **0.0190** | **Site** | **0.0075** |
| Sound | 0.0116 | System | 0.0132 | System | 0.0087 | **Symbol** | **0.0186** | **Pattern** | **0.0070** |
| System | 0.0103 | Valve | 0.0129 | Element | 0.0080 | **Plurality** | **0.0185** | **Registration** | **0.0070** |
| Power | 0.0092 | Water | 0.0121 | Nasal | 0.0073 | **Controller** | **0.0172** | **Respective** | **0.0068** |
| Control | 0.0090 | Inlet | 0.0099 | Adapted | 0.0072 | **Machine** | **0.0166** | **Lens** | **0.0067** |
| Electrical | 0.0088 | Pressure | 0.0097 | Frame | 0.0071 | **Player** | **0.0157** | **Symbol** | **0.0063** |
| Device | 0.0087 | Liquid | 0.0078 | Extending | 0.0066 | **Jackpot** | **0.0127** | **Image** | **0.0063** |
| *Topic 6* | | *Topic 7* | | *Topic 8* | | *Topic 9* | | *Topic 10* | |
| Time | 0.0112 | Material | 0.0196 | Portion | 0.0164 | System | 0.0202 | Substituted | 0.0204 |
| Determining | 0.0107 | Layer | 0.0119 | Apparatus | 0.0101 | Computer | 0.0202 | Optionally | 0.0190 |
| Signal | 0.0104 | Polymer | 0.0100 | Surface | 0.0101 | Memory | 0.0150 | Sequence | 0.0162 |
| Test | 0.0093 | Metal | 0.0093 | Device | 0.0098 | Device | 0.0139 | Compound | 0.0157 |
| Sensor | 0.0093 | Surface | 0.0092 | Body | 0.0088 | User | 0.0128 | Acid | 0.0151 |
| Flow | 0.0089 | Electrically | 0.0074 | Upper | 0.0088 | Plurality | 0.0081 | Seq | 0.0095 |
| Waveform | 0.0085 | Step | 0.0067 | Extending | 0.0087 | Coded | 0.0078 | Nucleic | 0.0084 |
| Pressure | 0.0085 | Conductive | 0.0064 | Lower | 0.0081 | Content | 0.0078 | Composition | 0.0079 |
| Predetermined | 0.0070 | Cell | 0.0057 | Container | 0.0081 | Printed | 0.0071 | Amino | 0.0072 |
| Plant | 0.0068 | Component | 0.0056 | Assembly | 0.0073 | Image | 0.0069 | Antibody | 0.0069 |

(continued)

**Table 11.3** (continued)

| Word | Probability | Word | Probability | Word | Probability | Word | Probability | Word | Probability |
|---|---|---|---|---|---|---|---|---|---|
| **Year 2013** | | | | | | | | | |
| *Topic 1* | | *Topic 2* | | *Topic 3* | | *Topic 4* | | *Topic 5* | |
| Portion | 0.0200 | Game | 0.0555 | Signal | 0.0206 | Cushion | 0.0345 | Composition | 0.0234 |
| Assembly | 0.0122 | Gaming | 0.0451 | Configured | 0.0181 | Mask | 0.0287 | Seq | 0.0184 |
| Body | 0.0107 | Symbol | 0.0322 | Apparatus | 0.0145 | Portion | 0.0285 | Acid | 0.0167 |
| Surface | 0.0091 | Plurality | 0.0274 | Device | 0.0139 | Assembly | 0.0191 | Sequence | 0.0158 |
| Extending | 0.0079 | Symbols | 0.0238 | Stimulation | 0.0105 | Frame | 0.0186 | Amino | 0.0102 |
| Wall | 0.0073 | Controller | 0.0226 | Signals | 0.0097 | Support | 0.0168 | Antibody | 0.0091 |
| Housing | 0.0072 | Player | 0.0189 | System | 0.0096 | Structure | 0.0154 | Cell | 0.0076 |
| Position | 0.0070 | System | 0.0177 | Power | 0.0096 | Full-face | 0.0124 | Nucleic | 0.0071 |
| Relative | 0.0063 | Arranged | 0.0152 | Flow | 0.0091 | Nasal | 0.0122 | Polypeptide | 0.0068 |
| Outer | 0.0062 | Machine | 0.0128 | Electrical | 0.0086 | Underlying | 0.0121 | Binding | 0.0066 |
| *Topic 6* | | *Topic 7* | | *Topic 8* | | *Topic 9* | | *Topic 10* | |
| Device | 0.0286 | Material | 0.0135 | Image | 0.0236 | **System** | **0.0272** | Substituted | 0.0583 |
| Wireless | 0.0132 | Layer | 0.0120 | Oligonucleotide | 0.0120 | **Computer** | **0.0260** | Optionally | 0.0513 |
| System | 0.0115 | Fluid | 0.0102 | Lens | 0.0098 | **User** | **0.0154** | Compound | 0.0160 |
| Plurality | 0.0112 | Gas | 0.0094 | Optical | 0.0095 | **Program** | **0.0112** | Alkyl | 0.0132 |
| Sensor | 0.0109 | Flow | 0.0084 | Antisense | 0.0086 | **Message** | **0.0103** | Independently | 0.0129 |
| Signal | 0.0092 | Water | 0.0083 | Light | 0.0085 | **Access** | **0.0088** | Formula | 0.0084 |
| Processing | 0.0088 | Liquid | 0.0081 | Plurality | 0.0077 | **Vehicle** | **0.0071** | Alkenyl | 0.0084 |
| Control | 0.0088 | Surface | 0.0075 | System | 0.007 | **Code** | **0.0061** | Salt | 0.0076 |
| Devices | 0.0087 | Step | 0.0067 | Laser | 0.0063 | **Storage** | **0.0060** | Alkynyl | 0.0066 |
| Component | 0.0082 | Electrode | 0.0066 | Step | 0.0062 | **Device** | **0.0059** | Acceptable | 0.0065 |

- The topics of year 2009 include printhead (0.0418) cartridge (0.0353), image (0.0217) device (0.0244), ink (0.0442) nozzle (0.0334), composition (0.0095) material (0.0065), portion (0.0246) assembly (0.0132), roller (0.0142) device (0.0122), alkyl (0.0109) compound (0.0183) formula (0.0111), computer (0.0079) gaming (0.0088), signal (0.0278) sensor (0.0108), and antibody (0.0379) sequence (0.0220).
- The topics of year 2010 contain portion (0.0217) assembly (0.0090), light (0.0131)/optical (0.0104) device (0.0104), ink (0.0518) printhead (0.0476), layer (0.0101) material (0.0144), computer (0.0191) memory (0.0253) plurality (0.0161), coded (0.0252) device (0.0269), antibody (0.0117) sequence (0.0172), pressure (0.0164) apparatus (0.0370), alkyl (0.0096) compound (0.0184), and electrode (0.0146) system (0.0175).
- The topics of year 2011 include layer (0.0166) material (0.0188), portion (0.0260) assembly (0.0202), ink (0.0579) printhead (0.0457), acid (0.0201) sequence (0.0234), alkyl (0.0142) compound (0.0159), pressure (0.0161) apparatus (0.0226), light (0.0133) device (0.0114), image (0.0170) print (0.0449), coded (0.0211) device (0.0207), and plurality (0.0084) apparatus (0.0096).
- The topics of year 2012 cover configured (0.0165) signal (0.0325), fluid (0.0209) chamber (0.0145), portion (0.0240) assembly (0.0213), gaming (0.0513) system (0.0205), light (0.0145) lens (0.0067), signal (0.0104) sensor (0.0093), layer (0.0119) material (0.0196), portion (0.0164) apparatus (0.0101), computer (0.0202) memory (0.0150), and acid (0.0151) sequence (0.0162).
- The topics of year 2013 comprise portion (0.0200) assembly (0.0122), gaming (0.0451) controller (0.0226), configured (0.0181) signal (0.0206), cushion (0.0345) mask (0.0287), acid (0.0167) sequence (0.0158), wireless (0.0132) signal (0.0092) sensor (0.0109), layer (0.0120) material (0.0135), optical (0.0095) lens (0.0098), message (0.0103) system (0.0272), and alkyl (0.0132) compound (0.0160).

### *11.4.3 Topic Change Identification*

After discovering main topics underlying in patent claims of each year's document collection, we then use the topic change model to identify the topic variation from years 2009 to 2013. For different groups of topics associated with two consecutive years, we conduct traversal comparison between the topics that belong to the later year with the topics related to the previous year. Topics that contain very similar words are considered as the same topic experiencing innovation; while topics that cannot match any existing ones count as new topics. Figure 11.6 illustrates the important topics that arose each year after 2009, by presenting the top 10 words for each topic using Pajek (Batagelj and Mrvar 2004).

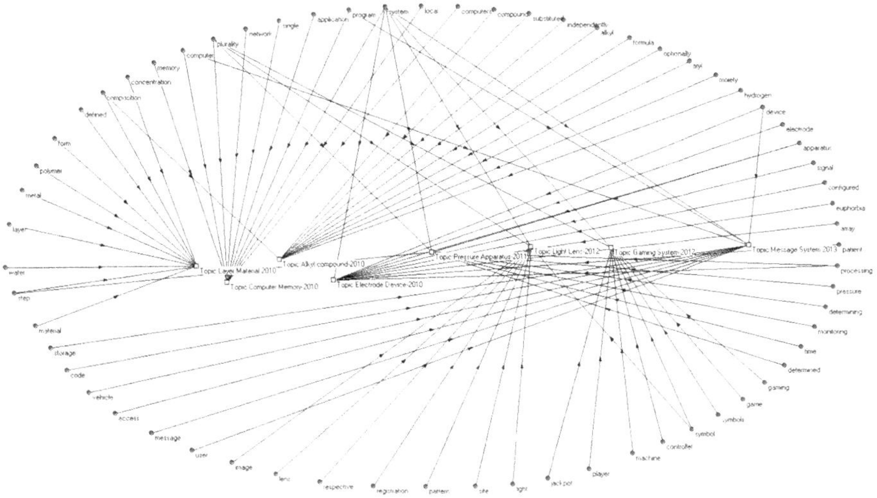


**Fig. 11.6** Topics became newly important in each year of 2010–2013 and topmost frequent words of each topic

In year 2010, there are four different topics appeared compared with year 2009, including layer material that related to metal and polymer composition, electrode device, computer memory, and alkyl compound. In year 2011, one newly important topic appeared, pressure apparatus. Then, year 2012 introduced two new topics including light lens and gaming system/controller compared with the previous year. Finally, for year 2013, computer system related to vehicle and message appeared as a new theme. All the topics above were identified without assistance of preset domain knowledge. The detailed words and their corresponding probabilities of the new topics mentioned above are highlighted in boldface in Table 11.3 of the Appendix.

### *11.4.4 Topic-Based Trend Estimation*

As mentioned, we can use the proposed approach to discover how the detailed content of a certain topic evolves from year to year and forecast the topic-based trend using historical status. In the case study, topic antibody fragment/sequence is chosen as an example. As shown in Fig. 11.7, we observe that the word distribution composing the topic develops over time. In year 2009, human and peptide were in the top words list, yet after this, the stress of the topic itself moved to plant, amino acid, nucleic acid, and polypeptide. The word "acid," instead of "antibody," ranked higher from year 2010 to 2013, which means they have larger probability of belonging to this topic as time goes on. The variation of the content of this topic may suggest that, in this area, the key point of technological research and development has shifted to amino/nucleic acid sequence.

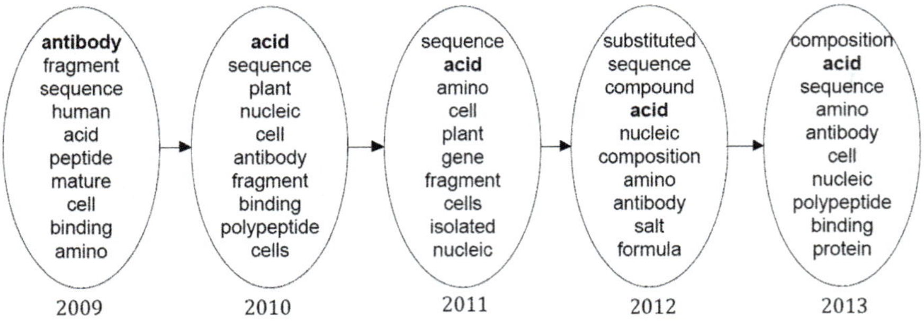


**Fig. 11.7** An example of the topic "antibody" evolving over time

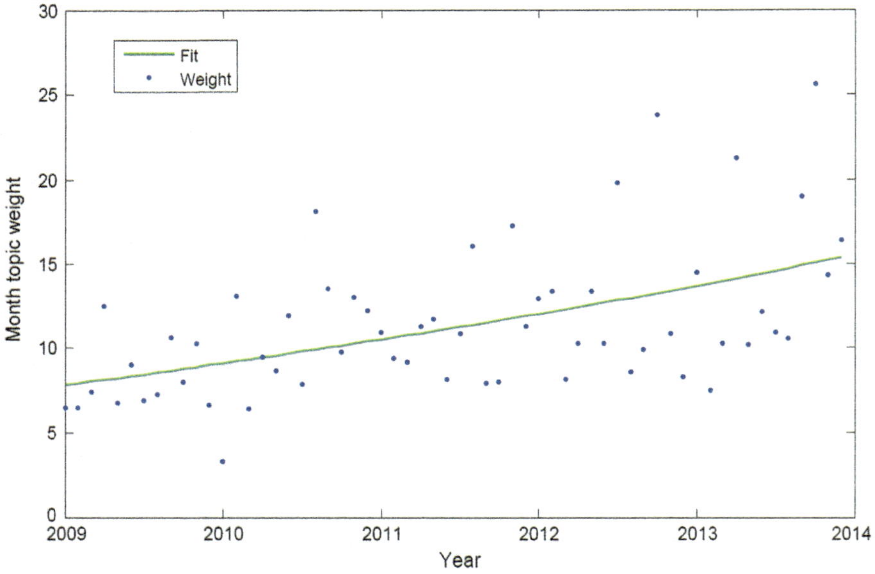


**Fig. 11.8** An example of the topic-based trend estimation of the theme "antibody"

To estimate topic-based trend of this topic, we then generate its temporal-weight matrix with one month as time interval. Each element in the matrix presents the weight of the topic in a corresponding time frame, from January 2009 to December 2013. We calculate the weight changes in a least-squares sense to estimate the general trend of the target topic. Figure 11.8 shows the final result of topic-based trend estimation of the theme "antibody." We can obverse directly that this topic appeared to have an upward trend. The significance of this topic kept growing continuously, from which we learn that the research and patenting for the topic of antibody is increasing over the past 5 years, and the importance of this topic has the potential to keep growing in future.

## 11.5 Conclusion and Future Work

This paper proposed an unsupervised topic change identification approach for patent mining using latent dirichlet allocation. Patent claims that embody the most significant technological terms are chosen as the main textual data source of our research. To improve the usage of LDA in patent topic extraction, we utilize USPC as a measurement of different estimations, to select the optimal model of topic modeling. Machine-identified topics are then placed into a topic change model to locate topic variation over time. Since there is no need to define any keywords in advance and all topics are automatically identified in an unsupervised way, this approach is able to set domain experts and analysts free from reading, understanding and summarizing massive technical documents and records. Finally, a case study, using USPTO patents published during the years 2009–2013 with Australia as their assignee country, is presented. The experimental results demonstrate that the proposed approach can be used as an automatic tool to extract topics and identify topic changes from a large volume of patent documents. From the application perspective, the discovered topic variations can be utilized to assist further decision making in R&D management, especially for newly created innovative enterprises, for example, to provide a full understanding of the topic structure of a certain industry, seek technological opportunities, and so on.

As patents and other technological indicators are generating and accumulating in an increasing rate, approaches for automatically identifying topic changes using data mining and machine learning methods will continue to be emphasized. In future work, we will keep focusing on locating topic changes that associate with more meaningful temporal segmentation, like trend-turning intervals (Chen et al. 2015), to identify and analyze the context that contributes to trend changing of patenting activities.

**Acknowledgments** The work presented in this paper is partly supported by the Australian Research Council (ARC) under Discovery Project DP140101366 and the National High Technology Research and Development Program of China (Grant No. 2014AA015105).

## Appendix

See Table 11.3.

## References

Abbas, A., Zhang, L., & Khan, S. U. (2014). A literature review on the state-of-the-art in patent analysis. *World Patent Inf. 37*, 3–13.

Batagelj, V., & Mrvar, A. (2004). *Pajek—Analysis and visualization of large networks*. Berlin: Springer.

Blei, D. M. (2012). Probabilistic topic models. *Communications of the ACM, 55*, 77–84.

Blei, D. M., Ng, A. Y., & Jordan, M. I. (2003). Latent dirichlet allocation. *The Journal of Machine Learning research, 3*, 993–1022.

Campbell, R. S. (1983). Patent trends as a technological forecasting tool. *World Patent Information, 5*, 137–143.

Camus, C., & Brancaleon, R. (2003). Intellectual assets management: From patents to knowledge. *World Patent Information, 25*, 155–159.

Chen, H., Zhang, G., Zhu, D., & Lu, J. (2015). A patent time series processing component for technology intelligence by trend identification functionality. *Neural Computing and Applications, 26*, 345–353.

Daim, T. U., Kocaoglu, D. F., & Anderson, T. R. (2011). Using technological intelligence for strategic decision making in high technology environments. *Technological Forecasting and Social Change, 78*, 197–198.

David, D., Lewis, Y. Y., Rose, T. G., Li, F. (2004). *SMART stopword list* [Online]. Cambridge: MIT Press. Available: http://jmlr.csail.mit.edu/papers/volume5/lewis04a/a11-smart-stop-list/english.stop

Ernst, H. (1997). The use of patent data for technological forecasting: The diffusion of CNC-technology in the machine tool industry. *Small Business Economics, 9*, 361–381.

Griffiths, T. L., & Steyvers, M. (2004). Finding scientific topics. *Proceedings of the National Academy of Sciences of the United States of America, 101*, 5228–5235.

Halkidi, M., Batistakis, Y., & Vazirgiannis, M. (2001). On clustering validation techniques. *Journal of Intelligent Information Systems, 17*, 107–145.

Haywood, S. (2003). *Academic vocabulary* [Online]. Nottingham: Nottingham University. Available: http://www.nottingham.ac.uk/alzsh3/acvocab/wordlists.htm, 2014

Heinrich, G. (2005). *Parameter estimation for text analysis*, version 2.9 ed. Darmstadt, Germany: Fraunhofer IGD.

Kim, D., & Oh, A. (2011). Topic chains for understanding a news corpus. In A. Gelbukh (Ed.), *Computational linguistics and intelligent text processing*. Berlin, Heidelberg: Springer.

Koltcov, S., Koltsova, O., & Nikolenko, S. (2014). Latent dirichlet allocation: stability and applications to studies of user-generated content. In *Proceedings of the 2014 ACM conference on Web science*. Bloomington, Indiana, USA: ACM.

Lai, K.-K., & Wu, S. J. (2005). Using the patent co-citation approach to establish a new patent classification system. *Information Processing and Management, 41*, 313–330.

Lukins, S. K., Kraft, N. A., & Etzkorn, L. H. (2010). Bug localization using latent Dirichlet allocation. *Information and Software Technology, 52*, 972–990.

Nishijima, Y., Anzai, T., & Sengoku, S. (2013). Application of bibliometric analysis to market analysis. In Proceedings of the 2013 Portland International Conference on Management of Engineering & Technology (pp. 2365–2377).

Noel, G. E., & Peterson, G. L. (2014). Applicability of Latent Dirichlet Allocation to multi-disk search. *Digital Investigation*.

Novelli, E. (2014). An examination of the antecedents and implications of patent scope. *Research Policy*.

Office U.S.P.A.T. (2015). *United States Patent and Trademark Office* [Online]. Available: http://www.uspto.gov/

Porter, L. A. (2005). QTIP: Quick technology intelligence processes. *Technological Forecasting and Social Change, 72*, 1070–1081.

Sheikh, N., Gomez, F. A., Yonghee, C., & Siddappa, J. (2011). Forecasting of advanced electronic packaging technologies using bibliometric analysis and Fisher-Pry diffusion model. In Proceedings of the 2011 Portland International Conference on Management of Engineering & Technology (pp. 1–20).

Sheldon, J. G. (1995). *How to write a patent application*. Practising Law Institute.

Steinbach, M., Karypis, G., & Kumar, V. (2000). A comparison of document clustering techniques. KDD workshop on text mining (pp. 525–526), Boston.

Steyvers, M., & Griffiths, T. (2007). Probabilistic topic models. In T. Landauer, D. S. McNamara, S. Dennis, & W. Kintsch (Ed.), *Latent semantic analysis: A road to meaning*. Laurence Erlbaum.

Tong, X., & Frame, J. D. (1994). Measuring national technological performance with patent claims data. *Research Policy, 23*, 133–141.

Trippe, A. J. (2003). Patinformatics: Tasks to tools. *World Patent Information, 25*, 211–221.

Tseng, Y.-H., Lin, C.-J., & Lin, Y.-I. (2007). Text mining techniques for patent analysis. *Information Processing and Management, 43*, 1216–1247.

USPTO. (2012). *Manual of patent examining procedure: Claim interpretation* [Online]. USPTO. Available: http://www.uspto.gov/web/offices/pac/mpep/s2111.html

Watts, R. J., & Porter, A. L. (1997). Innovation forecasting. *Technological Forecasting and Social Change, 56*, 25–47.

Wikipedia. (2014). *Transitional phrase* [Online]. Wikipedia. Available: http://en.wikipedia.org/wiki/Transitional_phrase, 2014.

WIPO. (2002). *Patent cooperation treaty (PCT) Article 6* [Online]. Washington: WIPO. Available: http://www.wipo.int/pct/en/texts/articles/a6.htm

WIPO. (2004). *WIPO intellectual property handbook: Policy, law and use*.

Xie, Z., & Miyazaki, K. (2013). Evaluating the effectiveness of keyword search strategy for patent identification. *World Patent Information, 35*, 20–30.

Yang, L., Qiu, M., Gottipati, S., Zhu, F., Jiang, J., Sun, H., & Chen, Z. (2013). Cqarank: Jointly model topics and expertise in community question answering. In *Proceedings of the 22nd ACM international conference on Conference on information & knowledge management* (pp. 99–108). ACM.

Yang, S., & Soo, V. (2012). Extract conceptual graphs from plain texts in patent claims. *Engineering Applications of Artificial Intelligence, 25*, 874–887.

Yoon, B. (2008). On the development of a technology intelligence tool for identifying technology opportunity. *Expert Systems with Applications, 35*, 124–135.

Yoon, B., & Park, Y. (2005). A systematic approach for identifying technology opportunities: Keyword-based morphology analysis. *Technological Forecasting and Social Change, 72*, 145–160.

Yoon, J., & Kim, K. (2012). TrendPerceptor: A property–function based technology intelligence system for identifying technology trends from patents. *Expert Systems with Applications, 39*, 2927–2938.

Zhang, Y., Porter, A. L., Hu, Z., Guo, Y., & Newman, N. C. (2014). "Term clumping" for technical intelligence: A case study on dye-sensitized solar cells. *Technological Forecasting and Social Change, 85*, 26–39.

Zhu, D., & Porter, A. L. (2002). Automated extraction and visualization of information for technological intelligence and forecasting. *Technological Forecasting and Social Change, 69*, 495–506.

# Chapter 12
# Semi-automatic Technology Roadmapping Composing Method for Multiple Science, Technology, and Innovation Data Incorporation

**Yi Zhang, Hongshu Chen and Donghua Zhu**

**Abstract** Since its first engagement with industry decades ago, technology roadmapping (TRM) is taking a more and more important role for competitive technical intelligence (CTI) in current R&D planning and innovation tracking. Important topics for both science policy and engineering management researchers involves with approaches that refer to real-world problems, explore value-added information from complex data sets, fuse analytic results and expert knowledge effectively and reasonable, and demonstrate to decision makers visually and understandably. The growing variety of Science, Technology, and Innovation (ST&I) data sources in the Big Data Age increases these challenges and opportunities. Addressing these concerns, this paper attempts to propose a semi-automatic TRM composing method to incorporate multiple ST&I data sources—we design an extendable interface for engaging diverse ST&I data sources and apply the fuzzy set to transfer vague expert knowledge to defined numeric values for automatic TRM generation. We focus on a case study on computer science-related R&D. Empirical data from the United States (US) National Science Foundation (NSF) Award data (innovative research ideas and proposals) and Derwent Innovation Index (DII) patent data source (technical and commercial information) affords vantage points at two stages of R&D process and also provide further capabilities for more ST&I data source incorporation. The understanding gained will also assist in description of computer science macro-trends for R&D decision makers.

Y. Zhang (✉) · H. Chen
Decision Systems & e-Service Intelligence Research Lab, Faculty of Engineering and Information Technology, Centre for Quantum Computation & Intelligent Systems, University of Technology Sydney, Ultimo, Australia
e-mail: yizhang.bit@gmail.com

Y. Zhang · H. Chen · D. Zhu
School of Management and Economics, Beijing Institute of Technology, Beijing, People's Republic of China



T.U. Daim et al. (eds.), *Anticipating Future Innovation Pathways Through Large Data Analysis*, Innovation, Technology, and Knowledge Management, DOI 10.1007/978-3-319-39056-7_12

**Keywords** Technology roadmapping · Bibliometrics · Fuzzy set · Competitive technical intelligence

## 12.1 Introduction

The growing variety of data sources in the Big Data Age not only increases the challenges and opportunities for competitive technical intelligence (CTI), but also leads the revolution of business management and decision making. Of significance is "data-driven," which uses Information Technology (IT) to support rigorous, constant experimentation that guides decisions and innovations (Bughin et al. 2010), and has being characterized as one of the most competitive features in various R&D planning and business models (McAfee et al. 2012).

Technology roadmapping (TRM) approaches, described as a representative, prominent, and flexible instrument for long-range technological forecasting and strategic planning, make good sense to actively incorporate business data into planning procedures (Phaal et al. 2004). Quantitative data, especially Science, Technology, and Innovation (ST&I) data, have already been largely involved in TRM models; however, since diverse emphases and possible time gap between different ST&I data sources, traditional TRM model usually only focuses on single ST&I data source, e.g., publications or patents. Undoubtedly, the booming new data sources, e.g., Twitter, news, customer comments, R&D project proposals, and product reports, also match the concept of ST&I researches perfectly, and the rapid engagement of multiple ST&I data with different formats and emphases introduces new challenges for current studies. Meanwhile, the way that current TRM study used to transfer vague human thoughts to defined numerical values (Lee et al. 2011) is still unfavorable, and the combination of qualitative and quantitative methodologies is also not as smart as what we imagine. At this stage, the emerging concerns are the approaches that refer to real-world problems, explore value-added information from complex data sets, fuse analytic results and expert knowledge effectively and reasonable, and demonstrate to decision makers visually and understandably.

Aiming to provide a solution for the questions—(1) how to incorporate multi-dimensional information from diverse ST&I data sources and (2) how to construct an intelligent TRM method, this paper develops a multiple ST&I data incorporation model and a fuzzy set-based semi-automatic TRM generation model. Based on a ST&I topic identification approach (Zhang et al. 2014b) and a traditional TRM composing method (Zhang et al. 2013), this paper designs a multilayer TRM method to arrange topics and related concepts (e.g., idea, technique, and product) to explore potential relationships. It is a challenge to build up an entire intelligent TRM composing method with novel IT techniques, e.g., training machine to discover potential linguistic relationships between technological components; however, this paper seeks approaches to introduce fuzzy set (Zadeh 1965) to transfer vague expert knowledge into defined numeric values and help automatically locate technological components for TRM composition. The empirical study selects the

United States (US) National Science Foundation (NSF) Award data (innovative research ideas and proposals) and the Derwent Innovation Index (DII) patent data source (technical products), which demonstrates the efficiency and feasibility of our methods and also provides vantage points at the top-bottom stages of R&D process and assists in description of computer science macro-trends for decision makers.

The main contributions of this paper are as follows: (1) We provide an approach to incorporate multiple ST&I data into TRM; (2) we construct a systematic approach to apply fuzzy set to the traditional TRM method and help compose TRM in a semi-automatic function; and (3) the process in which we think and solve problems emphasizes the combination of qualitative and quantitative methodologies and is also adaptive and transferrable to related ST&I researches.

This paper is organized as follows: The Related Works section reviews previous studies on TRM. In the Methodology section, we present the detailed research method for the semi-automatic TRM composing method, involving with a multiple ST&I data incorporation model and a fuzzy set-based semi-automatic TRM composing model. The Empirical Study section follows, using the US NSF Awards and DII patent data as the case. Finally, we conclude our current research and outline future work.

## 12.2 Related Works

This section reviews related literatures that include TRM and related ST&I data incorporation studies and also summarizes the limitations of these previous works.

Based on the significant work of Phaal et al. (2004), who summarized previous TRM methods and constructed an effective qualitative composing model, TRM research has already been extended from qualitative study only to a combination of qualitative and quantitative methodologies. Representatively, Huang et al. (2014) introduced a bibliometric technique-based four-dimensional TRM for the science and technology planning of China's solar cell industry; Zhang et al. (2014c) combined TRM model with Triple Helix innovation and Semantic TRIZ concepts and presented an empirical study on China's dye-sensitized solar cell industry; Lee et al. (2015) proposed a scenario-based TRM that involved with a plan assessment map and an activity assessment map for organizational plans, which also engaged the Bayesian network for topology and a causal relations definition. Geum et al. (2015) added the association rule mining to TRM to identify relationships between items on different layers of TRM.

The current TRM model has been combined with various concepts and methods, applying into real ST&I assessment, forecasting, and planning. According to previous studies, it is promising to conclude the benefits of current TRM studies as follows: (1) TRM is a visual model to present content, which enables easy understanding for both macro-, meso-, and micro-level problems (Zhang et al. 2013); (2) the hierarchical structure of TRM helps indicate potential relationships between items on different layers (Geum et al. 2015); (3) TRM is able to take dimensional impact factors into consideration, e.g., time, science policy, market pull, and technique push (Lee and Park 2005; Robinson and Propp 2008; Huang et al. 2014); and (4) previous TRM studies have perfect adaptability for general publication and patent data sources and match requests from different industrial domains.

Multiple ST&I data incorporation for TRM always holds great interests for ST&I analyzers. Data incorporation, known as data integration or data fusion, is one important branch of data mining and relies on strong computer science and IT background. Current research is to combine data residing at different sources and provide users with an unified view of these data (Lenzerini 2002). However, these data incorporation studies focus on data structure problem and ignore underlying insights and interactions of different data sources, which in contrast are the key factors for ST&I analyses. On the other hand, it has been a long time since the idea of multiple ST&I data incorporation generated, but until now related researches are rare. Robinson and Propp (2008) designed a multipath roadmapping framework where four layers—research lines, experimental integration, integrated platform/product, and application area—were used to emphasize the different phases of technology development. Zhang et al. (2013) engaged core terms derived from Web of Science (WoS) publications and International Classification Code (IPC) retrieved from the United States Patent Trademark Office (USPTO) into one TRM model to highlight the technological researches and applications.

Another tough task of current TRM study is the approach transferring vague human thoughts to defined numerical values (Lee et al. 2011), and this issue will heavily influence the efficiency and accuracy of TRM auto-generation process. In this context, fuzzy set would be a helpful instrument to minimize expert aid and consumed time, but maximize the usage of expert knowledge. Lu et al. (2011) constructed a novel group decision-making method with fuzzy hierarchical criteria for theme-based comprehensive evaluation in new product development, which calculated ranking results by fusing all assessment data from human beings and machines. Lee et al. (2011) introduced a fuzzy analytic hierarchy process to TRM model, which was definitely an innovative attempt for the combination of fuzzy concepts and TRM, although the empirical study only applied to a small-range data set (five sub-technologies of hydrogen energy) with expert ranking.

## 12.3 Methodology

On the purpose of providing solutions for constructing a multilayer TRM to reveal multidimensional information from emphasis-differed ST&I data sources and to explore insights for technical intelligence understandings, this paper constructs a TRM method for multiple ST&I data incorporation with the following steps:

1. Inputs: Grouped Topics—we focus on raw ST&I data and retrieve meaningful phrases and terms via a Term Clumping process (Zhang et al. 2014a) and then apply a K-means-based clustering approach (Zhang et al. 2014b) to identify hot research topics as key technological components;

2. Step 1: Multiple ST&I data incorporation model—we construct a designing process for multiple ST&I data incorporation that introduces Technology Readiness Level to analyze the emphases of ST&I data and proposes a questionnaire for expert consultation.
3. Step 2: Fuzzy set-based semi-automatic TRM generation model—in order to combine the qualitative and quantitative methodologies, we engage experts to evaluate topics by removing meaningless topics, consolidating duplicate topics, and highlighting significant ones, and we also evaluate each topic and group them into specified fuzzy sets, after which we generate the TRM in an automatic manner;
4. Outputs: TRM.

Note that the both steps would think about the combination of qualitative and quantitative methodologies. The step 1 is to consider the diverse emphases of applied ST&I data and data analyst would take an active role for the design process, while domain experts would dominate the step 2 for evaluating selected topic candidates. The framework of the TRM method for multiple ST&I data incorporation is given in Fig. 12.1.

### 12.3.1 Inputs: Grouped Topics

How to retrieve meaningful phrases and terms from raw ST&I data and identify valuable topics via clustering analysis are definitely interesting research questions, but they are not the foci of this paper. Briefly, we apply a revised version of the Term Clumping process (Zhang et al. 2014a) to retrieve phrases and terms from ST&I textual data by term removal, consolidation, and clustering, and then, a K-means-based clustering approach (Zhang et al. 2014b) is used to group related linguistic elements, e.g., phrases and terms, or records, into meaningful topics. We define these grouped topics as technological components, which reflect scientific or technical information of ST&I data.

### 12.3.2 Step 1: Multiple ST&I Data Integration

It is a general understanding that the emphases of different ST&I data are diverse. As shown in Fig. 12.2, we summarize the emphases of the selected mainstream ST&I data sources as below:

1. Academic proposal, e.g., NSF proposals, is usually granted by national government to support academic institutions for basic research, whose content focuses on new ideas, concepts, and unrealized innovative actions. Discoveries derived from academic proposal would be an express way to dive into innovation and promising for both academies and industries.

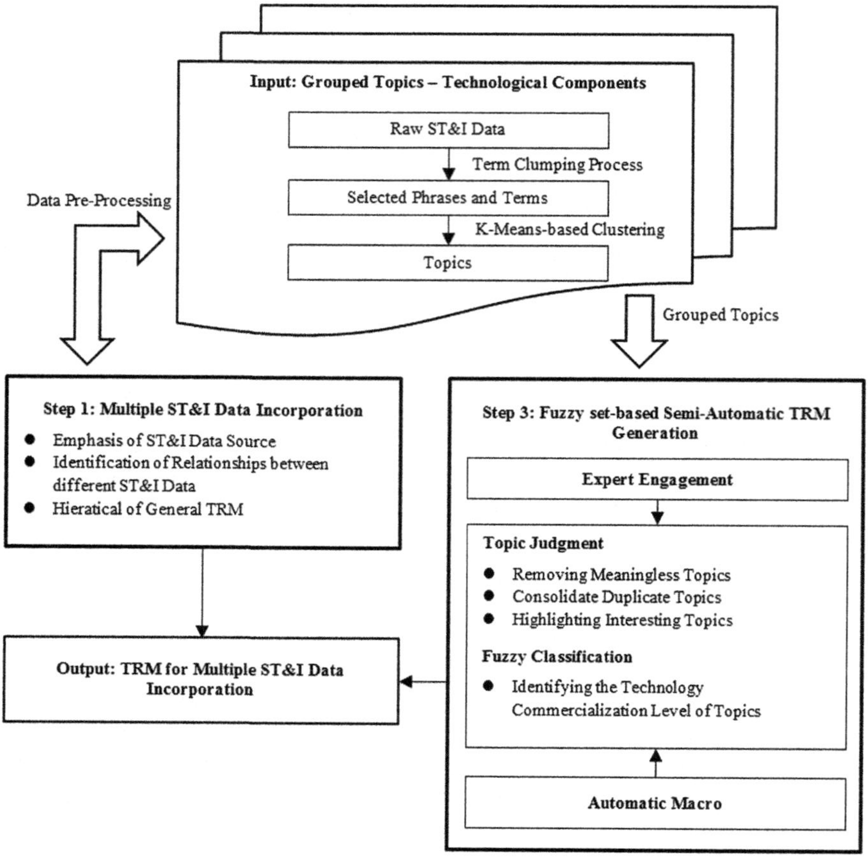


**Fig. 12.1** Framework of technology roadmapping method for multiple ST&I data incorporation

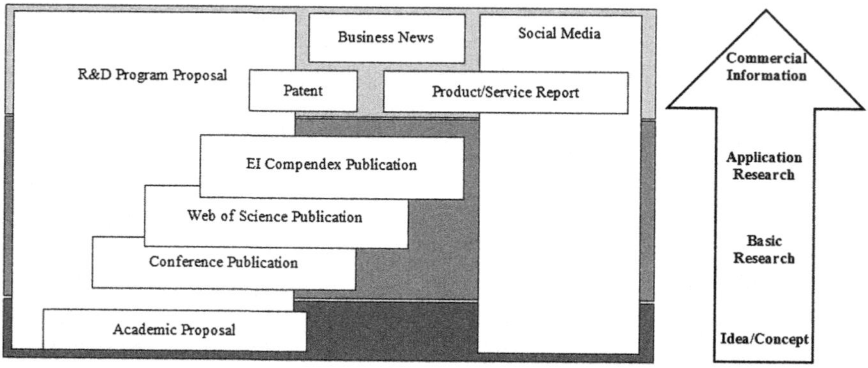


**Fig. 12.2** Emphasis of selected mainstream ST&I data sources

2. Publication contributes to both basic research and application research, but, in detail, conference paper, e.g., IEEE paper, mostly presents draft research frameworks, experimental results, or mature ideas, while publications in Web of Science including SCI and SSCI data, and the EI Compendex indexed papers emphasize fundamental research and application research, respectively.
3. Patent, e.g., DII patent, contributes to actual applications or products, the same as detailed technical report or guidebook of business services. In particular, patent terms should be vague and address legal effects (with patent barrier), and, comparably, academic terms are more clear and direct.
4. Business news, e.g., Factiva data, highlights social significances, where common technical terms and modifying adjectives will heavily influence the description.
5. Social media data, e.g., Twitter, are similar to the content of news, but it presents information in a more informal way. Sentimental analysis, sometimes, is applied to understand real meanings of these free texts.
6. National R&D program proposal, comparing with the five ST&I data above, is a complex one, which would include all aspects of ST&I emphases. Therefore, it would be not a smart option to incorporate national R&D programs with other ST&I data, but one feasible solution is to apply national R&D program proposal as a comparable case for other ST&I data.

The time gap problem is always the barrier that prevents incorporating multiple ST&I data effectively. It takes time to transfer an innovative idea into feasible plans and valid experiments, and related patents and mature products would be years or even decades after the idea firstly comes out. The time would be shorter and more unpredictable for emerging technologies. In this context, expert knowledge would be easier to deal with this situation than quantitative computation. Referring to the technology readiness levels (TRL) (Mankins 1995), we design a TRL scale for ST&I data sources, as shown in Table 12.1.

Although the definition of each TRL is still a fuzzy concept and its actual applications would depend on specified technological domains of ST&I data, Table 12.1 proposes a manner to transfer the emphases of ST&I data into an operable format for expert consultation. According to Table 12.1, we would recommend selecting ST&I data on neighbor TRLs, since the possible time gap would be tiny and could be ignored. However, if the time gap is able to be fully handled and considered, a comparison between the ST&I data sources at TRL 1 and TRL 5 would also make good sense to see an entire technological evolutionary pathway. Furthermore, based on the revised TRL scale, we design a questionnaire for multiple ST&I data incorporation in Table 12.2.

We attempt to provide a logical work flow to organize expert in a workshop process. The questions in Table 12.2 aim to lead the discussions and gain valuable information effectively. Note that what we list are only options for possible case studies, and we need to revise and refine this questionnaire to match actual requirements.

Table 12.1 Technology readiness level scale for ST&I data source

| TRL | Definition | Emphases of ST&I data |
|---|---|---|
| 1 | Technology concept and/or application formulated | Innovative ideas/concepts |
| 2 | Analytic and experimental critical function and/or characteristic proof-of-concept | Basic research |
| 3 | Component and/or breadboard and/or system/subsystem model and/or prototype demonstration in a laboratory or relevant environment | Application research |
| 4 | Actual system completed and "capability qualified" through test and demonstration | Application research/products |
| 5 | Actual system "capability proven" through successful mission operation | Products |

Table 12.2 The questionnaire for multiple ST&I data incorporation

| No. | Focus | Questions |
|---|---|---|
| 1 | ST&I data | Which ST&I data sources shall we use? |
| | | Is there any possible conflict between applied ST&I data? |
| 2 | Data structure | How to unify the structure of applied ST&I data? |
| 3 | Emphases | What are our emphases of this study? |
| | | Do we need to forecast? |
| | | How to rank these emphases and what is the priority one? |
| 4 | Time interval | Shall we look backward, forward, or both? |
| | | How long the time interval will be? |
| 5 | TRL | How the TRL distance between the emphases of applied ST&I data? |
| | | How to evaluate the possible time gap among applied ST&I data? |
| | | Is the time gap at an acceptable level? |
| 6 | Layer | Shall we divide applied ST&I data into separated layers or construct a multilayer TRM? |
| | | What is the criterion for classifying layers? |

Based on the hierarchical landscape and the expression form of components in our previous TRM models (Zhang et al. 2013; Zhou et al. 2014), we enrich the structure by the following: (1) distinguishing the scope of ST&I data sources with the shape of components, (2) softening the definition of *Y* axis to reserve an interface with the configuration of fuzzy set, and (3) defining linkages between components with multifactors, e.g., semantic similarity, time, science, policy. A sample of TRM for multiple ST&I data incorporation is given as in Fig. 12.3.

In Fig. 12.3, time is marked as the *X* axis, while *Y* axis is used for multilayer demonstration. Shaped components indicate topics derived from different ST&I data, and they are located among the multiple layers of TRM, and grouped components

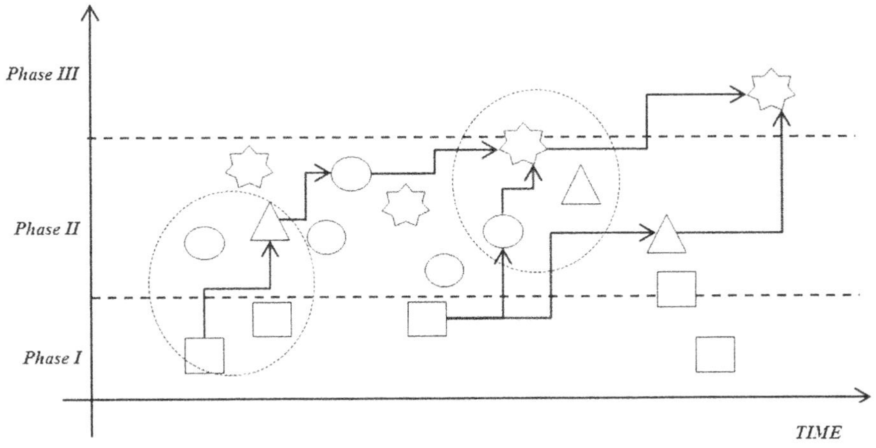


**Fig. 12.3** The sample of TRM for multiple ST&I data incorporation

would share similar topics with "some" possible linkages. It is also available to locate components on separated layers to highlight their origin ST&I data and relationships between different layers. In usual, exerts will help to identify the macro-level linkage between components to track possible evolutionary pathways.

### *12.3.3 Step 2: Fuzzy Set-Based Semi-automatic TRM Generation*

The balance between qualitative and quantitative methodologies in TRM composing model is definitely an intriguing and complicated research topic at the current time. Expert knowledge has been largely engaged with results refinement, component allocation, and the understanding of TRM for decision making (Zhang et al. 2015). At this stage, we aim to minimize the aid of experts and maximize the usage of expert knowledge in the limited time and scope, so that the fuzzy set can be considered as an effective tool to deal with this issue. As mentioned above, the emphases of ST&I data, the definition of TRL, the evaluation of the time gap, and the criterion for classifying the layers of TRM also depend on subjective expert knowledge. These vague elements definitely match the main concept of fuzzy set and afford possibilities for the engagement of fuzzy set.

As a general definition, we denote "all components" of TRM as the universe $X = \{x_1, x_2, \ldots, x_i, \ldots, x_{n-1}, x_n\}$ and "each phase/layer" as a fuzzy set $A_j$ defined on the $X$ where $j \in [1, m]$. The membership function $A_j(x_i)$ is considered as the degree that the component $x_i$ belongs to the phase/layer $A_j$ and will be decided depending on research purposes and empirical data. The detailed steps are outlined below:

- Considering specific case, to identify $X$, $A_j$ and $A_j(x_i)$;
- For each component $x_i$, experts would help classify into one of the fuzzy sets $A_j$ and mark a membership grade $A_j(x_i)$ for the selected fuzzy set;
- Based on $A_j(x_i)$, to calculate the $X(x_i)$ for each component and set as the $Y$ value;
- To generate TRM automatically via macros.

### 12.3.4 Outputs: TRM for Multiple ST&I Data Incorporation

After our 2-step method, we integrate multiple ST&I data sources, fuse the analytic results with expert knowledge, and generate the graphic TRM as our final output.

## 12.4 Empirical Study

Computer science would not be still considered as an emerging technology as what we did decades ago, but it has been integrated with IT techniques and various engineering applications and has become a fundamental instrument for multidisciplinary researches. This paper focused on the technology commercialization studies for computer science, and our purpose was to incorporate multiple ST&I data to track the evolutionary pathway of computer science technologies for commercialization. Since the limited condition for time and data sources, this paper only chose the NSF Award data (granted proposals) and the DII patent data for empirical study. Our consideration was the two data sources concentrated on the innovative ideas and mature technical products, respectively, and the contrast on the technology commercialization level would be better to indicate the importance of information fusion for multiple ST&I and also to demonstrate benefits of our TRM method. In addition, we built up our expert base with the help of twelve experts (Associate Professors, Lecturers, Researchers, and Ph.D. Candidates) from the Centre for Quantum Computation & Intelligent System, University of Technology Sydney, Australia, and the Knowledge Management and Data Analysis Laboratory, Beijing Institute of Technology, China.

### 12.4.1 Step 1: Incorporation of NSF Awards and DII Patents

The importance of NSF Awards and DII patens has been discussed separately in our previous studies (Zhang et al. 2014b, c), as shown in Table 12.2, and academic proposal is at the bottom of TRL, while patent belongs to the top of TRL; thus, the

**Table 12.3** Steps of term clumping processing

| | Step | NSF awards | | DII—2013[a] | |
|---|---|---|---|---|---|
| | | #T[b] | #A | #T | #A |
| 1 | Number of raw records | 12,915 | | 44,141 | |
| 2 | Natural language processing via VantagePoint (2015) | 254,992 | 17,859 | 706,739 | 154,791 |
| 3 | Basic cleaning with thesaurus | 214,172 | 16,208 | 679,736 | 131,577 |
| 4 | Fuzzy matching | 184,767 | 15,309 | – | – |
| 5 | Pruning[c] | 42,819 | 2470 | 19,926 | 19,930 |
| 6 | Extra fuzzy matching | 40,179 | 2395 | 15,510 | 16,603 |
| 7 | Computer science based common term cleaning | 30,015 | 2311 | 14,029 | 16,529 |

[a]Considering the larger amount of DII data, we divided it by year for the term clumping process, and therefore, we only present DII data in 2013 as a sample; also, we applied pruning to DII first and, then, fuzzy matching
[b]#$T$ = number of title terms and #$A$ = number of abstract terms
[c]In the pruning process, we removed terms appearing in only one record in the NSF awards and DII titles, but removed terms appearing in less than five records in DII abstracts

large distance between TRL scale introduces not only challenges but also promising insights, and the comparison would help identify the technology commercialization trend in a specified time interval. In this case, we aimed to track the rapid technological changes occurring with the coming of the Big Data Age—how innovative ideas boomed and how applicable techniques evolved; at this stage, we set the time interval from 2009 to 2013 to highlight Big Data-related techniques and their changes. Moreover, we consulted domain experts and decided to apply an integrated multilayer TRM to emphasize the contrast between NSF Awards and DII patents, and the layers included "basic research—TRL 1 and 2," "application research—TRL 3 and 4," and "products—TRL 5."

We grouped the topics of NSF Awards and DII patents separately. We selected 12,915 granted proposals under the Division of Computer and Communication Foundation in the NSF Awards (Zhang et al. 2014b) and 177,974 DII patents with the field Topic and Subject Category as "computer science," and the field Basic Patent Country and Priority Country as "US." The revised term Clumping steps were applied for feature extraction, and the process is given in Table 12.3.

We, then, applied the K-means-based clustering model (Zhang et al. 2014b) to group topics and acquire 54 topics from the NSF Awards and 44 ones from the DII patents.

### 12.4.2 Step 2: Fuzzy Set-Based Semi-automatic TRM Generation

According to the three layers for the technology commercialization study, we let the power set $A = \{A_1, A_2, A_3\}$ and introduced Gaussian distribution to define the

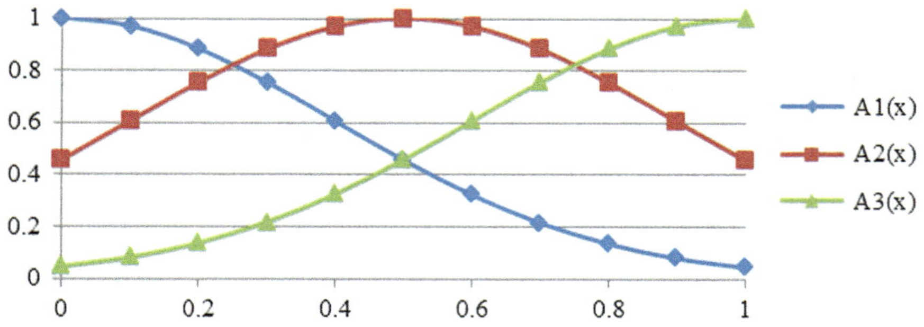


**Fig. 12.4** Distribution curves of member functions

membership functions. The three membership functions are provided below, and the distribution curves are shown in Fig. 12.4.

$$A_1(x): X \sim N\left(0, {}^1\!/_{2\pi}\right), x \in [0, 1]$$
$$A_2(x): X \sim N\left({}^1\!/_2, {}^1\!/_{2\pi}\right), x \in [0, 1]$$
$$A_3(x): X \sim N\left(1, {}^1\!/_{2\pi}\right), x \in [0, 1]$$

As shown in Fig. 12.4, we divided the universe into three intervals, where [0, 0.25] was mapped to "basic research," while [0.25, 0.75] and [0.75, 1] were for "application research" and "products," respectively. The experts firstly classified specified topic $x_i$ into one fuzzy set $A_j$ and marked the membership grade $A_j(x_i)$ to the fuzzy set. Then, we calculated the $X(x_i)$ and assigned the topic $x_i$ into one of the three fuzzy sets. As a sample, we list parts of the marked topics in Table 12.4, and the generated TRM is given in Fig. 12.5.

Although the experts only assigned each topic into one fuzzy set, we were able to calculate its membership grade for each fuzzy set. As an example, the topic "Video Frames" was assigned to $A_2(x)$ with the membership grade 0.81, and we got the $X(x)$ as 0.76, so according to the membership function, it was definitely possible to calculate the membership grade vector of "Video Frames" as (0.18, 0.81, 0.83). At this stage, we re-assigned the 98 topics with the following rules: (1) to classify the topic to the fuzzy set as the First Preference with the largest membership grade, if same membership grades occurred, we preferred to choose the fuzzy set for a lower TRL scale and (2) if the second largest membership grade was not less than 0.7, we set the related fuzzy set as the Second Preference. In this consideration, we found 51 topics in $A_2(x)$ (42 NSF topics) and 47 topics in $A_3(x)$ (35 DII topics) with First Preference and three topics in $A_1(x)$, all of which belonged to NSF Awards, 13 topics in $A_2(x)$ (8 NSF topics), and 20 topics in $A_3(x)$ (13 NSF topics) with Second Preference.

**Table 12.4** Big data-related topics with membership grades of three fuzzy sets

| Year | Topic | Topic description | Data | $X(x)$ | $A(x)$ |
|---|---|---|---|---|---|
| 2009 | Adaptive grasping | Adaptive grasping, automatic speech recognition, empirical mechanism design, hierarchical visual categorization, infinite Bayesian networks | NSF | 0.63 | $A_2(x) = 0.95$ |
| 2010 | Reading data | Reading data, RFID tag, tag memory, configuration data, service provider | DII | 0.88 | $A_3(x) = 0.96$ |
| 2011 | Solving large systems | Linear equations, parallel strategy, recursive divide, solving large systems | NSF | 0.67 | $A_2(x) = 0.91$ |
| 2011 | Remote location | Remote location, retail establishment, source code, information source, navigation database, road sign | DII | 0.74 | $A_2(x) = 0.83$ |
| 2012 | Real time | Real time, telecommunication network, advertisement server, mobile communication facility data, monetization platform | DII | 0.67 | $A_2(x) = 0.91$ |
| 2012 | Large asynchronous multichannel audio corpora | Large asynchronous multichannel audio corpora, novel speech processing advancements, robotic intelligence | NSF | 0.43 | $A_2(x) = 0.98$ |
| 2013 | Video frames | Video frames, encoding video frame, improving video quality, live multicast system, severe degradation | DII | 0.76 | $A_2(x) = 0.81$ |
| 2013 | Big data | Algorithm foundation, big data, parsimonious model, mathematical problems | NSF | 0.5 | $A_2(x) = 1$ |

### *12.4.3 Findings for the Commercialization of the Computer Science-Related Technologies*

In our previous studies, single ST&I data-based TRM took active role in technical intelligence studies, which hold more benefits on exploring inner features of related technologies and identifying technology development chains. However, the incorporation of multiple ST&I data sources makes possible to stand on a higher macro-level to understand technology evolutionary pathways and to discover the gaps during technology development and transfer. We attempted to understand the insights of Fig. 12.5 for R&D plan and technology management concerns, and our findings are given as below:

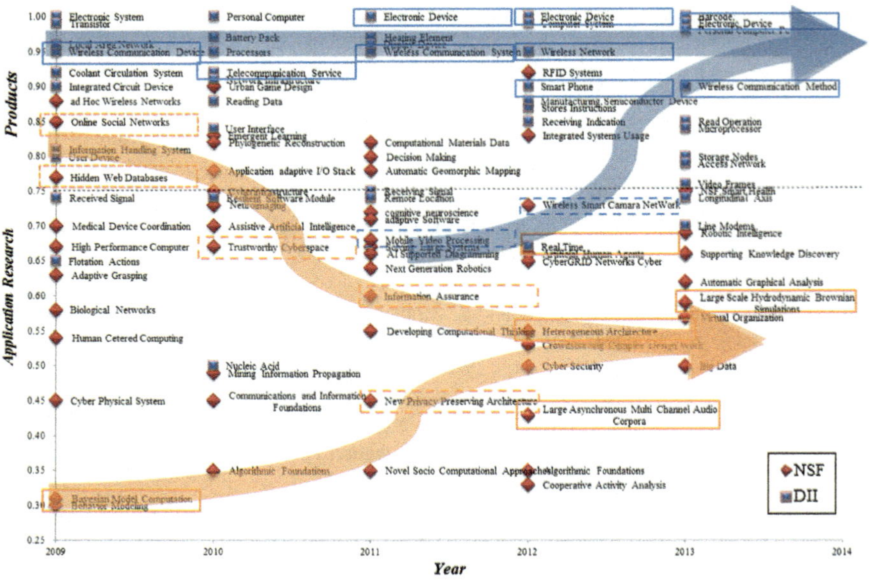


**Fig. 12.5** Technology roadmapping for computer science from 2009 to 2013 (based on the incorporation of NSF awards and DII patents)

**The "mobile device" and related techniques were, are, and still will be a hot commercial target in the near future.**

Obviously, the "mobile device" and related techniques (marked as the blue solid box) keep being identified as the hot topics from 2009 to 2013 in the DII patents, which means keen competitions occur or will occur in this filed, and the inventors (including the commercial firms) are seeking the intellectual property protection from the patents. At the same time, not only the mature products reflected in the DII patents, but also some undertaking researches in the NSF researches (marked as the blue dashed box) could be addressed in Fig. 12.5, e.g., "mobile video processing" and "wireless smart camera networks." Therefore, it is reasonable to conjecture that innovation and advanced techniques will also be engaged and transferred as a strong technical support for the follow-up developments.

**Big Data is not a creation, but a result of technology evolution and fusion, all related techniques of which are able to track down the origins.**

Big Data is an unavoidable topic in recent years. Social media, e.g., Twitter, Facebook, is more popular than any other periods in the history, and the boom of various new techniques, e.g., MapReduce, Hadoop, also illustrates revolutionary changes. In this situation, the voice that highlights the new creation of the Big Data-related techniques would be enough to have its supporters. However, as shown in Fig. 12.5, it is definite to declare that the Big Data could be considered as the results of technology evolution or fusion, and all related techniques are able to track down the origins. Extending the discussion by Zhang et al. (2014b), social media-leaded online social networks and web data (marked as the orange-dotted box) constitute parts of the foundation of Big Data, and the coming Big Data age also increases the concerns on the information security (marked as the orange-dashed box) rapidly. On the other hand, the efforts on the improvement of existed algorithms (marked as the orange solid box) have never been stopped, which compose the mainstream techniques of the Big Data Age.

In addition, if we narrow down the scope to technology commercialization, current Big Data-related research still concentrates on the NSF Awards and stands at the fundamental stages that include constructing concepts (e.g., Trust Worth Cyberspace, Real Time) and algorithms (e.g., Bayesian Network Computing, Large Asynchronous Multi-Channel Audio Corpora, Large Scale Hydrodynamic Brownian Simulations), and collecting data and applying it to experimental applications, while Big Data-related business models and real-world applications are crude, even there are no direct related topics in the DII patents. Thus, we should imagine that it will take time to transfer a new technique to commercial practices, and this kind of attempts would be an obvious trend in the near future.

**The process of technology commercialization is much faster than that several years ago.**

It is a common sense that the NSF interests include various fundamental researches that hold potential capabilities on further innovation, and the DII topics only concentrate on applicable techniques with any commercial benefits, e.g., software or hardware techniques. However, considering the three fuzzy sets for technology commercialization studies, only few topics belong to the set "basic research," and most of them are on the medium level between basic research and products. The possible explanation could be that the current process of technology commercialization is much faster than that several years ago, and new techniques could be used to solve real-world problems in a short time, or as we say, the experimental time is engaging into the commercialization process. In addition, more and more innovations are originated from real-world needs, which might be another strong driving force.

## 12.5 Discussion and Conclusions

Highlighting real-world needs and the engagement of emphasis-differed multiple ST&I data sources, this paper proposes an effective method to (1) incorporate multiple ST&I data to explore value-added information for R&D plans and technology innovation management and (2) introduce the fuzzy set concept to fuse the analytic results and expert knowledge smoothly and, then, help generate the TRM in a semi-automatic model. The thinking that combines qualitative and quantitative methodologies runs through the whole paper, the attempts on which provide great potential for related expert systems or decision-making processes.

We anticipate further study to look into the following directions: (1) to introduce novel IT techniques, e.g., machine learning, to enrich the semi-automatic model to an entire automatic composing model, and (2) to apply the multiple ST&I data incorporation for real-world applications. In addition, we will also consider the influences resulting from different empirical domains, e.g., emerging technology, social science, and mixed data with multidisciplinary, and address the concerns with more experiments.

**Acknowledgments** This work is supported by the Australian Research Council (ARC) under discovery grant DP140101366 and National High Technology Research and Development Program of China (Grant No. 2014AA015105).

## References

Bughin, J., Chui, M., & Manyika, J. (2010). Clouds, big data, and smart assets: Ten tech-enabled business trends to watch. *McKinsey Quarterly, 56*, 75–86.

Geum, Y., Lee, H., Lee, Y., & Park, Y. (2015). Development of data-driven technology roadmap considering dependency: An ARM-based technology roadmapping. *Technological Forecasting and Social Change, 91*, 264–279.

Huang, L., Zhang, Y., Guo, Y., Zhu, D., & Porter, A. L. (2014). Four dimensional science and technology planning: A new approach based on bibliometrics and technology roadmapping. *Technological Forecasting and Social Change, 81*, 39–48.

Lee, S., Mogi, G., Lee, S., & Kim, J. (2011). Prioritizing the weights of hydrogen energy technologies in the sector of the hydrogen economy by using a fuzzy AHP approach. *International Journal of Hydrogen Energy, 36*, 1897–1902.

Lee, S., & Park, Y. (2005). Customization of technology roadmaps according to roadmapping purposes: Overall process and detailed modules. *Technological Forecasting and Social Change, 72*, 567–583.

Lee, C., Song, B., & Park, Y. (2015). An instrument for scenario-based technology roadmapping: How to assess the impacts of future changes on organisational plans. *Technological Forecasting and Social Change, 90*, 285–301.

Lenzerini, M. (2002). Data integration: A theoretical perspective. In *Proceedings of the Twenty-First ACM SIGMOD-SIGACT-SIGART Symposium on Principles of Database Systems* (pp. 233–246). ACM.

Lu, J., Ma, J., Zhang, G., Zhu, Y., Zeng, X., & Koehl, L. (2011). Theme-based comprehensive evaluation in new product development using fuzzy hierarchical criteria group decision-making method. *IEEE Transactions on Industrial Electronics, 58*, 2236–2246.

Mankins, J. C. (1995). Technology readiness levels. *White Paper,* April 6.

McAfee, A., Brynjolfsson, E., Davenport, T. H., Patil, D., & Barton, D. (2012). Big data: The management revolution. *Harvard Business Review, 90*, 61–67.

Phaal, R., Farrukh, C. J., & Probert, D. R. (2004). Technology roadmapping—A planning framework for evolution and revolution. *Technological Forecasting and Social Change, 71*, 5–26.

Robinson, D. K., & Propp, T. (2008). Multi-path mapping for alignment strategies in emerging science and technologies. *Technological Forecasting and Social Change, 75*, 517–538.

VantagePoint. (2015). GA, USA: Search Tech Inc. Available: www.theVantagePoint.com. Accessed April 10, 2015.

Zadeh, L. A. (1965). Fuzzy sets. *Information and Control, 8*, 338–353.
Zhang, Y., Guo, Y., Wang, X., Zhu, D., & Porter, A. L. (2013). A hybrid visualisation model for technology roadmapping: Bibliometrics, qualitative methodology and empirical study. *Technology Analysis & Strategic Management, 25*, 707–724.
Zhang, Y., Porter, A. L., Hu, Z., Guo, Y., & Newman, N. C. (2014a). "Term clumping" for technical intelligence: A case study on dye-sensitized solar cells. *Technological Forecasting and Social Change, 85*, 26–39.
Zhang, Y., Robinson, D., Porter, A. L., Zhu, D., Zhang, G., & Lu, J. (2015). Technology roadmapping for competitive technical intelligence. *Technological Forecasting and Social Change* (to appear).
Zhang, Y., Zhang, G., Porter, A. L., Zhu, D., & Lu, J. (2014b). Science, technology & innovation textual data-oriented topic analysis and forecasting: Methodology and a case study. In *Proceedings of 5th International Conference on Future-Oriented Technology Analysis*. Brussels, Belgium.
Zhang, Y., Zhou, X., Porter, A. L., Gomila, J. M. V., & Yan, A. (2014c). Triple Helix innovation in China's dye-sensitized solar cell industry: Hybrid methods with semantic TRIZ and technology roadmapping. *Scientometrics, 99*, 55–75.
Zhou, X., Zhang, Y., Porter, A. L., Guo, Y., & Zhu, D. (2014). A patent analysis method to trace technology evolutionary pathways. *Scientometrics, 100*, 705–721.

# Chapter 13
# Generating Futures from Text—Scenario Development Using Text Mining

**Victoria Kayser and Erduana Shala**

**Abstract** Scenarios illustrate probable, plausible, and possible future developments and serve as a framework for strategic planning and decision making. They try to draw holistic images considering various aspects of today's world. Still, their development is complex and time-consuming. For example, at the beginning of the scenario development process, the literature needs to be screened in order to capture the state of the art and get an overview on influential aspects for the scenario stories. Here, this work concentrates on and proposes two alternative text mining approaches to improve this initial phase of scenario preparation. Text mining automatically processes texts and aggregates their content (scientific publications and reports in this case). This enables to summarize the topic and identify driving aspects. In order to draw a comparison, two different approaches are applied on two different cases. As the results show, the delimitation and structuring of the scenario field are supported and input for discussing the influences is delivered.

**Keywords** Scenario development · Text mining · Foresight · Topic modeling · Concept mapping

---

V. Kayser (✉) · E. Shala
Fraunhofer Institute for Systems and Innovation Research,
Breslauer Strasse 48, 76131 Karlsruhe, Germany
e-mail: Victoria.Kayser@isi.fraunhofer.de

E. Shala
e-mail: Erduana.Shala@isi.fraunhofer.de

V. Kayser
Chair of Innovation Economics, Technische Universität Berlin,
Müller-Breslau-Strasse 15 (VWS2), 10623 Berlin, Germany

E. Shala
Institute of Philosophy, Karlsruhe Institute of Technology,
Kaiserstr. 12, 76131 Karlsruhe, Germany



T.U. Daim et al. (eds.), *Anticipating Future Innovation Pathways Through Large Data Analysis*, Innovation, Technology, and Knowledge Management,
DOI 10.1007/978-3-319-39056-7_13

## 13.1 Introduction

Today, we are faced with a fast-changing world and a growing complexity. The impacts and interrelations of emerging technologies, environmental and social change challenge foresight processes and cause increasing uncertainty in decision making. However, an early debate of potential issues allows adjusting the future orientation. In this context, scenario planning is an established instrument (van der Heijden 2005; Mietzner 2009). Scenarios create multiple future images and thereby illustrate probable, plausible, and possible future developments which may be used as a framework for strategic decision making in the present (Kuosa 2012).

Basically, scenarios are developed in three phases: the preparation of the scenarios, the development of the scenario stories, and finally their usage (O'Brien and Meadows 2013). For the preparation of scenarios, a broad range of influences with a potential impact on the specific topic are taken into consideration such as social and environmental aspects, political arrangements, or technological progress. This work will concentrate on the first step of scenario planning where the purpose and focus of the exercise are clarified, and the scenario field needs to be delimited. Still, it is a challenging task which factors and dimensions of the topic to be included in the scenario development. This task strongly impacts the success of the whole process. At the moment, especially literature analysis, desk research, or surveys are used (Börjeson et al. 2006; Bradfield et al. 2005). In times of increasing data, an adaption of the scenario methodology is necessary for supporting a profound analysis of the scenario field. Besides, manual literature review and the identification of domain experts or interview partners are time-consuming. Here, hidden potential lays in text mining (Feldman and Sanger 2007; Manning et al. 2009), a technique to process text collections and extract influential aspects. Text mining automatically extracts relevant terms from texts and identifies patterns, trends, and dependencies within these data. This may reveal the additional information and delivers fundamental support for developing scenarios.

The objective of this work was to improve scenario planning by text mining and to capture the benefits of its integration. This work concentrates on the early phase of scenario planning, namely the structuring of the scenario field and identification of influential aspects. It is examined how to use text mining for this initial preparatory step, which text mining approach is appropriate and how complete and valuable the results are. Therefore, two comparative approaches are developed and evaluated by case studies.

The paper begins with a description of scenario planning and a consideration of the added value of text mining in the context of scenario planning (Sect. 13.2). In Sect. 13.3, the methodology is described, and the two comparative approaches are developed. These are tested and evaluated in Sect. 13.4 by conducting two case studies. The paper concludes with a discussion and highlighting points for future work in Sect. 13.5.

## 13.2 Scenario Planning: Is There a Demand for Text Mining?

Historically seen, scenario planning was not lanced as a fixed concept, but has evolved as a practitioners' strategic planning tool (Glenn 2009; Cuhls 2003). Having first been applied in the military field, scenarios are nowadays used in various contexts and purposes such as to support business planning (Bradfield et al. 2005), strategy development (O'Brien and Meadows 2013), technology assessment (Acatech 2012), or foresight activities (Ringland 2010). Over time, different typologies and process understandings have evolved ranging from narrative scenario writing to formalized concepts such as intuitive logics, trend impact analysis, or cross-impact analysis. Each of them follows its own understanding of scenario planning (Bradfield et al. 2005).

Principally, the process as applied in this article builds upon explorative, key factor-based scenario techniques (Fink et al. 2001; Reibnitz 1991). This process might be aggregated in three phases (see Fig. 13.1). Initially, the purpose and scope of the exercise are clarified. Here, it is framed on which level the scenarios will be located (e.g., corporate, national, international). Further, it is essential for the delimitation of the scenario field to get an overall picture of the regarded system and its driving variables (Godet and Durance 2011). In the following, a broad range of aspects with an impact on the specific topic are taken into consideration such as social changes, political arrangement, or technological progress. The summary of these aspects serves to describe the present situation. They are for example identified by experts in scenario workshops or by the literature analysis (Fink et al. 2001; Börjeson et al. 2006; van Notten et al. 2003). The initial phase is a crucial point in scenario development as the different future images are designed in the framework of the choices and prioritizations made here. The influence factors serve

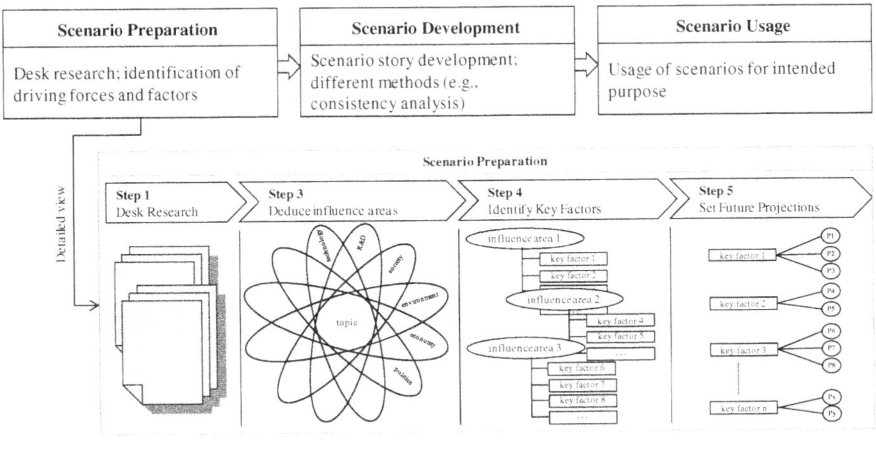


Fig. 13.1 Process of scenario planning

as a fundament from which alternative future projections are derived. The diversity of scenarios is reached by combining diverging future assumptions of the different influence factors. In the second step, these are bundled in logically sound scenario stories where each scenario represents one possible future. Techniques as consistency analysis or cross-impact analysis are frequently applied to reflect the given systemic interlinkages between the different influence areas and factors. In the third step, the created scenarios serve as a framework to think about future challenges and support decision making. Finally, the scenarios are used for various purposes as visioning or long-term strategic planning (Wilson 2000).

Scenarios are either created in an explorative or normative manner. The former explores future possibilities starting from today's knowledge base, whereas the latter presupposes certain future goals or desires before imagining the paths that lead to that point. In this work, an explorative approach is applied as illustrated in Fig. 13.2. When finally working with the scenarios, the process starts in the present with considering current challenges, tasks, and options. These are projected into the future where they are reflected and evaluated in the different future scenarios. Finally, these insights are reprojected into the present and thereby support decision making and future planning.

In the last 50 years, the purpose and the idea of forward looking activities (shift from forecasting to foresight) have frequently been discussed in the context of scenarios (Cuhls 2003). But the efficient design of the process itself is rarely addressed, in particular not in the face of increasing data and supportive IT-tools. Notwithstanding the effort needed to develop scenarios is assessed as problematic. As for example Raford (2014) states, critical points are the labor intensity (data collection, interviews, workshops, etc.) or the dependence on the skills of the workshop participants and scenario story writers, naturally leading to biases. According to Mietzner and Reger (2005), a deep knowledge and understanding of the examined scenario field are required. They emphasize that data and information from different sources need to be collected and interpreted which takes much time. Kuosa (2012) argues that database literature reviews usually consume many days in foresight processes. Further on, the initial phase requires a profound examination of the present developments because the final set of factors should construct a sound and holistic view of the considered field. As scenarios are built on these results, the quality of the whole process fully depends on this first step.

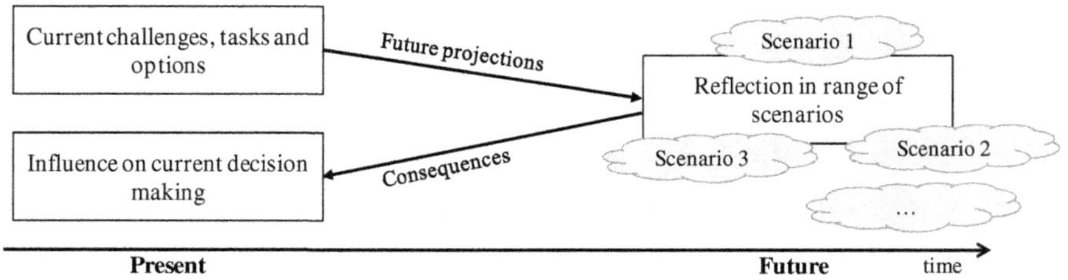


**Fig. 13.2** Explorative scenario planning

Other quantitative attempts to improve scenario planning primarily focus on later phases of the process. The existing software approaches in the scenario context concentrate on consistency analysis (e.g., INKA or Szenoplan) or address different (mathematical) challenges as for example to improve the efficiency of consistency analysis (Dönitz 2009) or to analyze impact networks (Weimer-Jehle 2009). Also, scenario development is largely a qualitative endeavor and strongly relies on judgments, expert opinions, or workshop results. As Martino (2003) summarizes "[…] *scenario generation is still a highly subjective art.*"

The aim of this article was not to change the process (our critique is not on the necessity of the conducted tasks), but to optimize it. There is still hidden potential to improve the initial phase, where a broad information base needs to be screened. Following the critiques described above, text mining might increase the process of knowledge generation and lead to new insights for scenario practitioners. To our knowledge, the intersection of scenario planning and text mining has not been tackled before in the scientific debate. With text mining, some of the above-mentioned points may be resolved. First, the reading effort during literature analysis might be reduced by aggregating the textual data. Instead of manual screening, automatic summaries and visual representations of the content can assist the planning process. Thereby, a differentiation of influences and factors is eased. This additionally reduces the time effort of scenario processes. In particular, it is expected to get a fast thematic overview and a fundament or entry point for further discussions when the influence factors have to be prioritized. Also, structuring the field and recognizing links between influence areas might be achieved.

## 13.3 Methodology—Text Mining-Based Scenario Development

The following describes the methodology to process text to support the identification of influential aspects for the scenarios. Related technical requirements are that full text is processable, and the results are adequately summarized to raise further discussions. As illustrated in Fig. 13.3, the process starts with converting the pdf files to a machine readable format. In this article, two different text mining

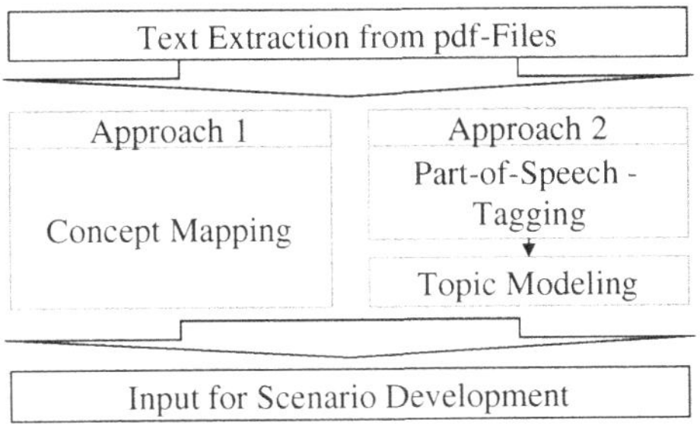


**Fig. 13.3** Process description—text mining

approaches are evaluated with regard to the technical requirements and the discussion in Sect. 13.2. These are the commercial software Leximancer (2011) and an own implementation developed in Python. Different algorithms for term extraction (stopwords-based vs. part of speech tagging) and the subsequent aggregation of information are applied (concept mapping vs. topic modeling). The results deliver a basis for deducing influence factors for the scenario development. This is further illustrated in Sect. 13.4 by two case studies. The following chapter focuses on the technical realization.

To begin with, the textual data are preprocessed and structured. For this, the texts in pdf format are transformed to plain text. Therefore, the package LAPDF is used (Ramakrishnan et al. 2012) which is especially designed for scientific texts, including figures, footnotes, or columns. For reports, a conversion using Adobe 9 Pro performed better because lapdftxt is not designed for this purpose (e.g., images, graphical elements). From the extracted text data, certain data fields were excluded, such as acknowledgements, figures, tables, and references due to uniformity and relevance for the further analysis.

### *13.3.1 Approach 1: Concept Mapping Using Leximancer*

Leximancer builds on concept mapping. The software aggregates documents to concepts based on the words they contain (Leximancer 2011). This approach is chosen because the program has an intuitively handling due to its graphical user interface. On the other hand, the results visualize the theme, and existing relationships between the thematic subgroup are clearly arranged. In other work, Leximancer is applied for quantitative summaries of thematic fields, exploration of unfamiliar domains, or theory building (Stockwell et al. 2009; Cameron et al. 2011) but is has not been applied to scenario planning before.

In technical terms, Leximancer builds on the experience of Stockwell et al. (2009) in the fast exploration of an unfamiliar domain and reduces complexity by aggregating words to concepts (Smith and Humphreys 2006). The algorithm underlying this program runs in two phases and is based on naive Bayes classification (Yarowsky 1995; Salton 1988). First, a classifier is built (semantic extraction) and applied in the second phase (relational extraction) (Smith and Humphreys 2006). The first step aims to learn the categorical coding scheme. It begins with omitting stopwords, merges word variants, and determines the frequency of single words and their co-occurrence. A concept as a group of related terms is built by a thesaurus as a term classifier. The concept bootstrapping algorithm results in concept seeds as start values for the concepts as clusters. The second step corresponds to coding as in content analysis. The text segments, normally 1 up to 3 sentences, are classified, and relations within and between the concepts are identified. Finally, the concepts (noted as nodes) are aggregated to themes (noted as circles) by a clustering approach. The closeness of the circles describes their contextual proximity. Related to scenario planning, the concepts indicate factors, while the themes are potential influence areas.

The Lexmiancer analysis in this work was conducted with its standard settings (automatic processing), but the stopword list was extended, word variants were merged, and the initial set of concepts was adapted (Leximancer 2011).

### *13.3.2 Approach 2: Part of Speech Tagging Based Approach*

The second approach is an individual solution especially implemented with respect to the scenario requirements described above and resolving some critiques on Leximancer (only extracts single terms and not phrases, difficult to track the intermediate process compared to own implementation). This approach is developed in Python (Bird et al. 2009) and opposed to Leximancer builds on noun phrase extraction. To begin with, the text is tokenized (broken into its single words). Next, part of speech tags is assigned to these tokens to characterize their grammatical instance. Then, chunk parsing is used based on regular expressions to extract noun phrases for further processing (e.g., *futures research* or *principle*). Certain mechanisms are implemented for term cleaning as lemmatization on plural forms or the removal of footnotes. Further on, a thesaurus matches varying spellings such as American and British English and replaces abbreviations.

Since frequent words are not necessarily the most relevant for mapping the content, a common weighting function is applied on the data set to compensate this, namely the tf-idf weighting scheme (Salton et al. 1983; Manning et al. 2009). Thereby, each term is scored per document. This is calculated by the following:

$$\text{tf - idf}_{\text{t,d}} = \text{tf}_{\text{t,d}} \times \text{idf}_{\text{t}}$$

The term frequency tf describes the frequency of a term $t$ in a document $d$. The inverse document frequency idf is a logarithmic measure scaling the importance of a term by $\log\frac{N}{n}$. The total number of documents $N$ is divided by the number of documents containing a certain term.

In the following, the content needs to be aggregated in such a manner which delivers influential areas for the scenarios. This requires an approach preferably running with few process interventions and easy to understand results. In this setting, topic modeling is suitable, realized using latent dirichlet allocation (LDA) (Blei et al. 2003). As an unsupervised approach, topic modeling affords no process interventions; nevertheless, the number of topics needs to be defined in advance. The results of topic modeling are mostly self-explanatory. The retrieved topics are related to influence areas, while the concepts represent potential factors. Compared to cluster analysis (Manning et al. 2009), for topic modeling the single documents are not only assigned to one cluster, but can belong to different topics.

Cluster analysis groups documents or words. So the topics can be interpreted as an additional layer between words and documents. Compared to the previously described concept mapping (see last section) topic modeling is closely related, but infers a statistical model during the generation process and applies soft clustering (words can belong to more than one topic) (Miner 2012). Topic models reveal the hidden thematic structure in texts. The underlying assumption is that documents are built from topics that draw words from word distributions. A topic is distributed over a fixed vocabulary. The topics are denoted by a probability mass function over each possible word. Topics have associated term probabilities, and for each document, topic proportions are computed (likelihood of a topic to appear in a document). For this implementation, the *gensim* package was used (Řehůřek and Sojka 2010). For a better performance, the stream of values is split into smaller chunks (500 words) for the training phase. The final topics are manually labeled.

## 13.4 Case Studies

To illustrate and test the procedure, two cases from a completed research project were taken. The EU FP7 *project European Security Trends and Threats in Society* (*ETTIS*) was conducted from 2011 to 2014 and developed a methodology for prioritizing societal security research topics.[1] Creating scenarios was part of this project. In the ETTIS scenario process, context, and threat scenarios in the domains *cyber*, *nuclear material* and *environment* were built and served as a framework for thinking about future societal security needs in R&D (Dönitz et al. 2013). The scenarios were built based on foresight studies and scientific literature. Initially, desk research was conducted, key stakeholders were interviewed, and lists of factors and future projections for the context and for the domains were built. The scenarios each focus on developments in the EU. In different focus group workshops, the lists were aggregated by prioritizing the factors. In preparation for the scenario writing, a consistency analysis and an influence analysis were applied.

For this work, the domains *nuclear material* and *environment* are chosen as examples. The same set of literature and foresight studies as in the project was used in the case studies to guarantee the comparability to the original project. In the following, the results are described and finally compared to the original set of influence areas and factors.

[1]For further details on the project please visit www.ettis-project.eu.

### 13.4.1 Case 1: Nuclear Scenarios

In concept maps, the circles represent themes, while the dots are concepts. The concepts are linked by lines as for example *risk* and *security* in Fig. 13.4. For the field of nuclear, the concept map (Fig. 13.4) shows seven concepts with varying likelihoods. *Energy* is the core theme containing concepts such as *oil*, *gas*, and *coal*. This theme has a huge overlap with *power* due to concepts such as *efficiency*. The *power* theme focuses on energy usage with concepts such as *cost*, *capacity*, and *investments*. To a certain point, on the left side of the map, there is the market perspective covering supply and demand (*energy* and *power*), while on the right side, research topics and the optimization of the fuel cycle are covered. The huge gap between these two might be related to a division of R&D and market realization. This is also underlined by *weapons* as a distinct theme, where also *safety* and *security* are covered concepts. This is close to the *fuel cycle* theme centering on the optimization of *uranium* usage and related *research*. In the middle, three further themes are mapped. *Countries* contain concepts such as *policy* and *economy* as regulatory aspects that have to be addressed on individual national levels. Further themes are *provisioning* or *storage* that is still not resolved in the field of energy.

Topic modeling showed the best results running on six topics (see Table 13.1). The first topic is mainly on indicators, materials, and their security. The second topic has a focus on nuclear weapons, attacks, and terrorists and sheds light on the side effects of nuclear technologies. The third topic deals with trust and safety issues. Also, the terms *attitude* and *belief* are contained indicating societal concerns. The fourth topic covers legal and political aspects. Fifth, $CO^2$ emissions and energy consumption are covered. Finally, a technical topic is about *fuel cycle* and reactors. Also, *costs* and *energy efficiency* are covered there.

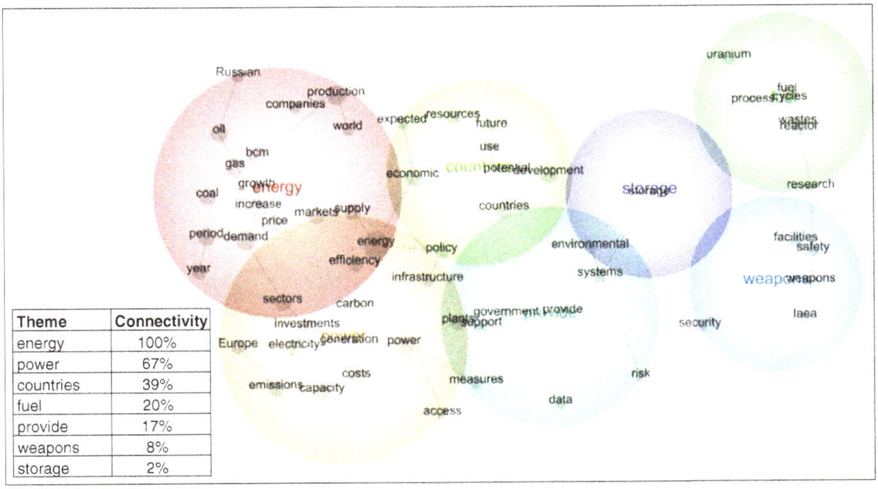


| Theme | Connectivity |
|---|---|
| energy | 100% |
| power | 67% |
| countries | 39% |
| fuel | 20% |
| provide | 17% |
| weapons | 8% |
| storage | 2% |

**Fig. 13.4** Concept map—nuclear scenarios

**Table 13.1** Topic model—nuclear scenarios

| No. | 10 most probable phrases per topic |
|---|---|
| 1: Indicators and material security | Indicator, index, energy chain, material security, material, score, category, material security condition, UCTE average, NTI index |
| 2: Weapons and attacks | Weapon, accident, convention, research reactor, attack, device, release, material, IAEA, terrorist |
| 3: Trust and safety concerns | Trust, safety series, wind power capacity, European wind energy association, VBN model, radiation protection, revision, French, attitude, belief |
| 4: Legal and political Issues | State party, reference case, IAEA, carbon value, hypothesis, convention, liberalization, pole, treaty, European union emission trading scheme |
| 5: $CO^2$ emissions | $CO^2$ emission, energy consumption, government engagement, coal product, accessibility, petroleum product, climate regime, shock, Latin America, ratio |
| 6: Fuel cycle | Fuel cycle, reactor, cost, pathway, repository, power sector, demand response, track case, option, energy efficiency |

The scenario stories developed in ETTIS are based on 15 different influence factors. For this work, they are aggregated to four main influence areas as illustrated in Fig. 13.5: *policy and regulation*, *society*, *nuclear security*, and *technology and R&D*.

As the comparison of the results shows, societal aspects are not directly covered in the concept map but in the topic modeling (topic 3). Policy and regulation are included among other aspects in topic 4 as well as in topic 1 (e.g., *indicators, material security*). Nuclear security is addressed in different topics and indirectly in the concept map as for example topic 2 covers many security relevant points (e.g., *attack, terrorist*). Technological aspects are included in both text mining results as for example *fuel cycle* or *weapons*. As these results already indicate, the text mining–solutions both show different results that are nevertheless equally contained in the factor lists. But there are issues that are not directly addressed in the scenario influence areas as for example nuclear weapons (own concept and included in topic 2).

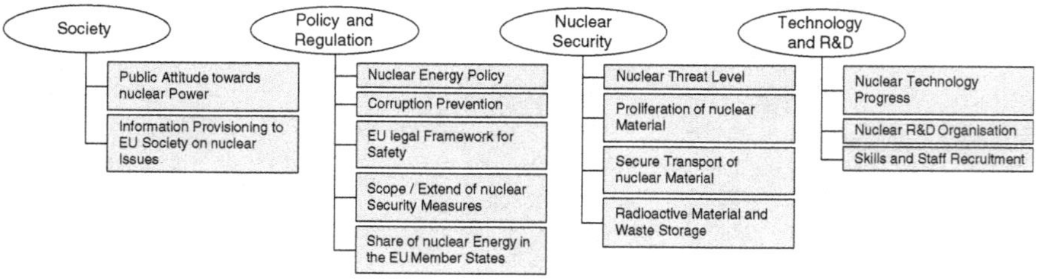


**Fig. 13.5** Influence factors of nuclear scenarios from the ETTIS project (Dönitz et al. 2013)

### 13.4.2 Case 2: Environment Scenarios

The concept map for environment shows eight themes ranging from *energy* to *values* (see Fig. 13.6). These themes show different ranges of connectivity. While *products* is a strongly connected theme (100 %), the *models* theme only shows a connectivity of 9 %. *Values* as an own theme underlines on the one hand the meaning of societal concerns and health issues in general and on the other hand the high impact of specific aspects as *willingness to pay*. The *energy* theme includes various possibilities of energy generation, such as *wind*, *PV*, or power plants. The *products* theme relates to *demand*, *prices*, or *growth* and summarizes the market perspective. The *data* theme covers concepts, such as *process*, *control*, and *analysis*. It is close to the *usage* theme. The themes *use* and *change* are closely related to the *models* theme which relates to the steady attempt to model usage and the changing external conditions, also with a future component.

The environment—topic model includes six topics (see Table 13.2). The first topic covers general (environmental) themes such as *biodiversity* or *photovoltaic*. The second has a more concrete focus on *IPR* and *patents* but also includes some energy and electricity aspects. Third, themes around energy savings and energy efficiency are addressed, and fourth, an own topic on futures research appears. The fifth topic deals with (social) values and risks. Sixth, food and nutrition are covered where also aspects such as *hunger reduction* are addressed.

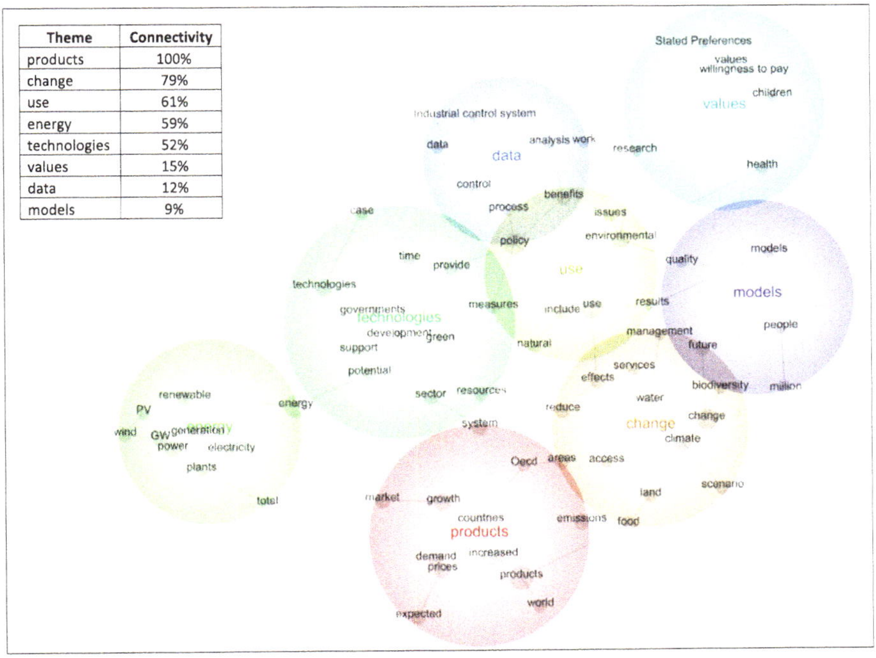


| Theme | Connectivity |
|---|---|
| products | 100% |
| change | 79% |
| use | 61% |
| energy | 59% |
| technologies | 52% |
| values | 15% |
| data | 12% |
| models | 9% |

**Fig. 13.6** Concept map—environment scenario

**Table 13.2** Topic model—environment scenarios

| No. | 10 most probable phrases per topic |
|---|---|
| 1: General aspects of environment | Island, water sector, photovoltaic, medium term, roadmap, science base, deployment, biodiversity, outlook period, outlook |
| 2: IPR and energy | Innovation, patent, electricity, inventor, energy consumption, Sweden, energy, energy foresight project, growth, carbon dioxide emission |
| 3: Energy efficiency | Megatrend, percent, energy efficiency, small-scale renewable energy scheme, storyline, energy saving, scenario, energy efficiency improvement, rebound effect, decommissioning |
| 4: Future research and ecology | Synthesis, future research, futurist, ecology, center, future research method, synthesis activity, delphi method, ecologist, scenario analysis |
| 5: Risks and values | Value of a statistical life, power plant fleet, child, hazard, habitat, risk reduction, risk change, value of a statistical life estimate, grazing marsh, adult |
| 6: Food and nutrition | Food system, food production, hunger, volatility, food producer, food price, hunger reduction, governance, food, food supply chain |

The list of the 12 influence factors of the environment scenario is illustrated in Fig. 13.7 (Dönitz et al. 2013). They were aggregated to five influence areas, such as *policy and regulation*, *environment and ecosystem*, *societal aspects*, *technology*, and *economy*.

As the comparison of the results shows, the policy and regulation area is not covered in such detail in the text mining results, only as a general *policy* term in the concept map. On the other hand, technology is covered in great detail by the text mining (own theme). Environment and ecosystem aspects are covered and spread over different topics, for example together with futures research in topic 4. All three solutions cover societal aspects such as *value system* or *environmental awareness*. Food and nutrition made up an own topic, but are not covered in the scenario factors. Both text mining approaches include energy themes that are not included in the factor lists.

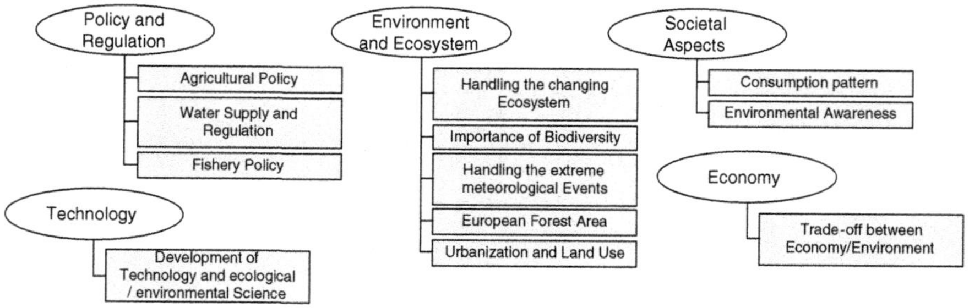


**Fig. 13.7** Influence factors of environment scenarios from the ETTIS project (Dönitz et al. 2013)

## 13.5 Discussion and Conclusion

This work contributes a first scientific approach to the integration of text mining in the initial phase of scenario processes. In the following, the results are discussed.

### *13.5.1 Comparison of Text Mining Approaches*

For the task of improving the initial phases in scenario processes, two text mining approaches were evaluated, a commercial software (Leximancer) and an own implementation. Both text mining–solutions introduced here are unsupervised approaches, and thereby, no background or domain knowledge is needed. This eases the application for non-domain experts. Leximancer is ideal to gain a fast overview of the considered field, also due to its fast algorithm. For the concept maps, the themes relate to key influential areas and the concepts relate to the factors in scenario planning. Further, these maps already hint toward interrelationships between the areas. This is most helpful for the later influence analysis. This software is very user friendly in its basic functions and delivers a solid foundation for the subsequent scenario process. Moreover, the detailed view (terms in context) offers further insights. Critical points are to intervene in the running process, but options are provided, and as for commercial software, the access is restricted. The main critique is that not noun phrases are extracted but single words. These three aspects were the main motivation to try a second approach based on noun phrase extraction, but also to have a comparable solution. The second solution affords a more in-depth knowledge of text mining and how to handle the software. In fact, programming skills are mandatory. The output of intermediate steps is possible, the process is fully traceable, and individual parameters can be set. Many options to intervene in the process exist. Additionally, PoS tagging for noun phrase extraction drastically reduces the effort for stopword removal as in Leximancer, but is meanwhile very computing intensive. Principally, the implemented topic modeling is an appropriate algorithm for the raised problem due to its assumption that text is arranged by different themes. What is currently missing for the second approach is an adequate visualization. Summarized, both approaches deliver input for discussing and prioritizing the influence factors, while Leximancer already indicates thematic dependencies and interrelations.

### *13.5.2 Text Mining and Its Contribution to Scenario Planning*

As described in Sect. 13.2, this article focuses on the preparatory step and intends to support the delimitation of the scenario field by a profound summary of the literature analyzed by text mining.

First results emphasize that the delimitation of the field is supported, and first interdependencies within the scenario field are illustrated even if there are partly huge differences between the text mining results and the project factor lists. One explanation is that topics not covered in the text corpus, that maybe arose in workshops or interviews, cannot be identified by text mining. Workshops and interviews strongly shape the direction of influence factor selection. Thus, a factor list is already directed toward the concrete objective of the project and adjusted to strategic goals. This explains why text mining results and the factor list of the project cannot be equal, but this was not claimed at any point. This rather underlines what is described in Fig. 13.3 that text mining delivers input or serves as a starting point for discussing the influence areas and factors. We argue that comprehensive foresight activities balance the strength of qualitative and quantitative approaches. The selection and discussion of the relevance of influence factors in workshops are still necessary. But text mining is very useful for bundling, structuring, and evaluating the scenario field. It delivers valuable input for discussing not only the influence factors but also their impact on each other. In practical terms, the text mining results give a first idea of what is covered in the scenario field. As the results of the two case studies indicate, this remarkably reduces the effort needed for a manual literature review. Further on, the dependence from the conceivably biased reading or the capabilities of the scenario team is reduced.

Due to these observations, the current scenario process understanding is extended by an additional text mining step (see Fig. 13.8). This figure also illustrates the weight qualitative and quantitative thinking has over the scenario process. This newly added text mining step enables to derive important variables and draw a future narrative varying the parameters in each scenario story. The here conducted analysis already delivers helpful information for the scenario story generation. Nevertheless, the concrete factors need to be manually refined to the exact objectives of the specific purpose of the scenario project. Special thematic requirements and foci are achieved by human interpretation and do not directly follow from data analysis.

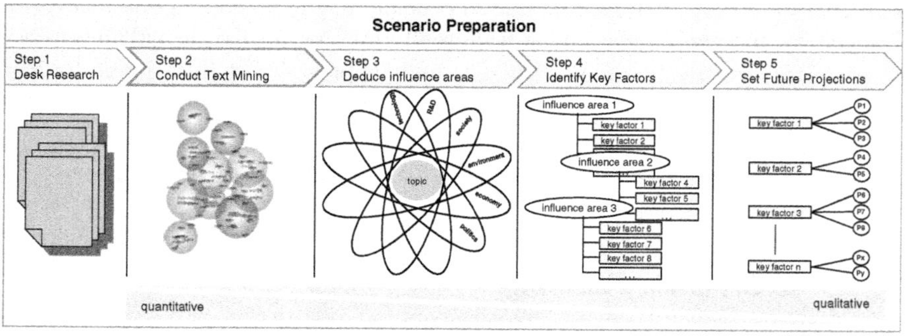


Fig. 13.8 Process of text mining-based scenario planning

For the two scenario cases presented in this paper, the same data as for the ETTIS scenarios were used, but also any other data source can be used in future (e.g., Twitter). Nevertheless, the quality of the output absolutely depends on the quality of the input data. This is a very important aspect of automatic analysis: If biased data are read in, the summary will also have these tendencies.

### 13.5.3 Final Conclusion and Future Work

This article is a first examination whether the combination of text mining and scenario planning is beneficial. In times of increasing data volumes, this approach is of high relevance. Reducing the reading effort is a crucial advantage, as we are nowadays faced with a growing amount of text to be screened. We argue that an advanced scenario building process needs technical extensions as text mining tools to cope with the new amounts and forms of data. Of course, the implemented approaches can be further improved (e.g., performance of algorithms, further visualizations), or further data sources can be added. Particularly, this methodology should be integrated from beginning to end in a scenario project. Then, its added value can be examined on another level as for example evaluating time savings. As discussed above, the results are promising, and the approach should be pursued and refined in future as it meets the requirements of foresight in the information age.

**Acknowledgments** The authors would like to thank Ewa Dönitz (Fraunhofer ISI) for her permission to work with the data of the ETTIS Scenarios. Furthermore, we want to thank a number of friends and colleagues for their comments on our work and their encouragement.

## References

Acatech. (2012). *Technikzukünfte: Vorausdenken - Erstellen - Bewerten*. Springer Vieweg; Acatech, [S.l.].

Bird, S., Klein, E., & Loper, E. (2009). *Natural language processing with python* (1st ed.). Beijing, Cambridge [Mass.]: O'Reilly.

Blei, D. M., Ng, A. Y., & Jordan, M. I. (2003). Latent dirichlet allocation. *The Journal of machine Learning research, 3*, 993–1022.

Börjeson, L., Höjer, M., Dreborg, K., Ekvall, T., & Finnveden, G. (2006). Scenario types and techniques: Towards a user's guide. *Futures, 38*(7), 723–739.

Bradfield, R., Wright, G., Burt, G., Cairns, G., & van der Heijden, K. (2005). The origins and evolution of scenario techniques in long range business planning. *Futures, 37*(8), 795–812.

Cameron, D., Finlayson, A., & Wotzko, R. (2011). Visualising social computing output: Mapping student blogs and tweets. In B. White, I. King, & P. Tsang (Eds.), *Social media tools and platforms in learning environments* (pp. 337–350). Berlin, Heidelberg: Springer.

Cuhls, K. (2003). From forecasting to foresight processes—New participative foresight activities in Germany. *J. Forecast., 22*(2–3), 93–111.

Dönitz, E. (2009). *Effizientere Szenariotechnik durch teilautomatische Generierung von Konsistenzmatrizen: Empirie, Konzeption, Fuzzy- und Neuro-Fuzzy-Ansätze*. Wiesbaden: Gabler Verlag/GWV Fachverlage GmbH.
Dönitz, E., Shala, E., Leimbach, T., Bierwisch, A., Grigoleitt, S., & Klerx, J. (2013) D4.4 catalogue of threat scenarios: ETTIS—European security trends and threats in society.
Feldman, R., & Sanger, J. (2007) *The text mining handbook: Advanced approaches in analyzing unstructured data*. Cambridge: Cambridge University Press.
Fink, A., Schlake, O., & Siebe, A. (2001). *Erfolg durch Szenario-Management: Prinzip und Werkzeuge der strategischen Vorausschau*. Frankfurt/Main, New York: Campus-Verlag.
Glenn, G. C. (2009) The futures group international scenarios. In J. C. Glenn & T. J. Gordon (Eds), *Futures research methodology* (pp. 1–25), Version 3.0. Millennium Project, Washington, DC.
Godet, M., & Durance, P. (2011). *Strategische Vorausschau: Für Unternehmen und Regionen*. Paris: Dunod.
Kuosa, T. (2012). *The Evolution of strategic foresight: Navigating public policy making*. Farnham: Ashgate Publishing Ltd.
Leximancer. (2011). Leximancer manual: Version 4.
Manning, C. D., Raghavan, P., & Schütze, H. (2009). *An introduction to information retrieval*. New York: Cambridge University Press.
Martino, J. P. (2003). A review of selected recent advances in technological forecasting. *Technological Forecasting and Social Change, 70*(8), 719–733.
Mietzner, D., & Reger, G. (2005). Advantages and disadvantages of scenario approaches for strategic foresight. *IJTIP, 1*(2), 220–239.
Mietzner, D. (2009). *Strategische Vorausschau und Szenarioanalysen: Methodenevaluation und neue Ansätze* (1st ed.). Wiesbaden: Gabler.
Miner, G. (2012). *Practical text mining and statistical analysis for non-structured text data applications* (1st ed.). Waltham, MA: Academic Press.
O'Brien, F. A., & Meadows, M. (2013). Scenario orientation and use to support strategy development. *Scenario Method: Current Developments in Theory and Practice, 80*(4), 643–656.
Raford, N. (2014). Online foresight platforms: Evidence for their impact on scenario planning & strategic foresight. Technological Forecasting and Social Change.
Ramakrishnan, C., Patnia, A., Hovy, E., & Burns, G. A. (2012). Layout-aware text extraction from full-text PDF of scientific articles. *Source Code for Biology and Medicine, 7*(1), 7.
Řehůřek, R, Sojka, P. (2010). Software framework for topic modelling with large corpora. In *Proceedings of LREC 2010 Workshop 2010* (pp. 46–50).
Reibnitz, U. V. (1991). *Szenario-Technik: Instrumente für die unternehmerische und persönliche Erfolgsplanung*. Wiesbaden: Gabler.
Ringland, G. (2010). The role of scenarios in strategic foresight. *Technological Forecasting and Social Change, 77*(9), 1493–1498.
Salton, G., Fox, E. A., & Wu, H. (1983). Extended Boolean information retrieval. *Communications of the ACM, 26*(11), 1022–1036.
Salton, G. (1988). *Automatic text processing: The transformation, analysis, and retrieval of information by computer*. Reading, Mass: Addison-Wesley.
Smith, A. E., & Humphreys, M. S. (2006). Evaluation of unsupervised semantic mapping of natural language with Leximancer concept mapping. *Behavior Research Methods, 38*(2), 262–279.
Stockwell, P., Colomb, R. M., Smith, A. E., & Wiles, J. (2009). Use of an automatic content analysis tool: A technique for seeing both local and global scope. *International Journal of Human-Computer Studies, 67*(5), 424–436.
van der Heijden, K. (2005). *Scenarios: The art of strategic conversation* (2nd ed.). Chichester, England, Hoboken, NJ: Wiley.
van Notten, P. W. F., Rotmans, J., van Asselt, M. B. A., & Rothman, D. S. (2003). An updated scenario typology. *Futures, 35*(5), 423–443.

Weimer-Jehle, W. (2009). Szenarienentwicklung mit der Cross-Impact-Bilanzanalyse. In: J. Gausemeier (Ed.), *Vorausschau und Technologieplanung: 5. Symposium für Vorausschau und Technologieplanung* (pp. 435–454). Heinz-Nixdorf-Institut, 19. und 20. November 2009 in der Berlin-Brandenburgischen Akademie der Wissenschaften. HNI, Paderborn.
Wilson, I. (2000). From scenario thinking to strategic action. *Technological Forecasting and Social Change, 65*(1), 23–29.
Yarowsky, D. (1995) Unsupervised word sense disambiguation rivaling supervised methods. In *Proceedings of the 33rd Annual Meeting on Association for Computational Linguistics* (pp. 189–196).

# Part III
# Anticipating the Future—Cases and Frameworks

# Chapter 14
# Additive Manufacturing: Importance and Challenges for Latin America

**Marisela Rodríguez Salvador, Ana Marcela Hernández de Menéndez and David Alfredo Arcos Novillo**

**Abstract** In this chapter, Rodríguez et al. (Res Educ 11(3):165–173, 2002) Competitive Technical Intelligence methodology is adapted and applied to an analysis of additive manufacturing technology. The global current stage of this technology and an assessment of its future potential are presented and compared with actual development in Latin America. Market information is also evaluated. Later, a scientometric patent and scientific literature analysis are used to compare global and regional circumstances in regard to this technology. The insights obtained from these assessments reveal that patent and research activity regarding additive manufacturing throughout Latin America is behind that of developed countries. However, some companies are making use of additive manufacturing in their current processes. A major adoption of this technology is expected to occur in Latin American as a result of the advancements in additive manufacturing.

**Keywords** Additive manufacturing · 3D printing · Additive fabrication · Latin America · Scientometrics · Patent analysis · Scientific analysis

## 14.1 Introduction

Additive manufacturing is the process of joining materials to make three-dimensional objects from digital models. Unlike traditional subtractive manufacturing methods, this process is usually developed layer upon layer (Scott et al. 2012). The term "3D printing" is more commonly used as a synonym for additive manufacturing (Wohlers Associates 2013). Consequently, both "additive manufacturing" and "3D printing" are used interchangeably in this chapter.

---

M. Rodríguez Salvador (✉) · A.M. Hernández de Menéndez · D.A. Arcos Novillo
Escuela de Ingeniería y Ciencias, Tecnológico de Monterrey,
Avenida Eugenio Garza Sada 2501 Sur, Colonia Tecnológico,
64849 Monterrey, Nuevo León, Mexico
e-mail: marisrod@itesm.mx



T.U. Daim et al. (eds.), *Anticipating Future Innovation Pathways Through Large Data Analysis*, Innovation, Technology, and Knowledge Management,
DOI 10.1007/978-3-319-39056-7_14

The first patent related to this technology was granted on March 11, 1986 in the USA under the registration number 4,575,330. This patent, invented by (Hull 1986), was for the development of an "apparatus for production of three-dimensional objects by stereolithography." The latest of the former patents only recently expired (Basiliere 2014b).

The growth of this technology was slow during the first two decades; however, the 3D printing market has expanded dramatically since 2012. The participation of independent creators, hobbyists, and early adopters has begun to heavily publicize the subject (Basiliere 2014a). As an example of this great interest, 3D printing publications in specialized journals have grown from 1600 to 16,000 articles from 2011 to 2012, an increase of 10 times in a single year (Wohlers Associates 2013). Industries that are taking advantage of this technology include education, aerospace, defense, architecture, automotive, consumer products, electronics, and medical devices (Basiliere et al. 2014).

At present, the impact of additive manufacturing continues to grow in terms of commercial and scholarly activities (Bourell et al. 2009). Some commercially available products already use this technology; for example, a hip joint for general use manufactured by Arcam AB has been approved by the China Food and Drug Administration (Chen 2015). The US Food and Drug Administration (FDA) also certificated a robotic arm developed by DEKA Integrated Solutions Corporation to be manufactured, marketed, and made available to the US Veteran Affairs health system, though its release date has yet to be defined (U.S. Department of Veterans Affairs 2014). These advancements are expected to reduce surgery costs and improve patient quality of life. The importance of this technology is comparable to that of the development of the semiconductor, the computer, and the Internet (Wohlers Associates 2013).

This chapter presents a Competitive Technological Intelligence research on additive manufacturing technology. Market insights regarding this technology will be presented, and a scientometric patent and scientific literature analysis will be used to compare the state of the technology at both regional (Latin America) and global levels. The importance of this technology and the main challenges facing its development and adoption in Latin American countries will also be discussed.

This work is organized as follows: it starts with the definition of additive manufacturing and the different types that involves, after that the methodology that was developed to analyze this technology is presented, followed by a discussion of the results obtained concerning additive manufacturing market and the impact of the technology. The incursion of this technology in Latin America will then be discussed, and the scientific and patent productions at global and regional levels will be compared. The main challenges facing the implementation of this technology will be described, after which conclusions will be provided.

## 14.2 Additive Manufacturing Technology

The American Society for Testing and Materials (ASTM) defines additive manufacturing as the process of joining materials to make objects from 3D model data, usually layer upon layer, as opposed to subtractive manufacturing methods (Mahamood et al. 2014). This process is also known as additive fabrication, additive processes, freeform fabrication, rapid prototyping, and 3D printing (Wohlers Associates 2013). The current study uses "3D printing" to refer to additive manufacturing, as this version of the term is most popular (Beer 2013; Shah and Basiliere 2014). Additive manufacturing can also be classified according to the technology used to create the final product. This will be discussed in the section below.

### *14.2.1 Types of Additive Manufacturing Technologies*

This emerging technology is based on several methods that have been classified by the ASTM regarding seven different technologies, as shown in Table 14.1.

The use of these technologies depends on the desired product characteristics. Directed energy deposition technology, for example, is used during metal processing to produce high-quality metal parts for the military and aerospace industries. Powder bed fusion is well known for its production of complex geometries, which represents a big advantage for medical manufacturers in products like dental devices and knee implants (Basiliere and Shanler 2014).

**Table 14.1** Additive manufacturing technologies

| Technology | Definition |
|---|---|
| Binder jetting | Particles of powdered material are selectively joined using a liquid bonding agent |
| Directed energy deposition | Materials are fused via melting while they are being deposited; the fusion is achieved using a "focused thermal energy" such as a laser, electron beam, or plasma arc |
| Material extrusion | Material is dispensed through a nozzle or orifice in order to be selectively joined |
| Material jetting | Material in fine droplets is deposited to be selectively joined |
| Powder bed fusion | Particles of powdered material deposited in a bed are selectively fused using a thermal energy |
| Sheet lamination | Sheets of material are bonded to build an object |
| Stereolithography | Liquid photopolymer in a vat is selectively cured via light-activated polymerization |

Adapted from: Basiliere (2015), Shah and Basiliere (2014)

## 14.3 Methodology

The Competitive Technology Intelligence methodology applied in the present study is based on a previous study by Rodríguez et al. (2002). A graphic description of each stage involved in the methodology is presented in Fig. 14.1.

### *14.3.1 Planning*

The competitive technological intelligence process was developed in this stage to determine the requirements and set the main goals, objectives, and activities needed to carry out the current project. The research strategy was determined considering experts from academia and industry and relevant keywords found in databases and software search engines. Next, a scientific analysis of the existing literature on additive manufacturing and competitive technical intelligence was conducted. The purpose of this study was to refine and adapt the previous competitive technological intelligence methodology developed by the first author of the current research, as well as to compare the potential of additive manufacturing with its actual state in Latin America.

### *14.3.2 Information Evaluation*

First, the primary and secondary sources of information at an international level were identified and evaluated. Experts[1] on additive manufacturing and competitive technical intelligence field were consulted, and databases related to the market, science, and technology were identified using meta-Internet searches. Additional sources of information, such as global reports from non-governmental organizations, were also included during the development of the project.

### *14.3.3 Information Collection*

This stage consisted of gathering information from the previously identified secondary sources. The scope of this activity included analysis of global and regional market information regarding additive manufacturing, as well as of scientific and technological information at an international level. For market data, the latest information from specialized databases such as Gartner was consulted using previously determined keywords. Gartner is the world's leading information technology research and advisory

[1]From Tecnológico de Monterrey (Mexico) and Manchester University (UK).

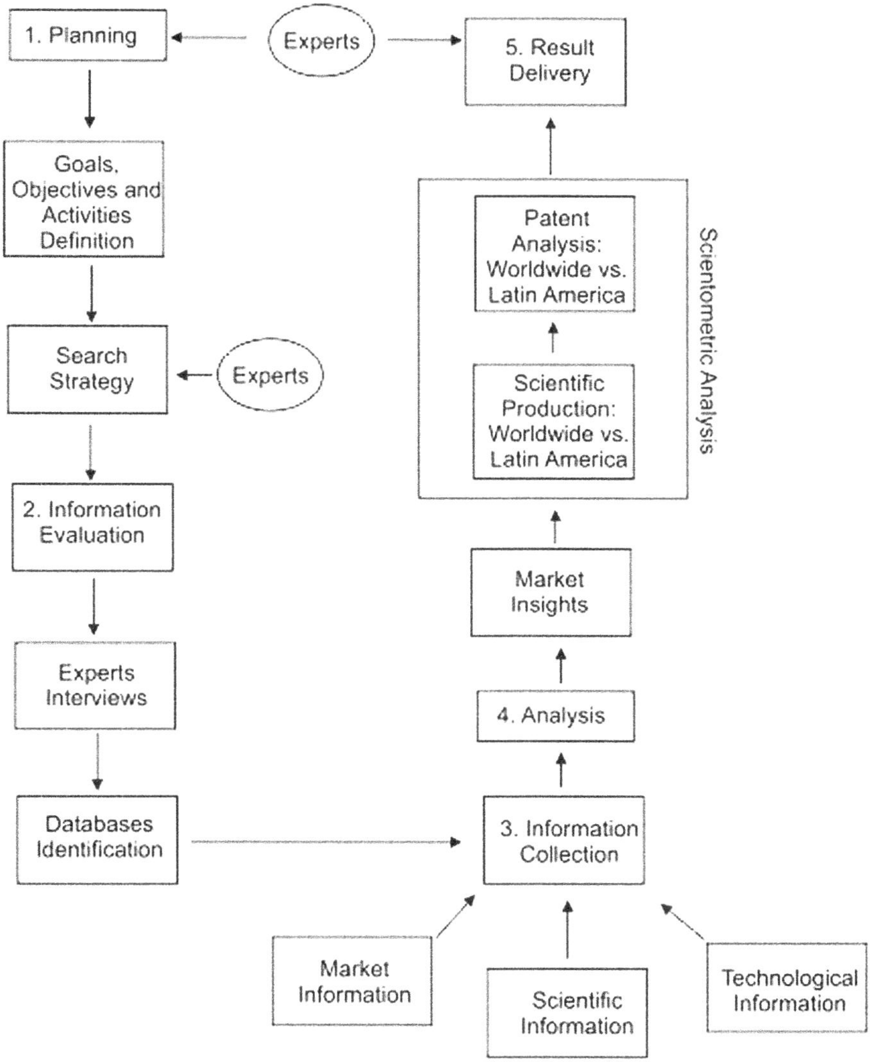


**Fig. 14.1** Competitive technological intelligence methodology (Adapted from: Rodríguez et al. 2002)

company. It offers technology-related insight for strategic decision making (Gartner 2015). Next, roadmaps of advanced manufacturing were identified to support the determination of future development for additive manufacturing and definition of the challenges that the Latin American region could face.

The Scopus database was used to determine the dynamic and focus of the additive manufacturing scientific research in Latin America, covering from 1984 (when the first patent on additive manufacturing technology was applied for) to May 5, 2015 (when this stage of research ended). Scopus is the largest abstract and citation database of peer-reviewed literature. It brings an overview of the world's research output in the fields of science and technology, among other areas (Elsevier 2015). The Patent Insight Pro software was used to perform a network analysis of Latin American organizations developing additive manufacturing research. This software is a patent search and analytics platform that uses advanced text mining algorithms to analyze patents and scientific literature (Gridlogics 2015).

Regarding technological information, a scientometric patent analysis was developed by applying Patseer software. This is a global patent database and research platform with integrated analytic tools covering more than 92 million records from major authorities worldwide (Sinha and Pandurangi 2015). A time period from 1984 to June 17, 2015 was defined to perform the analysis. The search strategy for developing the current research is described below:

- Scientific information
  - TITLE-ABS-KEY ("Additive manufacturing" OR "Additive manufacture")
  - TITLE-ABS-KEY ("3D printing" OR "3D printer" OR "3D print")
  - TITLE-ABS-KEY ("Rapid prototyping" OR "Rapid prototype")
  - TITLE-ABS-KEY ("Additive fabrication")
  - TITLE-ABS-KEY ("Rapid manufacturing" OR "Rapid manufacture")
- Technological information
  - TAC: (Additive Manufacturing) OR (Additively Manufacturing) OR (Additive Manufacture)
  - TAC: (3D printing) OR (Three-dimensional Printing) OR (3D Printer)
  - TAC: (Rapid Prototyping) OR (Deposition Modeling)
  - TAC: (Additive Fabrication) OR (Rapid Manufacturing) OR (Dimensional Printer)

This stage included expert validation and a complementary review with additional information that was found during the study.

### *14.3.4 Analysis*

All data were integrated before the gathered information was analyzed. Because different types of sources were involved, the information first had to be classified to standardize it and obtain a general perspective of the research. Irrelevant information and duplicates were removed. Market data were then analyzed to identify insights related to the research objectives. Scientific information was analyzed using the Scopus (classification by country) and Patent Insight Pro (correlation map) analytical tools. The patent analysis was performed with the aid of Patseer advanced software to determine the patent activity regarding additive manufacturing in Latin America and compare it with global statistics.

### 14.3.5 *Results Delivery*

This stage consisted of the dissemination of results via two methods. First, a comparison of the global standard for additive manufacturing against that currently in place in Latin America was presented to a Mexican research group that focuses on this technology. Second, the most important identified insights were presented in a specialized conference attended by researchers of additive manufacturing and technology mining.

## 14.4 Results

This section presents the results obtained from the research regarding the market and scientific and technological information about additive manufacturing. To begin, a general overview of the technology is provided below.

### 14.4.1 *Prices in the Market*

Insights obtained revealed that 3D printing is a relatively expensive technology as we can see in the following table. However, reductions in price are expected as use of the technology increases. The range of prices of the different additive manufacturing technologies available up to 2014 is presented in Table 14.2.

From Table 14.2, it can be observed that material extrusion is the most affordable technology, followed by stereolithography. Directed energy deposition and fusion bed of powder are the most expensive.

**Table 14.2** Additive manufacturing price range

| Technology | Price lower bound (USD $) | Price upper bound (USD $) |
|---|---|---|
| Material extrusion | 500 | 400,000 |
| Stereolithography | 3000 | 800,000 |
| Inkjet adhesive | 5000 | 800,000 |
| Inkjet material | 20,000 | 600,000 |
| Directed energy deposition | 200,000 | 5,000,000 |
| Fusion bed of powder | 19,800 | 2,000,000 |
| Lamination of sheets | 37,000 | 1,000,000 |

Adapted from: Basiliere (2014a)

#### 14.4.1.1 Cross-Industry Additive Manufacturing

Initially, additive manufacturing was only used for the development of prototypes (Mahamood et al. 2014). Currently, it is utilized in a wide range of industries, especially in areas such as consumer products and electronics (22 % of users of additive manufacturing technology worldwide), the automotive industry (19 %), medical and dental applications (16 %), industrial and business machines (13 %), and aerospace (10 %) (Wohlers Associates 2013). In addition, additive manufacturing is starting to be used to develop products of high added value such as human tissue, food, and airplane wings, and NASA recognizes 3D printing as an important technology for space exploration (Atlantic Council 2013).

Companies' motivators for using 3D printing devices are directly related to the new generation or improvement of products; however, companies also consider other factors like supply chains efficiency (Basiliere 2014a). The adoption of additive manufacturing accelerates the process of product commercialization; it pushes production to the customer and allows products to be more innovative in terms of design, faster production, etc. (Rodríguez et al. 2014).

#### 14.4.1.2 Impact of Additive Manufacturing Technology in Industry

3D printing enables the creation of products in a faster and, in some cases, more affordable way. It is even possible to manufacture complex geometries and minimize inventory as production is pushed to the point of consumption (Cearley et al. 2015). At present, the range of applications for additive manufacturing among companies is varied and still growing as the technology becomes more specialized (Plummer et al. 2014). In addition, the emergence of new types of businesses as a consequence of the advantages of the additive manufacturing's special capabilities is expected. There are already successful cases of companies that could not have arisen without the prior existence of additive manufacturing; for example, Align Technology uses this technology to manufacture plastic aligners to replace metal orthodontia brackets (Wohlers Associates 2013). Moreover, global manufacturers such as General Electric, Boeing, and Ford are already using 3D printing machines to produce critical parts for airplanes, automobiles, and wind turbines (Atlantic Council 2013).

Additive manufacturing allows the reinvention of many old products and is expected to lead to the development of innovative new ones. In the future, there will not be limitations in design as additive manufacturing will enable people to print anything that can be modeled by a computer. In fact, it is expected that this innovative technology will create new industries and professions (Campbell et al. 2011). Despite this being an increasingly global technology, its use has yet been limited in developing countries such as those in Latin America, as will be discussed below.

#### 14.4.1.3 Additive Manufacturing Incursion in Latin America

3D printing can be of great use, particularly for regions that lack significant production capacity and must depend on imports even for basic consumer goods. The cost of establishing a complete 3D printing facility is approximately USD $10,000, a much more feasible amount than what is required to set up a conventional factory (Atlantic Council 2013). In Latin America, interest in additive manufacturing technology is increasing. However, its adoption is in an emerging phase. A recent report grouped Latin American countries in the "other countries" section of an account of the additive manufacturing systems installed from 1988 through to the end of 2012; this entire "other" section represented only 12 % of the global total (Wohlers Associates 2013). The following are some examples of Latin American companies that installed additive manufacturing systems during this time frame:

- Thinker Thing Company, based in Santiago, Chile, developed an innovative process that allows consumers to design real-world objects using their own preferences. After the design stage, clients simply press the "print" button and their object, made of the material of their choice, arrives in the post. However, the company does not utilize its own 3D printers; it uses the 3D printing services of a US-based company (Thinker Thing 2015).
- Brazil is gaining experience with 3D printing mainly in the automotive sector. Although important multinational additive manufacturing companies have entered the Brazilian market, there are two local companies that manufacture their own devices using 3D printing (Wohlers Associates 2013):
  - Metamaquina is a Brazilian company that develops and produces low-cost 3D printers and 3D printing materials for its national market. The company also has laboratories where clients can learn to model and prototype objects (Metamaquina 2015).
  - Cliever also develops and produces 3D printers and 3D printing materials. In addition, the company has developed a software for modeling objects, and it offers other related accessories (Cliever 2015).
- Kikai Labs is an Argentine company that develops 3D printers and offers printing services. It also sells 3D printing materials produced in association with a regional material manufacturer. This company has won national awards for its entrepreneurial efforts (Kikai Labs 2015).

Additive manufacturing is one of the technologies with the greatest potential for improving Latin American education. A major adoption of this technology in the education field is expected to take place in the next three years, mainly for the development of educational prototypes. For example, students of mechanical and electrical engineering from Universidad de Piura, Peru utilized an affordable 3D printer to build car prototypes with educational purposes (Johnson et al. 2013). Furthermore, the Advanced Manufacturing Group of the Tecnológico de Monterrey, Mexico is devoting important research efforts to additive manufacturing technology (Cantú 2015). They also offer 3D printing services to companies looking for customized prototypes.

Important actions have also been taken by the Latin American healthcare industry. As an example, Brazil has already used 3D printing to produce patient-specific cranial adaptive prostheses for skull injuries. Uruguay is involved in a partnership with the Department of Capital and Technology-Intensive Sectors of Brazil for innovation in this field (United Nations 2015). In both cases, important efforts to develop this technology are being carried out.

At present, Latin America does not have a significant presence in the additive manufacturing field. However, while the presence of this technology is still low compared to that in other regions, its influence has begun to permeate the development of a variety of start-ups.

### *14.4.2 Additive Manufacturing Scientific Production in Latin America*

Scientific production related to additive manufacturing technology in Latin America is in its infancy. This is demonstrated by the small number of papers published by individuals or researchers assigned to Latin American organizations. The results of the analysis of these papers are presented in Table 14.3.

**Table 14.3** Additive manufacturing scientific research developed by Latin American individuals or organizations (own elaboration based on Scopus analysis)

| Keyword | Number of papers (global) | Number of papers (Latin America) | Latin America (%) | Number of papers (Brazil) | Number of papers (Mexico) | Number of papers (other Latin American countries) |
|---|---|---|---|---|---|---|
| Additive manufacturing OR additive manufacture | 2885 | 50 | 1.7 | 41 | 6 | 3 |
| 3D printing OR 3D printer OR 3D print | 3914 | 81 | 2.1 | 55 | 13 | 13 |
| Rapid prototyping OR rapid prototype | 16,584 | 376 | 2.3 | 267 | 40 | 69 |
| Additive fabrication | 367 | 3 | 0.8 | 2 | 1 | 0 |
| Rapid manufacturing OR Rapid manufacture | 1499 | 26 | 1.7 | 22 | 4 | 0 |
| Total | 25,249 | 536 | 2.1 | 387 | 64 | 85 |

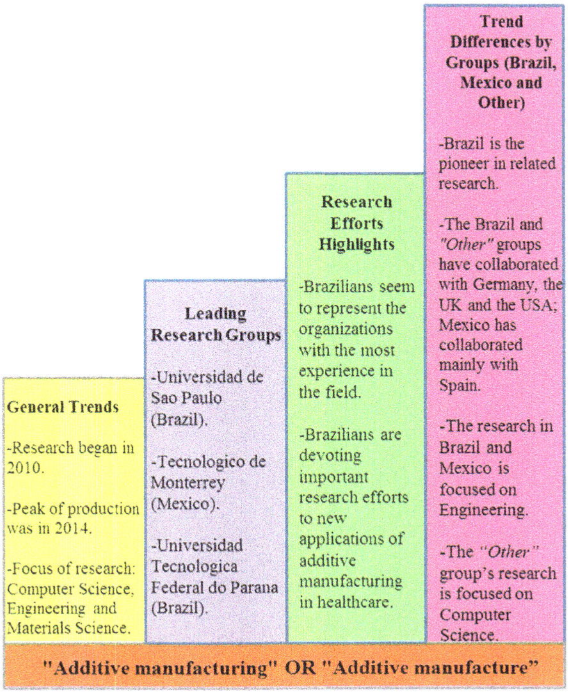


**Fig. 14.2** Latin American scientific production of additive manufacturing based on "additive manufacturing" OR "additive manufacture" queries (own elaboration based on Scopus analysis)

This information demonstrates that only 2 % of the global scientific production related to additive manufacturing occurs in Latin American organizations, of which rapid prototyping is the field with the highest number of publications. As shown, 536 additive manufacturing papers were identified within the Latin American region; deeper analyses of these results are presented in Figs. 14.2, 14.3, 14.4, 14.5, and 14.6. These data were analyzed using the advanced tools of the Scopus database. Each figure presents the general trends, key research collaborations (country level), main research efforts, and differences by group (Brazil, Mexico, and Other) for each search strategy.

An analysis of the collaborations with Latin American organizations was made with consideration of the results of Table 14.3 (Brazil, Mexico, and Other). The following results were obtained (Table 14.4).

Now we drive our attention on the collaboration dynamics regarding the highest number of publications developed by Latin American organizations. In this respect, Fig. 14.7 illustrates a collaboration network that was established using the "rapid prototyping" results from the previous section (376 papers in total). The Renato Archer Center in Brazil conducts joint research with organizations such as the Imaging Department of Sao Paulo State University (Brazil) and Pontific Catholic University of Rio Grande do Sul (Brazil), as shown in the yellow circle in Fig. 14.7.

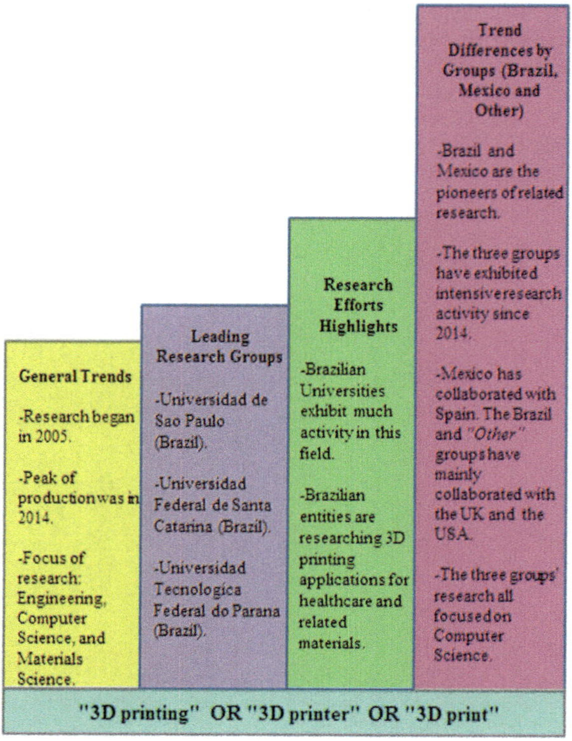


**Fig. 14.3** Latin American scientific production of additive manufacturing based on "3D printing" OR "3D printer" OR "3D print" queries (own elaboration based on Scopus analysis)

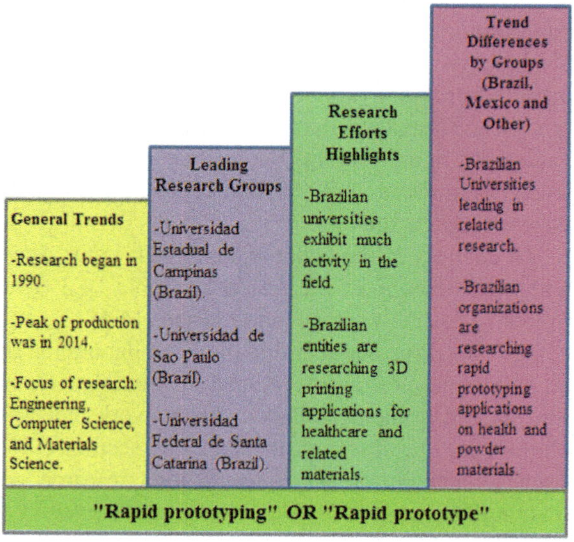


**Fig. 14.4** Latin American scientific production of additive manufacturing based on "rapid prototyping" OR "rapid prototype" queries (own elaboration based on Scopus analysis)

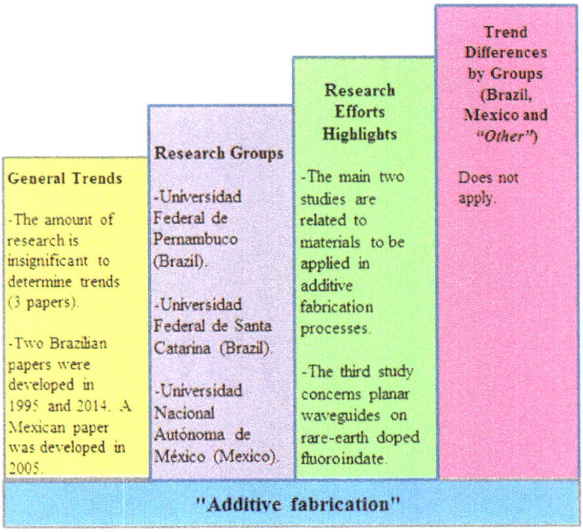


**Fig. 14.5** Latin American scientific production of additive manufacturing based on "additive fabrication" query (own elaboration based on Scopus analysis)

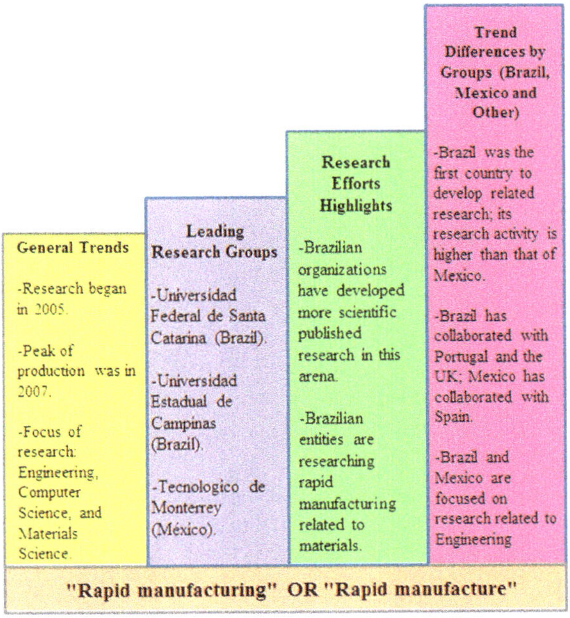


**Fig. 14.6** Latin American scientific production of additive manufacturing based on "rapid manufacturing" OR "rapid manufacture" queries (own elaboration based on Scopus analysis)

Shown in the purple circles are Tsinghua University (China)'s research network that includes various Latin American organizations, such as Universidad Federal de Santa Catarina (Brazil), Universidad de Brasilia (Brazil), Instituto Médico ENERI (Argentina), and the Department of Surgery at Pontificia Universidad Católica Do Río Grande (Brazil). The green circle indicates a dense collaboration between organizations that include the University of Campinas (Brazil), Brazilian Telecom

**Table 14.4** Examples of global organizations collaborating with each Latin American group (Brazil, Mexico, and Other) (own elaboration based on Patent Insight Pro analysis)

| Keyword | Brazil | | Mexico | | Other Latin American countries | |
|---|---|---|---|---|---|---|
| | Brazilian organization | Collaborator | Mexican organization | Collaborator | "Other" organization | Collaborator |
| "Additive manufacturing" OR "additive manufacture" | Federal Institute of Santa Catarina | Aachen University, Laser Technology (Germany) | Tecnológico de Monterrey | Universitat de Girona (Spain) | University of Antioquia-GIB-Eafit (Colombia) | University of Buffalo (US) |
| | Escola de Matemática Aplicada FGV/EMAp | University of Brighton (UK) | | | Universidad Autonoma de Occidente (Colombia) | Rochester Institute of Technology (US) |
| | Federal University of Rio de Janeiro-UFRJ/COPPE | New York University (US) | | | | |
| "3D printing" OR "3D printer" OR "3D print" | Federal University of Santa Catarina | Hewlett Packard Laboratories (US) | Lab. de Tecnologias de Información, CINVESTAV | Universite de Poitiers (France) | Instituto Tecnológico de Buenos Aires (Argentina) | Massachusetts Institute of Technology (US) |
| | Universidad Federal de Bahía (Brazil) and Escola de Matemática Aplicada | University of Brighton (UK) | Tecnológico de Monterrey | Universitat de Girona (Spain) | Universidad de Antioquia (Colombia) | University College London (UK) |
| "Rapid prototyping" OR "Rapid prototype" | Universidade Federal Do Rio De Janeiro | University of Maryland (US) | Universidad de las Americas | Texas A and M University (US) and Sandia National Laboratories (US) | Pontifi cia Universidad Catolica de Chile (Chile) | University of Texas at Austin (US) |
| | São Paulo State University | Instituto Politécnico de Leiria (Portugal) | Metalsa SA | University of Girona (Spain) | Central University of Venezuela (Venezuela) | Polytechnic University of Catalonia (Spain) |

(continued)

**Table 14.4** (continued)

| Keyword | Brazil | | Mexico | | Other Latin American countries | |
|---|---|---|---|---|---|---|
| | Brazilian organization | Collaborator | Mexican organization | Collaborator | "Other" organization | Collaborator |
| "Additive fabrication" | Universidad Federal de Pernambuco | Without collaboration | Universidad Nacional Autonoma de Mexico | University of California at San Diego (US) and Osram Sylvania Central Research (US) | Does not apply | Does not apply |
| | Universidade Federal de Santa Catarina and Instituto Federal de Santa Catarina | Without International collaboration | | | | |
| "Rapid manufacturing" OR "Rapid manufacture" | Universidade Federal de Uberlândia | Instituto Superior Tecnico, ICEMS (Portugal) | Tecnológico de Monterrey | University of Girona (Spain) | Does not apply | Does not apply |
| | Universidade Federal de Santa Catarina | Loughborough University (UK) | | | | |

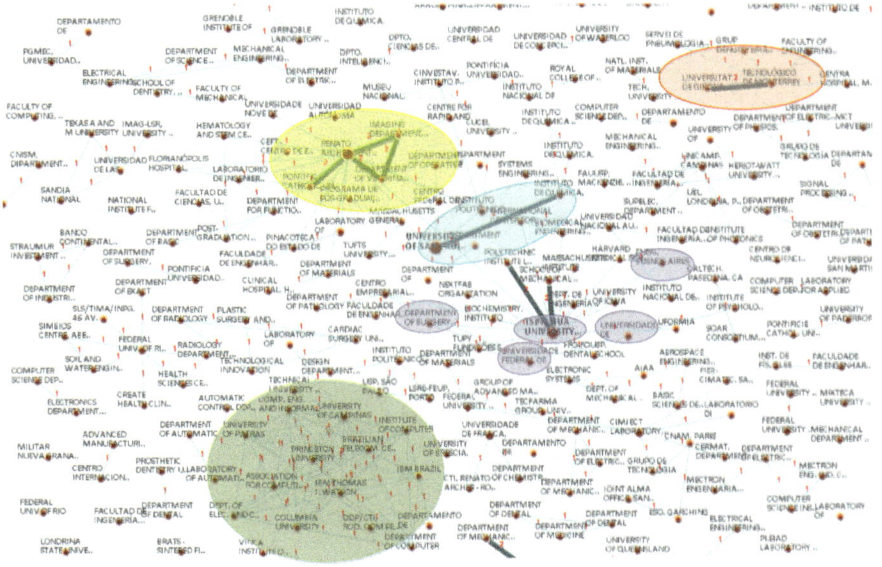

**Fig. 14.7** Key research network on the Latin American scientific production of additive manufacturing. *Circles* indicate research networks, and lines show the number of joint papers (own elaboration based on Patent Insight Pro analysis)

(Brazil), IBM Brazil (Brazil), Columbia University (USA), IBM Thomas Watson Research Center (USA), and the University of Crete (Greece). The orange circle indicates the relationship between Tecnologico de Monterrey (México) and Universidad de Girona (Spain). Finally, the partnership between the University of Sao Paulo (Brazil) and the Instituto de Química of Universidad de Campinias (Brazil) is presented in the light blue circle.

Using previous analyses, research trends regarding additive manufacturing were determined as follows:

- Research on additive manufacturing and its related terms began in Latin America in 1990. Production has spiked in recent years, particularly since 2014.
- Research is focused mainly in the engineering field, followed by the computer (process optimization) and material sciences.
- Brazil has pioneered developing additive manufacturing research since 1990, exhibiting intensive activity.
- Some collaboration efforts have developed between Latin America and countries including Spain, the UK, and the USA. However, their joint research production is low compared to that by Latin American organizations alone.
- Most research leaders are from Brazilian universities, such as Universidad de Sao Paulo, Universidad Estadual de Campinas, and Universidad Federal de Santa Catarina.

- Brazil has conducted intensive research of the technology compared to work conducted in other Latin American countries. The strong national collaboration among Brazilian organizations could be the result of the key research network analysis that has been developed. The main innovative research focus of the collaborating organizations is on healthcare applications (e.g., bio-printing, scaffolds, dental devices, and implants).

### 14.4.3 Additive Manufacturing Patent Production in Latin America

Although there is growing interest in additive manufacturing in Latin America, the rate of related patent production is not significant. This appreciation remains in the low percentage (0.02 %) of additive manufacturing inventions developed and patented by Latin American organizations or individuals compared to worldwide production rates. These figures are shown in Table 14.5.

Of the three obtained patents detailed in Table 14.5, the Mexican invention involves the use of 3D printing to manufacture centrifuge support; the Brazilian patent focuses on using 3D printing to develop bone scaffolds (patented by two different authorities, Brazil and WIPO); and the Chilean patent involves the development of a 3D printing support for 2D printing. This last invention has been patented by 14 different patent authorities in Mexico, Russia, Argentina, Canada, Europe, Japan, and others.

### 14.4.4 Additive Manufacturing and Challenges for Latin America

Countries around the world are placing additive manufacturing at the core of their strategies for development. Important investments are being made in the research and development of additive manufacturing technologies as part of the governmental strategies in the USA, Japan, China, Singapore, South Africa, Australia, Belgium, France, Germany, the Netherlands, Poland, Portugal, Spain, Sweden, the UK, and others (European Commission 2014). The European Factories of the Future Research Association (2013) developed a roadmap by which to determine the ongoing importance of manufacturing in the European economy and develop a vision for this sector up to the year 2030. To respond to world challenges, it is expected that factories will have to be green and sustainable, use small amounts of resources, consume little energy, and produce zero emissions and waste. In fact, the Association previously mentioned specifies the importance of using new manufacturing technologies, such as additive manufacturing, to develop sustainable business.

**Table 14.5** Additive manufacturing patent production in Latin America (own elaboration based on Patseer analysis)

| Keyword | Number of patents (global) | Number of patents (Latin America) | Latin America (%) | Number of patents (Brazil) | Number of patents (Mexico) | Number of patents (Chile) | Number of patents (other Latin American countries) |
|---|---|---|---|---|---|---|---|
| Additive Manufacturing OR additively manufacturing OR additive manufacture | 1957 | 0 | 0.00 | 0 | 0 | 0 | 0 |
| 3D printing OR Three-dimensional Printing OR 3D Printer | 5417 | 2 | 0.04 | 0 | 1 | 1 | 0 |
| Rapid prototyping OR deposition modeling | 4468 | 1 | 0.02 | 1 | 0 | 0 | 0 |
| Additive fabrication OR rapid manufacturing OR dimensional printer | 2022 | 0 | 0.00 | 0 | 0 | 0 | 0 |
| Total | 13,864 | 3 | 0.02 | 1 | 1 | 1 | 0 |

The aerospace, automotive, and electronic sectors are the most propitious European industries to develop additive manufacturing business in the future. By 2020, these industries are expected to be characterized by customization; this will foster the utilization of additive manufacturing technology, mainly to develop more innovative products. However, some requirements should be considered to ensure the correct adoption of additive manufacturing in producing such innovations; these include process stability, certifications, design rules, and the ability to control part quality during the production process (Gausemeier 2013).

Experts in the additive manufacturing field predict a promising future for this technology. In the next 10–15 years, it is expected that this technology will become an integral component of available manufacturing processes, especially for metallic parts; machines producing 3D mechanical and electrical components are expected to become commonplace. Moreover, the mass production of low-cost 3D printers is anticipated to meet the projected home and office demands of consumers. Bio-printing also represents a highly promising application of additive manufacturing. Functional tissue and organs may be able to be fabricated using these methods within the next 10–15 years. Designers will not have limitations; they will be able to meet every product requirement in terms of design and materials. It is expected that there will be large factories, regional assembly centers, and local 3D printing shops to support the growing industry. To help these predictions come true, governments should create research programs that support such developments. Experts have determined that in addition to focusing research efforts on new additive manufacturing materials and processes, it will be necessary to investigate design and implementation aspects (Bourell et al. 2009).

As the global acceptance of additive manufacturing technology becomes a reality, Latin America will have to adopt the technology as well. However, some challenges must be addressed for this region to enter this worldwide dynamic. At this moment, additive manufacturing is characterized by low volume production and high prices. The cost of machines, materials, and maintenance inhibits the wider adoption of the technology (Bourell et al. 2009). In this regard, Latin America is at more of a disadvantage than other regions, partly because Latin American countries are characterized by low purchasing power. In 2014, the Latin American Gross Domestic Product (GDP) per capita based on purchasing power parity (PPP) was 12,443 USD (Comisión Económica para América Latina y el Caribe 2014). This is low compared to countries such as the USA, France, and Japan, which have GDP per capita based on PPPs of 53,000, 37,500, and 36,000 USD, respectively (The World Bank 2013).

For Latin America to develop sustainably, investment in the production and export of advanced technologies is necessary (Atilano et al. 2015); this includes additive manufacturing. With the exception of Brazil, Chile, and Mexico, Latin American economies exhibit low technological development due to several factors, including the restricted access to knowledge with high added value, low productivity, consumerism (high imports of technological goods and services), unqualified human resources, low public and private investment in research and development activities, and a lack of policies to improve the environment for dissemination and technological innovation. Patent production is relatively low and focuses on the traditional sectors of manufacturing, chemicals, petroleum, and steel. This could be the result of a lack of national policies for the management of intellectual property systems (Serrano 2014).

In the more advanced countries of Latin America, such as Mexico, there is an awareness of the importance of developing advanced manufacturing technologies to achieve sustainable economic growth. The Mexican government has already developed advanced manufacturing roadmaps that focus on talent management and on boosting the capabilities of the design, development, and engineering of processes, products, and materials produced in Mexico (ProMéxico 2011). It is hoped that most other Latin American countries will follow the example of the more advanced ones; however, such progress takes time. In the meantime, there is a need for strategy if these countries are to face the outlined challenges successfully.

The results of this research offer valuable information about the state of additive manufacturing in Latin America. There is a growing interest in this technology in the region. Global additive manufacturing developers might find the insights achieved in this study useful; they might, for example, use this study's results to identify the leading organizations researching additive manufacturing technology, their research focus, key research networks, organizations that are already commercializing these processes, and related patent inventions. Such information, as well as information regarding what potential competitors are already doing, could be useful to developers who wish to devise strategies by which to introduce and implement this technology in Latin America. In summary, the results of this research are of great strategic value in the additive manufacturing industry.

## 14.5 Conclusions

Additive manufacturing or 3D printing is an innovative technology that is changing conventional production processes and gaining ground in worldwide markets. The most valuable characteristic of this technology is its capacity to produce high-quality objects with complex geometries. Because of these advantages, successful and innovative business has been developed around additive manufacturing.

There are high expectations for the acceptance of this technology, not only in developed countries but also in developing ones. However, Latin America seems to be accepting the technology at a slower pace, as demonstrated by the low amount of scientific literature and patent production related to additive manufacturing in these countries. Results obtained from the present study's scientific literature production analysis indicate that while 25,249 papers related to additive manufacturing were developed worldwide between 1984 and 2015 (through May 5), only 536 were produced by individuals or researchers assigned to Latin American organizations (i.e., 2 %, the majority of which were from Brazil and Mexico). Further, the results of the patent analysis conducted in this review reveal no significant patent production related to additive manufacturing currently occurring in Latin America. Of the worldwide data regarding patent production related to this technology (13,864 records) from 1984 to 2015 (through June 17), only 0.02 % correspond to Latin America.

There is an observable global trend toward the development and adoption of technologies that make manufacturing processes more sustainable. Advanced countries, such as the USA, Japan, and France, have assigned great importance to this subject by developing long-term technology strategies that will enable them to maintain and increase their competitiveness. Results of this research show that Latin America is far behind that which is happening globally. There is an emerging interest in researching and developing new technologies like additive manufacturing; however, no strategic plans in the form of roadmaps have been identified for this technology in the Latin American region. Latin American countries face significant challenges regarding the adoption of advanced technologies such as additive manufacturing; such obstacles include the scant access to high-value knowledge and inadequate governmental policies to stimulate strong technological innovation. However, countries such as Brazil and Mexico demonstrate the greatest progress in the additive manufacturing adoption process; thus, it may still be that other countries of the region will follow their example.

**Acknowledgments** We thank all of those who directly and indirectly provided academic support for the development of this research. This research was supported by Tecnológico de Monterrey through Centro de Innovación en Diseño y tecnología and its research group in Advanced Manufacturing. In addition, we would like to thank Andres Avila (research assistant) for his valuable time and support.

## References

Atilano, A., Mercado, J., & Casanova, H. (2015). Indicadores de innovación tecnológica de los países de américa latina y el caribe. Banco de Desarrollo de América Latina. http://scioteca.caf.com/handle/123456789/724. Accessed 11 June 2015.

Atlantic Council. (2013). Envisioning 2030: US strategy for the coming technology revolution. The Atlantic Council of the United States. http://www.atlanticcouncil.org/images/publications/Envisioning_2030_US_Strategy_for_the_Coming_Tech_Revolution_web.pdf. Accessed 01 Apr 2015.

Basiliere, P. (2014a). Market guide for 3D printing. Gartner. http://www.gartner.com/document/2934020?ref=QuickSearch. Accessed 25 Mar 2015.

Basiliere, P. (2014b). Technology overview for stereolithography 3D printing. Gartner. http://www.gartner.com/document/2755618?ref=QuickSearch. Accessed 25 Mar 2015.

Basiliere, P. (2015). Technology overview for binder jet 3D printing. http://www.gartner.com/document/2981218?ref=QuickSearch. Accessed 25 Mar 2015.

Basiliere, P., Halpern, M., Burt, M., & Shanler, M. (2014). Cool vendors in 3D printing. Gartner. http://www.gartner.com/document/2726917?ref=QuickSearch. Accessed 25 Mar 2015.

Basiliere, P., & Shanler, M. (2014). Hype cycle for 3D printing. Gartner. http://www.gartner.com/document/2803426?ref=QuickSearch. Accessed 25 Mar 2015.

Beer, N. (2013). Additive manufacturing. Turning mind into matter. Sierra College Center for Applied Competitive Technologies (CACT). http://sierracollegetraining.com/uploads/201307/sierra-college-cact-additive-manufacturing-report-and-recommendations-may2013.pdf. Accessed 11 June 2015.

Bourell, D., Leu, M., & Rosen, D. (2009). Roadmap for additive manufacturing identifying the future of freeform processing. The University of Texas at Austin, Laboratory for Freeform Fabrication Advanced Manufacturing Center, United States. https://wohlersassociates.com/roadmap2009.pdf. Accessed 11 June 2015.

Campbell, T., Williams, C., Ivanova, O., & Garrett, B. (2011). Could 3D printing change the world? Technologies, potential, and implications of additive manufacturing. Atlantic Council. http://3dprintingindustry.com/wp-content/uploads/2013/05/Atlantis-Report-on-3D-printing.pdf. Accessed 11 June 2015.

Cantú, F. (Ed.). (2015). Strategic research groups. Tecnológico de Monterrey. http://www.sitios.itesm.mx/webtools/research/ITESMResearchGroupsBrochure2015.pdf. Accessed 11 June 2015.

Cearley, D., Walker, M., & Blösch, M. (2015). The top 10 strategic technology trends for 2015. Gartner. http://www.gartner.com/document/2964518?ref=QuickSearch. Accessed 25 Mar 2015.

Chen, S. (2015). This is just the beginning: China approves world's first 3D-printed hip joint for general use. Science and Research. http://www.scmp.com/tech/science-research/article/1854369/just-beginning-china-approves-worlds-first-3d-printed-hip. Accessed 08 Dec 2015.

Cliever. (2015). Productos. http://www.cliever.com.br/. Accessed 08 June 2015.

Comisión Económica para América Latina y el Caribe. (2014). CEPAL publica estimaciones de las paridades de poder adquisitivo de los países de la región. http://www.cepal.org/es/noticias/cepal-publica-estimaciones-de-las-paridades-de-poder-adquisitivo-de-los-paises-de-la-region. Accessed 09 June 2015.

Elsevier. (2015). Scopus. http://www.elsevier.com/solutions/scopus. Accessed 08 June 2015.

European Comission. (2014). Additive manufacturing in FP7 and horizon 2020. Report from the EC workshop on additive manufacturing held on 18 June 2014. http://www.rm-platform.com/linkdoc/EC%20AM%20Workshop%20Report%202014.pdf. Accessed 11 June 2015.

European Factories of the Future Research Association. (2013). Factories of the future. Multi-annual roadmap for the contractual PPP under horizon 2020. http://www.effra.eu/. Accessed 11 June 2015.

Gartner. (2015). About Gartner. http://www.gartner.com/technology/about.jsp. Accessed 08 June 2015.

Gausemeier, J. (2013). Thinking ahead the future of additive manufacturing. Innovation roadmapping of required advancements. Direct Manufacturing Research Center, University of Paderborn., Germany. http://dmrc.uni-paderborn.de/fileadmin/dmrc/Download/data/DMRC_Studien/DMRC_Study_Part_3.pdf. Accessed 11 June 2015.

Gridlogics. (2015). Patent insight pro. http://gridlogics.com/?portfolio=patent-insight-pro-by-gridlogics. Accessed 07 Oct 2015.

Hull, C. (1986). Apparatus for production of three-dimensional objects by stereolithography. US patent 4575330.

Johnson, L., Adams Becker, S., Gago, D. Garcia, E., & Martín, S. (2013). NMC perspectivas tecnológicas: educación superior en américa latina 2013–2018. Un análisis regional del informe horizon del NMC. Austin, Texas: The New Media Consortium. http://www.oei.es/noticias/spip.php?article13253. Accessed 11 June 2015.

Kikai Labs. (2015). Servicios. http://kikailabs.com.ar/servicios/. Accessed 08 June 2015.

Mahamood, R., Akinlabi, E., Shukla, M., & Pityana, S. (2014). Revolutionary additive manufacturing: An overview. *Lasers in Engineering (Old City Publishing), 27*, 161–178.

Metamaquina. (2015). Metamaquina. http://metamaquina.com.br/. Accessed 09 June 2015.

Plummer, D., Fiering, L., Dulaney, K., McGuire, M., Da Rold, C., Sarner, A., et al. (2014). Top 10 strategic predictions for 2015 and beyond: Digital business is driving 'big change'. Gartner. http://www.gartner.com/document/2864817?ref=lib. Accessed 25 Feb 2015.

ProMéxico. (2011). Diseñado en México. Mapa de ruta de diseño, ingeniería y manufactura avanzada, ProMéxico: México DF. http://www.promexico.gob.mx/documentos/mapas-de-ruta/MRT-Manufactura-Avanzada.pdf. Accessed 11 June 2015.

Rodríguez, M., Cruz, P., Avila, A., Olivares, E., & Arellano, B. (2014). Strategic foresight: determining patent trends in additive manufacturing. *Journal of Intelligence Studies in Business, 4*(3).

Rodríguez, M., Eddy, A., & Garza, R. (2002). Industry/university cooperative research in competitive technical intelligence: a case of identifying technological trends for a Mexican steel manufacturer. *Research Evaluation, 11*(3), 165–173.

Scott, J., Gupta, N., Weber, C., Newsome, S., Wohlers, T., & Caffrey, T. (2012). Additive manufacturing: Status and opportunities. *Science and Technology Police Institute.*

Serrano, E. (2014). Desarrollo tecnológico y Brecha tecnológica entre países de América Latina. *Ánfora, 21*(36), 41–65.

Shah, Z., & Basiliere, P. (2014). Technology overview for powder bed fusion. Gartner. http://www.gartner.com/document/2830619?ref=QuickSearch. Accessed 25 Mar 2015.

Sinha, M., & Pandurangi, A. (2015). Guide to practical patent searching and how to use PatSeer for patent search and analysis. Gridlogics Technologies Pvt. Ltd. http://patseer.com/2015/04/e-book-patent-searching/. Accessed 29 Apr 2015.

The World Bank. (2013). GDP per capita, PPP (current international $). http://data.worldbank.org/indicator/NY.GDP.PCAP.PP.CD. Accessed 09 June 2015.

Thinker Thing. (2015). About. http://thinkerthing.com/about/. Accessed 08 June 2015.

United Nations. (2015). Exploring advanced technologies in Latin America. http://www.unido.org/news/press/exploring-advanced-t.html. Accessed 09 June 2015.

U.S. Department of Veterans Affairs. (2014). DEKA advanced prosthetic arm gains FDA approval. Office of research and development. http://www.research.va.gov/currents/spring2014/spring2014-34.cfm. Accessed 08 Dec 2015.

Wohlers Associates. (2013). Wohlers report 2013, additive manufacturing and 3D printing state of the industry, annual worldwide progress report. Fort Collins, Colorado, USA: Wohlers Associates, Inc. https://wohlersassociates.com/2013report.htm. Accessed 11 June 2015.

# Chapter 15
# The Application of Social Network Analysis: Case of Smart Roofing

**Tugrul U. Daim, Monticha Khammuang and Edwin Garces**

**Abstract** The use of social network analysis (SNA) becomes popular in social science research in the recent years. It is a practical application because it helps organizations to have better conceptualized and new understandings of the interactions. It could help organizations interpret and understand complexity, systems, pattern of changes, and structure of interactions. Moreover, SNA applications have been applied in many complicated fields to identify knowledge leaders in organizations, measure collaboration of teams, illustrate the hidden patterns of structure, and exploring the paths of interactions. In addition, many software programs were developed for personal or limited distribution by mathematicians, sociologists, graph theorists, and information technology specialists enabling SNA to facilitate the analysis of data and the creation of sociograms easier than before. Applying SNA in organizations could benefit many internal activities. It could help organizations to identify the group of experts for technology roadmapping (TRM) or R&D related activities, to know who the most appropriate expert for future collaboration may be, and to see the pattern of the interactions for future R&D planning. This chapter proposes an analysis of smart roofing using SNA to identify the group of experts, the interactions among experts, and the patterns of these interactions to help researchers to gain a better understanding of the current situation of smart roofing research and development programs and also to help them to prepare related future plans in order to promote the progress of smart roofing research and development programs.

**Keywords** Social network analysis · SNA · Practical approach · The application of SNA · The use of SNA

---

T.U. Daim (✉) · M. Khammuang · E. Garces
Department of Engineering and Technology Management, Portland State University, 900 SW 4th Avenue, Lower Level, Suite 50-02, Portland, OR 97201, USA
e-mail: tugrul.u.daim@pdx.edu


T.U. Daim et al. (eds.), *Anticipating Future Innovation Pathways Through Large Data Analysis*, Innovation, Technology, and Knowledge Management,
DOI 10.1007/978-3-319-39056-7_15

## 15.1 Introduction

The use of social network analysis (SNA) is new for the evaluation in social science. It has yet to be completely studied in this field. The use of SNA has gradually risen over the past ten to fifteen years (Durland and Fredericks 2005).

There are three factors related to the increasing usage of SNA. Firstly, SNA is a powerful tool that helps organization to have a clear understanding of interactions. After the Dot Com boom in 1990s, online networking tools for individuals to create and explore their personal and business networks grew up rapidly. Some companies mine data and sell data back to other business. Many companies such as IBM, Accenture, and Mars are also using SNA to determine the influencers, the relationships among teams/projects, and patterns of interaction among teams (Durland and Fredericks 2005).

SNA was developed for understanding the complexity and system of the networks so it could help organizations to do the evaluation of designs and program development effectively because it could explain the complexity and interactional nature of structures (Durland and Fredericks 2005).

## 15.2 Basic Concepts in SNA and Their Relation to Expert Identification

### *15.2.1 What Is Social Network Analysis?*

Social network analysis (SNA) is a general approach for investigating social structures or networks and the relationship among them (Wellman and Berkowitz 1988). It represents a concept of social structure in terms of a network connecting members and channeling resources together. Moreover, it focuses on the characteristics of the network rather than on the characteristics of the individuals and point out a group of networks as personal communities (Wetherell et al. 1994). It also focuses on individual actors making alternatives without considering the behavior of others. This approach neglects the social context of the actor. SNA considers the relationships between actors as the first priority, and individual properties are second priority (Knoke and Kuklinski 1982). Moreover, another important function of SNA is to study how structural regularities influence actors' behavior (Knoke and Kuklinski 1982). It is clear that original application of SNA is to investigate the interactions and we could classify the investigation into two major patterns: the ego network and the global network. The first one is to analyze the network of one person. The second one is to analyze all relations among the participants in that network (White 2000).

SNA has two very unique characteristics which differentiates it from other analysis tools. Firstly, it helps to describe and understand relational data better than others because of its own set of measures and analysis tools. Relational data represent a relationship between two components and also the value of that relationship. SNA focuses on the social context and behavior of relationships among actors rather than on the rational choices individual actors make so this characteristic differentiates SNA from other methodologies (Durland and Fredericks 2005). Another unique character is that other methodologies result in an understanding of importance or the significance of the correlations whereas SNA gives a path into the complexity that often starts with a small thing that opens up into something much bigger. This does not mean that SNA has a never-ending path into an analysis, but it provides many more hidden points into the function of programs (Durland and Fredericks 2005). By the way, the results from SNA are more complex, and the significance from what we found cannot be measured with one statistic.

### *15.2.2 What Are the Components of the Network?*

The main components of the networks are actors and their relations. Actors can be called nodes, vertices, or points. Relations can be called as arcs, edges, or ties (Scott 1988). The picture of components of the network can be seen in Fig. 15.1.

Points in Fig. 15.1 are used to represent the actors, and lines are used to represent relations.

### *15.2.3 How Many Types of Relations Are There Within a Network?*

Generally, relations can be divided into two groups: directed or undirected. In a directed relation (see in Fig. 15.2), the edges have their own direction or arrow. This kind of edges is similar to arcs and can be used as ordered pairs of vertices. We

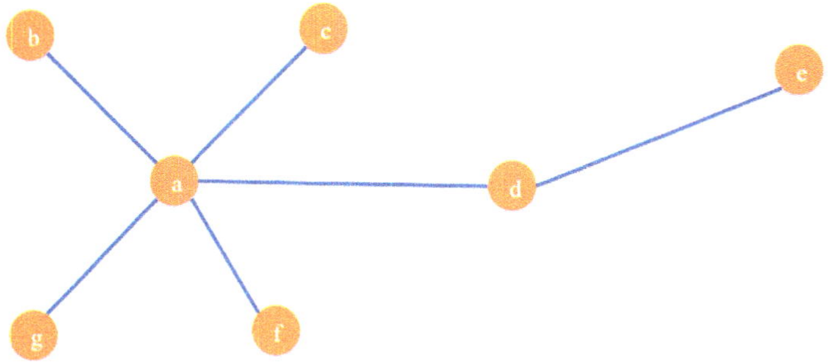


**Fig. 15.1** Components in a network

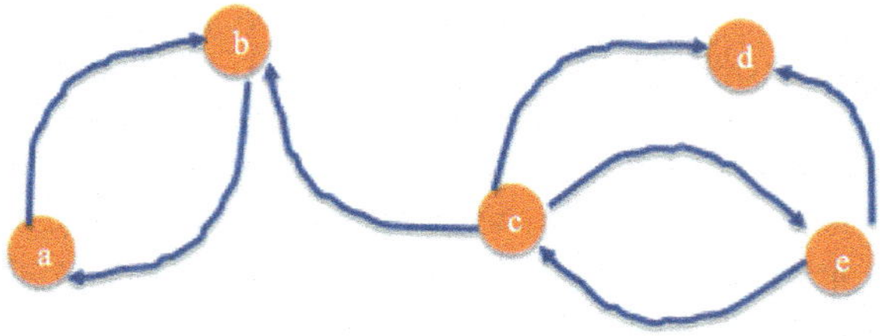


**Fig. 15.2** Directed relation

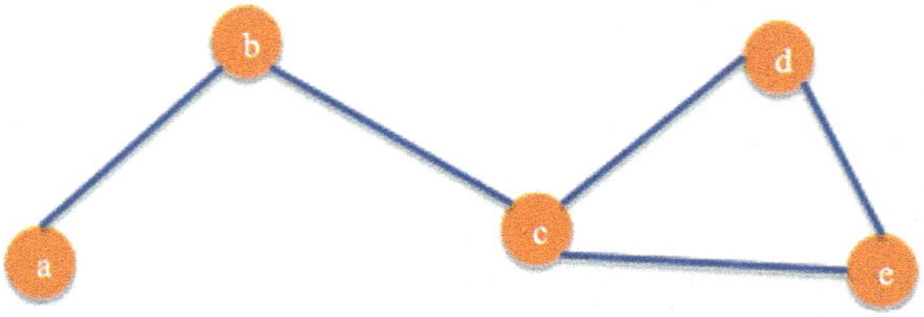


**Fig. 15.3** The undirected relation

use directed relations to show relational phenomena that has a sense of direction (Borgatti and Everett 1997).

In an undirected relation (see in Fig. 15.3), the edges have no order pairs. This kind of relation is used where there is no direction or the direction does not make sense or it is not clear about exactly the direction (Borgatti and Everett 1997).

For general research, we expect that the networks have different structures and have their unique form of relations.

### *15.2.4 What Are the Types of Network Data?*

Data used in SNA consist of an array/table of measurements. The rows of the array could represent the cases, subjects, or observations. Each cell of the array shows a relationship between the actors. We could see the example of network data in Table 15.1, which describes the network of friendship relations among four people (Hanneman and Riddle 2005).

The difference between conventional data and network data is that conventional data focus on actors and attributes, whereas network data focus on actors and

**Table 15.1** Example of social network data

| Who reports liking whom? | | | | |
|---|---|---|---|---|
| Chooser | Choice | | | |
| | Bob | Carol | Ted | Alice |
| Bob | – | 0 | 1 | 1 |
| Carol | 1 | – | 0 | 1 |
| Ted | 0 | 1 | – | 1 |
| Alice | 1 | 0 | 0 | – |

relations. This difference causes the researchers to design the research before collecting data in order to conduct sampling, develop measurement, and handle the resulting data.

We can separate social network data into two groups: ***1-mode and 2-mode***. The first group shows edges based on directed contact between actors in the network, and all of the nodes are of the same type such as people, organizations, and ideas, whereas the 2-mode data show nodes from two different classes and ties are across classes (Borgatti and Everett 1997; Hanneman and Riddle 2005).

### 15.2.5 How to Classify Level of Network for Investigation?

We can classify the level of networks into three levels: ego network, partial network, or global network.

***Ego network***: This level of network focuses on the individual, rather than on the whole network (see in Fig. 15.5). At this level, we collect information from the connections where the actors are connected to each focal ego. This information is useful for researchers because it could enable them to see the incomplete picture of the whole network and to understand how networks affect individuals. However, researchers can obtain only some information from ego network level. In ego networks, we cannot measure the overall density of the population. If we have some reasonable explanation to explain about alters in terms of their social roles, rather than as individuals, ego networks can tell us more about their local social structures such as alters connected to an ego by a friendship relation as kin, co-worker, member of the same church, co-author, etc., see in Fig. 15.5 (Hanneman and Riddle 2005).

***Partial network***: There are some cases that we cannot track down the full network. This partial network is an alternative approach to begin with a selection of focal egos and identify connected egos. Then, we determine the first stage of egos connected to one another. This partial network approach is suitable for collecting data from very large populations. For instance, we collect data of female university students about their close friends and ask them to identify which of their friends know one another. This partial network approach could give us a clear and reliable overall picture of networks in which individuals are embedded. This kind of data could be very useful to help us understand the opportunities and constraints ego has as a result of the path they are related in their networks. Moreover, this kind of network also gives us some information about the network as a whole. It represents micro-network data sets or a sampling of local areas of larger networks (Hanneman and Riddle 2005).

***Global network or complete network***: This kind of network (see in Fig. 15.4) focuses on multiple attributes of actors and also multiple kinds of ties that connect

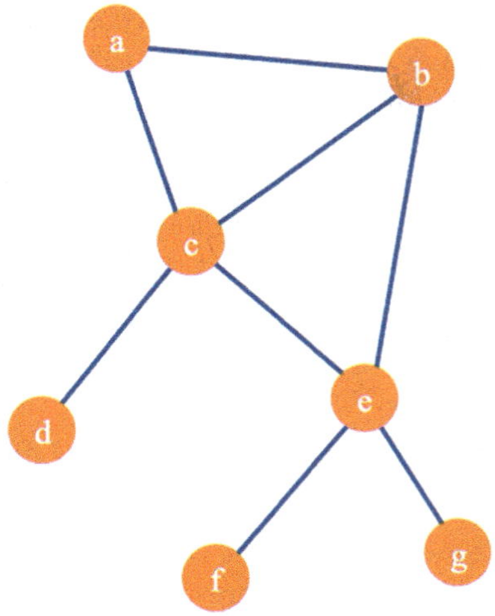


**Fig. 15.4** The global network

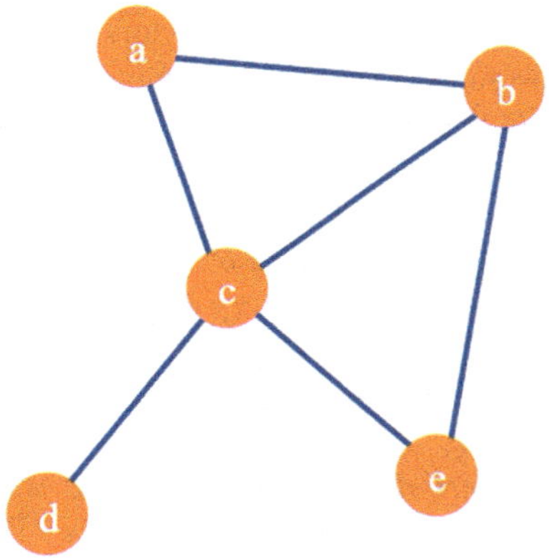


**Fig. 15.5** The ego network of "ego c"

actors in a network. For instance, we might want to know which faculty have the same group of students, serve on the same community, and have one or more fields of expertise and co-author in common. These actors might be tied together closely in one relational network; however, they might be quite far from one another in a different relational network. The establishment of actors in multi-relational networks and the topology of networks combined of multiple relations are the most interesting part of SNA. When researchers collect data about relations among actors, we are trying to sample from a population of possible relations. Network correlation, multi-dimensional scaling and clustering, and role algebras are related to the study of global network or complete network data (Fig. 15.6) (Hanneman and Riddle 2005).

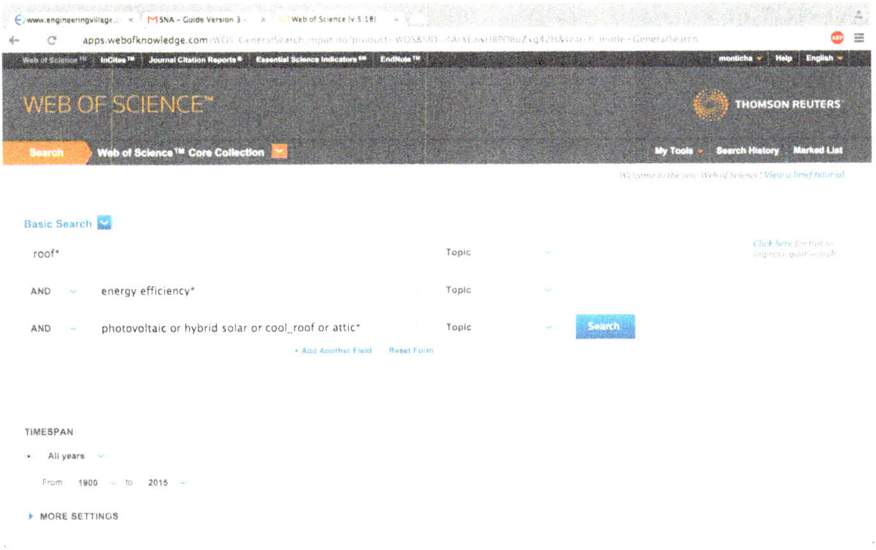


**Fig. 15.6** Web of Science search screen with specific keywords

## *15.2.6 How to Identify Network Structure?*

Researchers need to identify network structure because it shows the characteristic and the cohesion of that network. We could use "***Connectivity***" to identify the structure of the network (Hanneman and Riddle 2005).

***Connectivity***: This term is used to explain how actors in one part of the network are connected to another actor of other part of the network or how two actors are connected to each other in undirected network data. Actors and their connections play important role in SNA so it is necessary to begin to investigate the networks by examining these very connectivity properties. Firstly, we should look at the whole network and then focus on the number of actors, the number of possible connections, and the number of actual connections. The differences in the size of networks and how the actors are connected could tell us about human populations. Population size is one of the most critical factors in sociological analysis. The connection of a small group is different from a large group in many ways. The ways they are connected to each other could be a key indicator of the cohesion, solidarity, moral density, and complexity of the network. Individuals and networks have different basic demographic features. Individual actors might have many or few ties. Individuals might be the source of ties or might be the actors that receive ties, but do not send them, or might be both. The number and kinds of ties that actors have are critical factors to determine how much they embed in the network, what constraints related to their behavior exist, what the range of opportunities is, or how much influence and power they have (Hanneman and Riddle 2005).

In order to analyze connectivity, researchers could use "reachability, density, distance, a path/a walk/a cycle, number of walks, and flow":

*Reachability*. It is used when a target actor is reachable by another. In general, if the data are directed, it is possible that actor A might be able to reach actor B, but actor B might be unable to reach actor A. In undirected data, each pair of actors might be able or unable to reach to one another. If there is the case of unreachability in a network, there might be the potential of a division of the network or it could be interpreted that the target population is composed of more than one subpopulation (Hanneman and Riddle 2005).

*Density*. The density is used to indicate the level of connectedness of a network. It is calculated by using the number of lines in a graph divided by the maximum number of lines (in case that every author is connected to every other one). Consequently, its value is between 0 and 1. For example, if the value of the density of the central network is 0.05, this network is very loose and is not a dense network at all (Otte and Rousseau 2002).

*Distance*. It is the distance between an actor and others. It is used to capture how individuals are embedded in networks. Knowing number of actors stay at various distances from each actor is very important in order to understand the differences among actors in terms of their limitations and advantages. Sometimes there are multiple paths between two actors. Multiple connections may indicate a stronger connection between two actors than a single one. The distances among actors in a network could be a critical macro-characteristic of the whole network. If the distances are great, it may take a long time for information to diffuse across population or it could be that some actors are quite unaware of, and influenced by others even if they are technically reachable. The variance across the actors in the length that they have from other actors could be a basis for differentiation and stratification. Actors who are closer to more others may be able to put more power than those who are more distant (Hanneman and Riddle 2005).

*A path/a walk/a cycle*. A path is a sequence of nodes and edges: starting with one node and ending with another node. It also represents the tracing of the indirect connection between the two nodes. On a path, it is impossible to go backward or revisit the same node twice, whereas *a walk can be* any sequence of nodes and edges and it is possible to go backwards on a walk. A path which starts and ends at the same node is named *a cycle* (Otte and Rousseau 2002).

*Number of walks/paths*. This metric is used to count how many linkage actors have been compared to one another. These data provide a way of thinking about the strength of ties or relations. Actors connected at short distances might have stronger connections if they are connected many times or even if they have many more number of paths. The numbers of walks/paths could be found by raising the matrix to that power. These differences help researchers to understand how information moves in the network, which actors have stronger power, and also other important properties (Hanneman and Riddle 2005).

*Flow*. This metric is used to identify the movement of information from actors to actors. It is used to answer how many different actors in the neighborhood lead to a target. Flow also helps to assess the strength of the ties (Otte and Rousseau 2002).

### 15.2.7 How to Identify Key/Central Nodes in Network?

Researchers need to identify key/central nodes in network of the networks because these nodes show the key players of that network. The term used to identify key nodes in a network is called "***Centrality***" (Freeman 1979).

***Centrality*** refers to location, indicating where an actor resides in a network. This term could help to formalize intuitive notions about the distinction between insiders and outsiders. In order to analyze connectivity, researchers could use "degree centrality, closeness, betweenness centrality" as basic centrality elements (Freeman 1979; Borgatti 2005).

- *Degree centrality*. It is the number of connections that a node has with other nodes. For example, having higher degree of centrality means that this scientist has collaborated with many colleagues. Moreover, we could measure the degree centrality of the whole network. Lower degree of centrality of the whole network indicates that many authors in this network are not connected to each other (Freeman 1979).
- *Closeness*. Another way of analyzing centrality is using the closeness factor. This closeness indicator is more general than the degree centrality, because it includes the structural position of actors in the whole network. A high value of closeness for an actor means that actor is related to all others through a small number of paths (Freeman 1979).
- *Betweenness*. This indicator relied on the number of shortest paths passing through an actor. Actors who have a high value of betweenness seem to play important role of connecting different groups or they might have higher power in communication, communication control, and communication flow than others (Borgatti 2005).

### 15.2.8 What Is SNA Process?

According to Otte and Rousseau (2002), Hansen (2009), the process of SNA can be separated into 3 steps.

**Step 1: Designing of the analysis**

Researchers need to define the objective and clarify the scope of the analysis. They also need to determine what kinds of networks and what kinds of relations they want to study. Moreover, they need to formulate the hypotheses and research questions to set the right path to the analysis (Otte and Rousseau 2002).

**Step 2: Collecting network data**

There are two main methods to collect data for SNA. Firstly, we could use questionnaires and interviews to collect data about the relationships within a specific group. In this case, researchers need to gather background information such as using interviewing senior managers and key staff to understand specific needs and hidden issues. This way is suitable for organizations to identify the relationship among teams/projects, the flow of information among teams/project, or the most influencer in the network. After that, researchers need to develop the survey/interview methodology, design the questionnaire, and survey/interview the individuals/teams/and units in the target network. Secondly, researchers could gather data from academic websites such as for Web of Science or Compendex. They could get the data about the authors and the co-authors in order to identify relationship between authors/coauthors. This method is suitable for knowledge management, collaboration, or other academic purposes. After that, researchers could go directly to that website and search information by using the specific keywords in order to collect target data. These keywords are very important because website/program will gather the article/journal from giving keywords. If keywords are not that specific, the result might be too large and difficult to use for analysis, but if keywords are too specific, the result might be too small and we cannot find the relationships in the target network (Otte and Rousseau 2002; Hansen 2009).

**Step 3: Measuring network data and analyzing network data**

Researchers could use SNA software to calculate the basic terms used for SNA and create the visualization of the network. The examples of software are (Huisman and Van Duijn 2005);

- Pajek (Windows, free)
- UCInet (Windows, shareware)
- Netdraw (Windows, free)
- Mage (Windows, free)
- GUESS (with all platforms, freeware)
- R packages for SNA (with all platforms, freeware)
- Gephi (with all platforms, freeware).

### *15.2.9 Application of SNA in Organizations*

There are many advantages of applying SNA. One of the biggest advantages of SNA is that it can be visualized using the appropriate tools very clearly. This leads to a deeper understanding of the structures and relationships of a network. The analysis of social networks focused on the fact that a relation between persons and a relation between organizations are important, because they make and display attitudes, communication, and information flow of products. SNA provides the methods to investigate these relationships, to represent graphically, evaluating and building on that to develop it further (Krause and Croft 2007). Organizations could apply SNA in many business activities such as they could apply SNA in merger acquisition assessment in order to control the network management with strategic alliances and collaborations. In this field, they could measure how successful the integration after mergers is. They also can apply SNA to identify how good the network is connected, how stable the network is, and whether or not there are any holes in the network (Krause and Croft 2007).

We can apply SNA to find the experts in order to build communities of practice, tighten the internal knowledge, and build up the knowledge management system by survey and analyze leaders' opinions to identify who has the most influence, what the organizational causes of conflicts are, or how efficient the information is (Fritsch and Kauffeld-Mon 2010). Last but not least, viral marketing like word of mouth (Staab 2005), procurement, supply chain management, and human resource development also can use the application of SNA as well (Staab 2005).

### *15.2.10 The Importance of Expert Identification*

It is critically important for organizations to identify experts within and outside the organization as they need their help to manage tacit knowledge within organization in an effective way. This kind of knowledge is usually transferred better through mentoring or face-to-face interactions among those internal experts. In a very highly competitive business environment, managing explicit knowledge is also important. Organizations need this kind of knowledge to create innovation within organization by transferring knowledge from external experts. With this combination between tacit and explicit knowledge, organizations could gain the true value added intellectual assets of an organization and they could maintain the organization's core competency and innovate (Yang and Huh 2008). It is also important to identify the right experts for collaboration purposes or to develop strategic technology road maps (Müller et al. 2012; Daim and Oliver 2008).

## 15.3 Methodology

In this research, the methodology was divided into three steps: data collection using web of science, expert identification using RStudio and Gephi, and expert data analysis.

***Step 1: Data collection using Web of Science***. Because we need expert data in smart roofing field, we need to collect expert data from reliable multi-databases and easy-to-access sources so we chose to use Web of Science as our database. Web of science is an online scientific citation database which provides a comprehensive citation search. It gives access to multiple databases and allows us to do in-depth exploration of academic fields.

After choosing the source, then we have to create "Keywords" used to search for smart roofing data. The size of the data is very important for the expert analysis. If data are too large, the analysis might be too rough. We might be unable to analyze the whole network, but if data are too small, we might be unable to get any significant results and may not find the right network of experts. That is why choosing the right size of data is very important for this kind of research.

The strategy we used to identify keywords was to "***search with generic keywords and then narrow down using specific keywords.***" When we searched by using generic keywords, we got more than 1000 data points, and then, we narrowed down by using specific names. Finally, we got down to 80 data points, which was suitable for our analysis. For example in the case of Web of Science, system will run the query by starting to look for articles that contain "roof with any following word" in article's topic, and then, it will start to search in article's title and collect articles that contain photovoltaic or hybrid solar or cool roof or attic in their title from a time period which in this research we selected as being from 1900 to 2015. The search screen we used can be seen in Figs. 15.6, and 15.7.

After that, we have to go directly to each article and check whether the article is exactly related to our topic or not by clicking on each topic, to see the detail and delete the unrelated articles. After we check and clean all data, we could save the data with full record and cited references and be ready to move to the next step.

***Step 2: Expert identification using RStudio and Gephi***. In this step, we will interpret the data we got from previous step by using RStudio with Shiny package. RStudio is a free and open source programming language used for statistical computing and graphics, whereas Shiny package is a package used to calculate

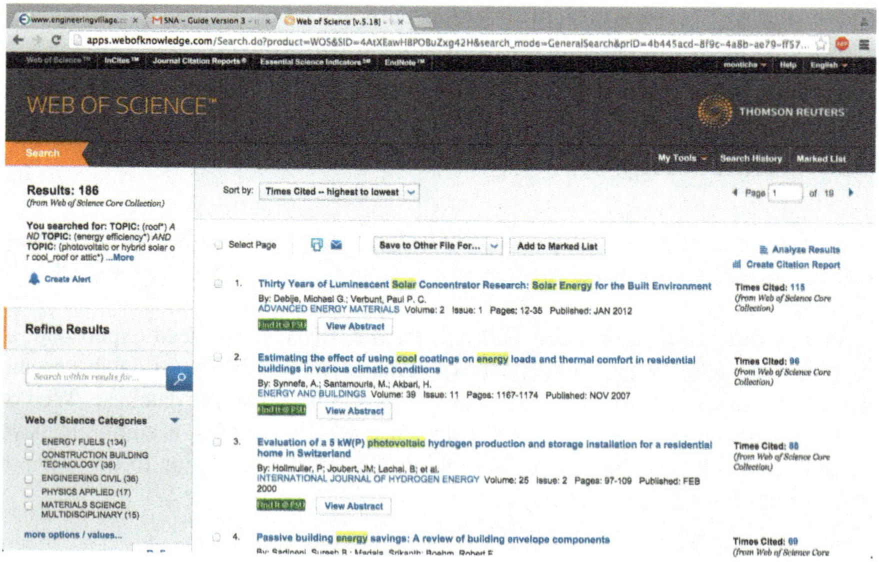


**Fig. 15.7** Web of Science using specific keywords

SNA basic elements such as betweenness, degree, closeness, and centrality. This package is developed in house at the authors' institution, and it supports for databases from Web of Science/Compendex. The way the RStudio with Shiny package was used can be seen in Figs. 15.8, 15.9, and 15.10.

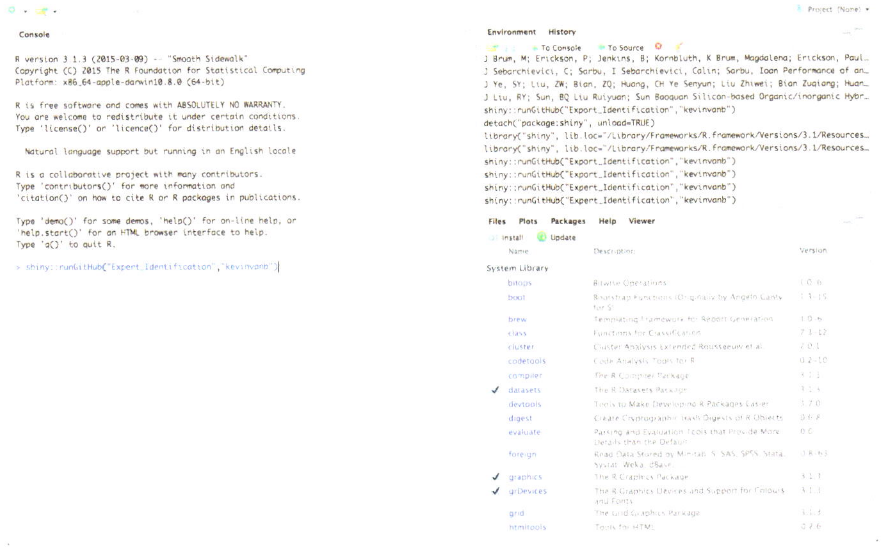


**Fig. 15.8** RStudio with shiny package

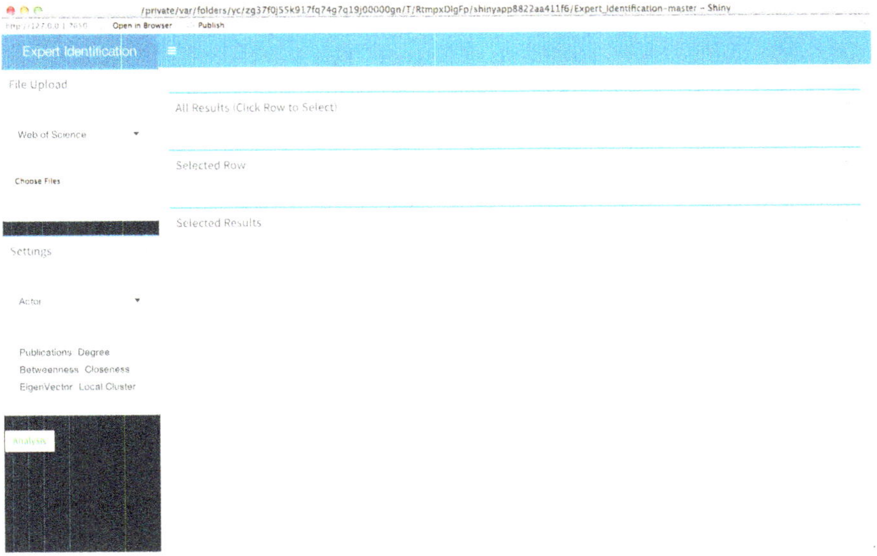


**Fig. 15.9** RStudio with shiny package

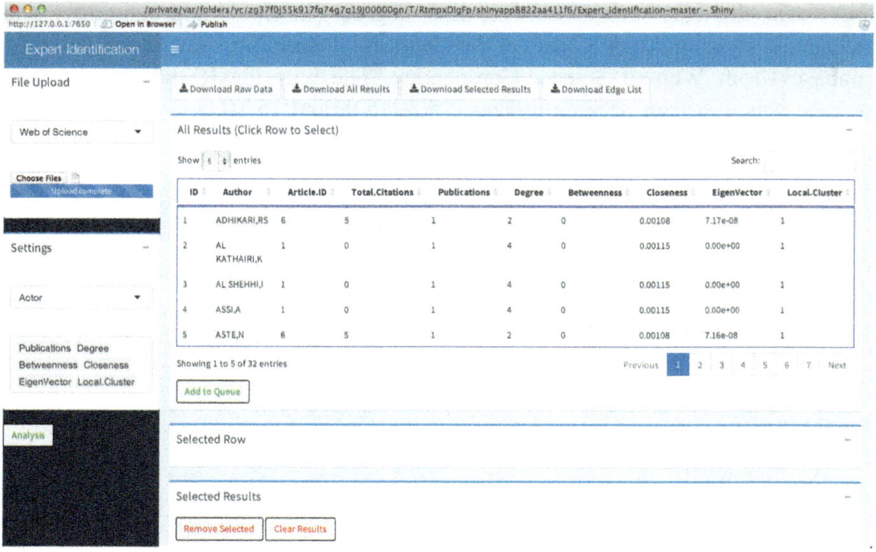

**Fig. 15.10** RStudio with shiny package

At the end of this step, we could calculate important terms used for SNA analysis such as betweenness, degree, closeness, and centrality. The result of this step can be seen in Appendix A.

After we got this result, we did the visual analysis using Gephi, which is a free program used to analyze and visualize network data. There are two data sets we used for Gephi. One data set is the one with all the data (186 entries) and the other one is the data set with only the most top ten experts in this field. The way the Gephi was used can be seen in Figs. 15.11 and 15.12.

***Step 3: Expert data analysis***. After completing steps 1 and 2, we analyze the results to identify the groups of experts for smart roofing, the relations between experts, the structure of this network, and also the centrality of the experts.

## 15.4 Results and Discussion

In this section, we will analyze two groups of results to identify the experts in smart roofing field. The first group of results is the centrality-related metrics including degree centrality, closeness, number of citations, and betweenness. These results will help us to understand more about the key experts in the network. The second group of results is the expert network pictures or visualizations. These results will help us to better understand the structure of the network, structure of the sub-networks, and the network flow.

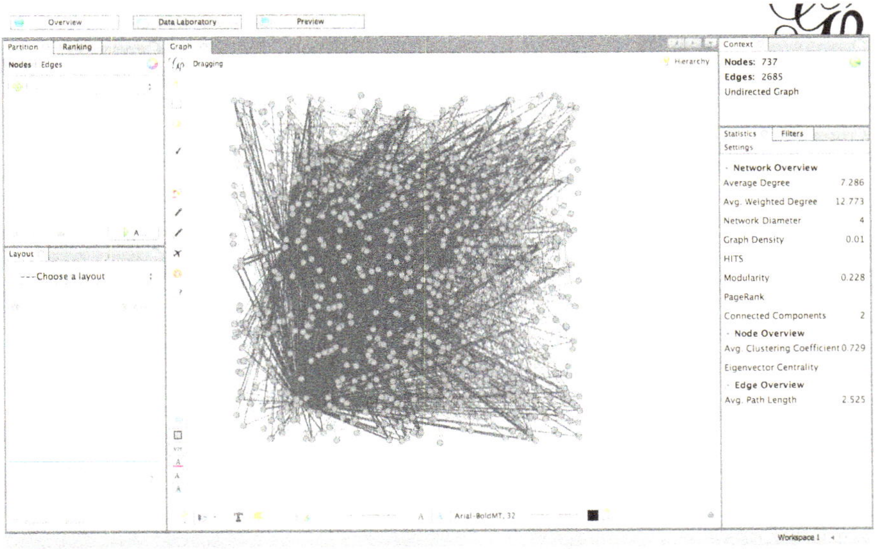


Fig. 15.11 Working screen of gephi program

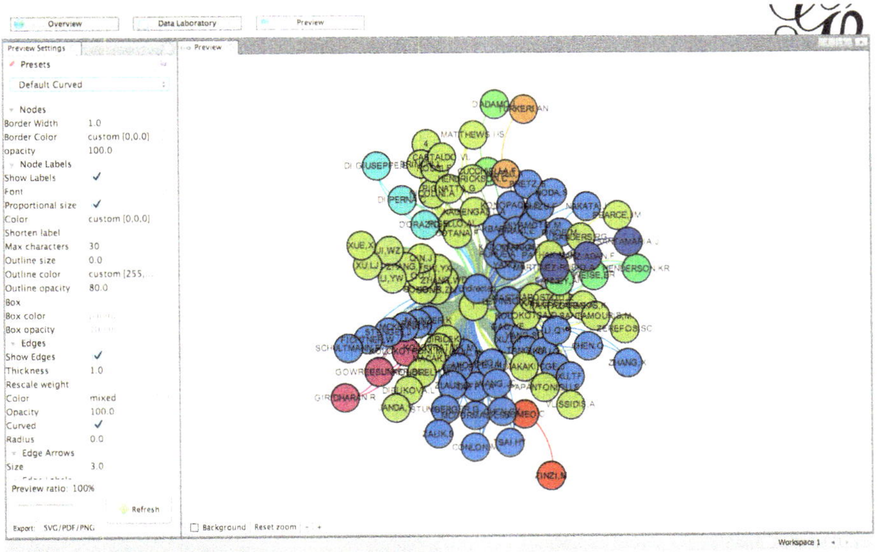


Fig. 15.12 Visualization data created by gephi

## 15.4.1 *Centrality Analysis*

### 15.4.1.1 Degree Centrality

Table 15.2 shows the degree of centrality from Web of Science database. This data set is ranked from the largest value of degree centrality to the lowest one. According to these data, we could see that LEVINSON-R, JI-WZ, LI-YW, QIN-J QU-J, SHI-YX, SONG-JR, SONG-ZN, XU-LJ, XUE-X, ZHANG-T, and ZHANG-WD have the highest values of degree centrality. It means that these people have stronger collaborations than the others. They might be collaborating with many colleagues. We might assume that they may know many experts in this field or they might have had more opportunity to work, share, or communicate with other experts compared to those who have lower values of degree centrality.

### 15.4.1.2 Closeness

Table 15.3 shows the closeness values from Web of Science database. This data set is ranked from the highest value of closeness to the lowest one. According to these

**Table 15.2** Degree of centrality results (top 20) from Web of Science database

| ID | Author | Closeness |
|---|---|---|
| 1 | LEVINSON, R | 0.000103 |
| 2 | JI, WZ | 0.000103 |
| 3 | LI, YW | 0.000103 |
| 4 | QIN, J | 0.000103 |
| 5 | QU, J | 0.000103 |
| 6 | SHI, YX | 0.000103 |
| 7 | SONG, JR | 0.000103 |
| 8 | SONG, ZN | 0.000103 |
| 9 | XU, LJ | 0.000103 |
| 10 | XUE, X | 0.000103 |
| 11 | ZHANG, T | 0.000103 |
| 12 | ZHANG, WD | 0.000103 |
| 13 | GAO, YF | 0.000103 |
| 14 | GE, J | 0.000103 |
| 15 | TANG, XM | 0.000103 |
| 16 | XU, JM | 0.000103 |
| 17 | XU, TF | 0.000103 |
| 18 | YANG, SC | 0.000103 |
| 19 | ZHOU, Q | 0.000103 |
| 20 | AKBARI, H | 0.000103 |
| 21 | BRETZ, S | 0.000103 |
| 22 | KONOPACKI, S | 0.000103 |

**Table 15.3** Closeness (top 22) from Web of Science database

| ID | Author | Closeness |
|---|---|---|
| 1 | LEVINSON, R | 0.000103 |
| 2 | JI, WZ | 0.000103 |
| 3 | LI, YW | 0.000103 |
| 4 | QIN, J | 0.000103 |
| 5 | QU, J | 0.000103 |
| 6 | SHI, YX | 0.000103 |
| 7 | SONG, JR | 0.000103 |
| 8 | SONG, ZN | 0.000103 |
| 9 | XU, LJ | 0.000103 |
| 10 | XUE, X | 0.000103 |
| 11 | ZHANG, T | 0.000103 |
| 12 | ZHANG, WD | 0.000103 |
| 13 | GAO, YF | 0.000103 |
| 14 | GE, J | 0.000103 |
| 15 | TANG, XM | 0.000103 |
| 16 | XU, JM | 0.000103 |
| 17 | XU, TF | 0.000103 |
| 18 | YANG, SC | 0.000103 |
| 19 | ZHOU, Q | 0.000103 |
| 20 | AKBARI, H | 0.000103 |
| 21 | BRETZ, S | 0.000103 |
| 22 | KONOPACKI, S | 0.000103 |

data, we could see that all of these 22 people have the highest value of closeness (0.000103). While this value is not so high, these people would be the ones to connect to establish collaborations.

#### 15.4.1.3 Betweenness

Table 15.4 shows the betweenness values from Web of Science database. This data set is ranked from the highest value of betweenness to the lowest one. According to these data, we could see that *LEVINSON-R, KOLOKOTSA-D, YANO-A, COTANA-F, PISELLO-AL* have the highest value of closeness (21, 20, 14, 4, 4). It means that they have number of shortest paths passing through them. These people seem to play an important role of connecting different groups or they might have higher power in communication, communication control, and communication flow than others in the same network.

By comparison of Tables 15.2, 15.3, and 15.4, it can be seen that *LEVINSON-R* is the only one that has the highest rank in every table. According to these data, we could assume that he might be the most influential expert in this network because he stays close to other experts, he knows many experts, and he also has the shortest

**Table 15.4** Betweenness results (top 16) from Web of Science database

| ID | Author | Betweenness |
|---|---|---|
| 1 | LEVINSON, R | 21 |
| 2 | KOLOKOTSA, D | 20 |
| 3 | YANO, A | 14 |
| 4 | COTANA, F | 4 |
| 5 | PISELLO, AL | 4 |
| 6 | SANTAMOURIS, M | 2 |
| 7 | GOBAKIS, K | 0.667 |
| 8 | KARLESSI, T | 0.667 |
| 9 | MASTRAPOSTOLI, E | 0.667 |
| 10 | PANTAZARAS, A | 0.667 |
| 11 | ZEREFOS, SC | 0.667 |
| 12 | BRINCHI, L | 0.2 |
| 13 | NICOLINI, A | 0.2 |
| 14 | CASTALDO, VL | 0.2 |
| 15 | PIGNATTA, G | 0.2 |
| 16 | ROSSI, F | 0.2 |

**Table 15.5** Comparing data among degree centrality, closeness, and betweenness

| ID | Author | Degree | ID | Author | Closeness | ID | Author | Betweenness |
|---|---|---|---|---|---|---|---|---|
| **1** | **LEVINSON, R** | 10 | **1** | **LEVINSON, R** | 0.000103 | **1** | **LEVINSON, R** | 21 |
| 2 | JI, WZ | 10 | 2 | JI, WZ | 0.000103 | **2** | **KOLOKOTSA, D** | 20 |
| 3 | LI, YW | 10 | 3 | LI, YW | 0.000103 | **3** | **YANO, A** | 14 |
| 4 | QIN, J | 10 | 4 | QIN, J | 0.000103 | 4 | COTANA, F | 4 |
| 5 | QU, J | 10 | 5 | QU, J | 0.000103 | 5 | PISELLO, AL | 4 |
| 6 | SHI, YX | 10 | 6 | SHI, YX | 0.000103 | 6 | SANTAMOURIS, M | 2 |
| 7 | SONG, JR | 10 | 7 | SONG, JR | 0.000103 | 7 | GOBAKIS, K | 0.667 |
| 8 | SONG, ZN | 10 | 8 | SONG, ZN | 0.000103 | 8 | KARLESSI, T | 0.667 |
| 9 | XU, LJ | 10 | 9 | XU, LJ | 0.000103 | 9 | MASTRAPOSTOLI, E | 0.667 |
| 10 | XUE, X | 10 | 10 | XUE, X | 0.000103 | 10 | PANTAZARAS, A | 0.667 |
| 11 | ZHANG, T | 10 | 11 | ZHANG, T | 0.000103 | 11 | ZEREFOS, SC | 0.667 |
| 12 | ZHANG, WD | 10 | 12 | ZHANG, WD | 0.000103 | 12 | BRINCHI, L | 0.2 |
| **13** | **KOLOKOTSA, D** | 9 | 13 | GAO, YF | 0.000103 | 13 | NICOLINI, A | 0.2 |
| **14** | **YANO, A** | 9 | 14 | GE, J | 0.000103 | 14 | CASTALDO, VL | 0.2 |
| 15 | FURUE, A | 7 | 15 | TANG, XM | 0.000103 | 15 | PIGNATTA, G | 0.2 |
| 16 | HIRAKI, E | 7 | 16 | XU, JM | 0.000103 | 16 | ROSSI, F | 0.2 |

paths passing. Moreover, KOLOKOTSA-D and YANO-A are other influential experts, because they got higher values of betweenness and degree centrality. This can be seen in Table 15.5.

#### 15.4.1.4 Number of Citations

Table 15.6 shows the number of citations from Web of Science database. This data set is ranked from the highest value of citation number to the lowest one. According to these data, we could see that *COTANA-F and PISELLO-AL* have the highest values of citations number (41). It means that their articles have been cited by other experts for many times. We can assume that their work is significant and also related to other experts' work. They might be the ones developing the general theory and could be used as the basic knowledge or reference for other experts; or their work might be easy to apply to different fields.

By comparison of Tables 15.5 and 15.6, it can be seen that *LEVINSON-R* seems to be the most important export in this network too because his work has been cited by many experts (Table 15.7).

#### 15.4.1.5 Overall

According to all information we got (degree centrality, closeness, betweenness, and citations number), we could come up with the ranking in Table 15.8.

**Table 15.6** Number of citations from Web of Science database

| ID | Author | Total citations |
|---|---|---|
| 1 | COTANA, F | 41 |
| 2 | PISELLO, AL | 41 |
| 3 | AYOMPE, LM | 38 |
| 4 | CONLON, M | 38 |
| 5 | DUFFY, A | 38 |
| 6 | MCCORMACK, SJ | 38 |
| 7 | LEVINSON, R | 33 |
| 8 | AKBARI, H | 32 |
| 9 | BRETZ, S | 32 |
| 10 | KONOPACKI, S | 32 |
| 11 | KOLOKOTSA, D | 31 |
| 12 | CUCCHIELLA, F | 29 |
| 13 | D'ADAMO, I | 29 |
| 14 | SANTAMOURIS, M | 17 |
| 15 | GIRIDHARAN, R | 15 |
| 16 | GOWREESUNKER, BL | 15 |
| 17 | KOLOKOTRONI, M | 15 |
| 18 | DIAKAKI, C | 14 |
| 19 | PAPANTONIOU, S | 14 |
| 20 | VLISSIDIS, A | 14 |

**Table 15.7** Comparison of data among degree centrality, closeness, betweenness, and citations number

| ID | Author | Degree | ID | Author | Closeness | ID | Author | Betweenness | ID | Author | # Cited |
|---|---|---|---|---|---|---|---|---|---|---|---|
| **1** | **LEVINSON, R** | 10 | **1** | **LEVINSON, R** | 0.000103 | **1** | **LEVINSON, R** | 21 | 1 | COTANA, F | 41 |
| 2 | JI, WZ | 10 | 2 | JI, WZ | 0.000103 | 2 | KOLOKOTSA, D | 20 | 2 | PISELLO, AL | 41 |
| 3 | LI, YW | 10 | 3 | LI, YW | 0.000103 | 3 | YANO, A | 14 | 3 | AYOMPE, LM | 38 |
| 4 | QIN, J | 10 | 4 | QIN, J | 0.000103 | 4 | COTANA, F | 4 | 4 | CONLON, M | 38 |
| 5 | QU, J | 10 | 5 | QU, J | 0.000103 | 5 | PISELLO, AL | 4 | 5 | DUFFY, A | 38 |
| 6 | SHI, YX | 10 | 6 | SHI, YX | 0.000103 | 6 | SANTAMOURIS, M | 2 | 6 | MCCORMACK, SJ | 38 |
| 7 | SONG, JR | 10 | 7 | SONG, JR | 0.000103 | 7 | GOBAKIS, K | 0.667 | 7 | **LEVINSON, R** | 33 |
| 8 | SONG, ZN | 10 | 8 | SONG, ZN | 0.000103 | 8 | KARLESSI, T | 0.667 | 8 | AKBARI, H | 32 |
| 9 | XU, LJ | 10 | 9 | XU, LJ | 0.000103 | 9 | MASTRAPOSTOLI, E | 0.667 | 9 | BRETZ, S | 32 |
| 10 | XUE, X | 10 | 10 | XUE, X | 0.000103 | 10 | PANTAZARAS, A | 0.667 | 10 | KONOPACKI, S | 32 |
| 11 | ZHANG, T | 10 | 11 | ZHANG, T | 0.000103 | 11 | ZEREFOS, SC | 0.667 | 11 | KOLOKOTSA, D | 31 |
| 12 | ZHANG, WD | 10 | 12 | ZHANG, WD | 0.000103 | 12 | BRINCHI, L | 0.2 | 12 | CUCCHIELLA, F | 29 |
| 13 | KOLOKOTSA, D | 9 | 13 | GAO, YF | 0.000103 | 13 | NICOLINI, A | 0.2 | 13 | D'ADAMO, I | 29 |
| 14 | YANO, A | 9 | 14 | GE, J | 0.000103 | 14 | CASTALDO, VL | 0.2 | 14 | SANTAMOURIS, M | 17 |
| 15 | FURUE, A | 7 | 15 | TANG, XM | 0.000103 | 15 | PIGNATTA, G | 0.2 | 15 | GIRIDHARAN, R | 15 |
| 16 | HIRAKI, E | 7 | 16 | XU, JM | 0.000103 | 16 | ROSSI, F | 0.2 | 16 | GOWREESUNKER, BL | 15 |
| 17 | ISHIZU, F | 7 | 17 | XU, TF | 0.000103 | | | | 17 | KOLOKOTRONI, M | 15 |
| 18 | KADOWAKI, M | 7 | 18 | YANG, SC | 0.000103 | | | | 18 | DIAKAKI, C | 14 |
| 19 | MIYAMOTO, M | 7 | 19 | ZHOU, Q | 0.000103 | | | | 19 | PAPANTONIOU, S | 14 |
| 20 | NODA, S | 7 | 20 | AKBARI, H | 0.000103 | | | | 20 | VLISSIDIS, A | 14 |
| | | | 21 | BRETZ, S | 0.000103 | | | | | | |
| | | | 22 | KONOPACKI, S | 0.000103 | | | | | | |

Basic centrality elements show that Levinson-R is the expert who has the highest value of every centrality element except total citations number. From this analysis, we could say that Levinson-R is the most important expert in this network. He might be the key of this network. Moreover, Kolokotsa-D and Yano-A could be the other two of the most important and influential experts in this network due to the high values of degree centrality, citations number, and betweenness.

## 15.4.2 Visualization

### 15.4.2.1 Structure of the Whole Network

The overall picture of this network with 186 data entries is shown in Fig. 15.13. It is an undirected graph with 575 nodes and 1950 edges. The average degree centrality of this network is 3.391 (max 10, min 0), the network diameter is 3, the density of the network is 0.006, and the average path length is 1.084.

It is difficult to see the relationships among specific experts in Fig. 15.13. So we narrow data down by focusing only on the group of specific experts we identified through the previous step and used that data to generate the visualization again with Gephi. We did this because we wanted to understand relationships among the specific experts better. The resulting visualization is shown in Fig. 15.14.

Figure 15.14 is an undirected graph with 62 nodes and 88 edges. The average degree centrality of this network is 2.839 (max 10, min 0), the network diameter is 4, the density of the network is 0.047, and the average path length is 3.198.

From the above graph, we could see that the nodes have four different colors. These colors are determined by modularity, which is another measure of the structure of networks. It is used to represent the strength of division of a network into nodes. We could call them clusters or communities of the network. We could see that there are four clusters or modularity within this network: blue, red, yellow, and green. The green one is the biggest cluster, whereas the yellow one is the smallest cluster in this network. It means that there are many experts in the green cluster and may be these people are connected to more people than experts in other clusters.

Moreover, we could see that there are six nodes that have strong edges (*Cotana-F, Pisello-AL, Yano-A, Kolokatsa-D, Levenson-R, and Zhang-WD*). These people seem to have higher power than others, and they might be able to influence other people because they are in the center or core of the network. They are connected to too many experts, and also they stay close to other strong nodes. This could imply that the most important experts in this network are the nodes that have strong edges. In order to see the picture clearly, we use Fig. 15.15.

**Table 15.8** The most important experts in this network analyzed by author

| ID | Author | Degree | Total citations | Betweenness | Local cluster |
|---|---|---|---|---|---|
| 1 | LEVINSON, R | 10 | 33 | 21 | 0.533 |
| 2 | KOLOKOTSA, D | 9 | 31 | 20 | 0.389 |
| 3 | YANO, A | 9 | 13 | 14 | 0.611 |
| 4 | COTANA, F | 6 | 41 | 4 | 0.467 |
| 5 | PISELLO, AL | 6 | 41 | 4 | 0.467 |
| 6 | SANTAMOURIS, M | 6 | 17 | 2 | 0.733 |
| 7 | ZHANG, T | 10 | 8 | 0 | 1 |
| 8 | ZHANG, WD | 10 | 8 | 0 | 1 |
| 9 | AKBARI, H | 3 | 32 | 0 | 1 |
| 10 | BRETZ, S | 3 | 32 | 0 | 1 |
| 11 | KONOPACKI, S | 3 | 32 | 0 | 1 |

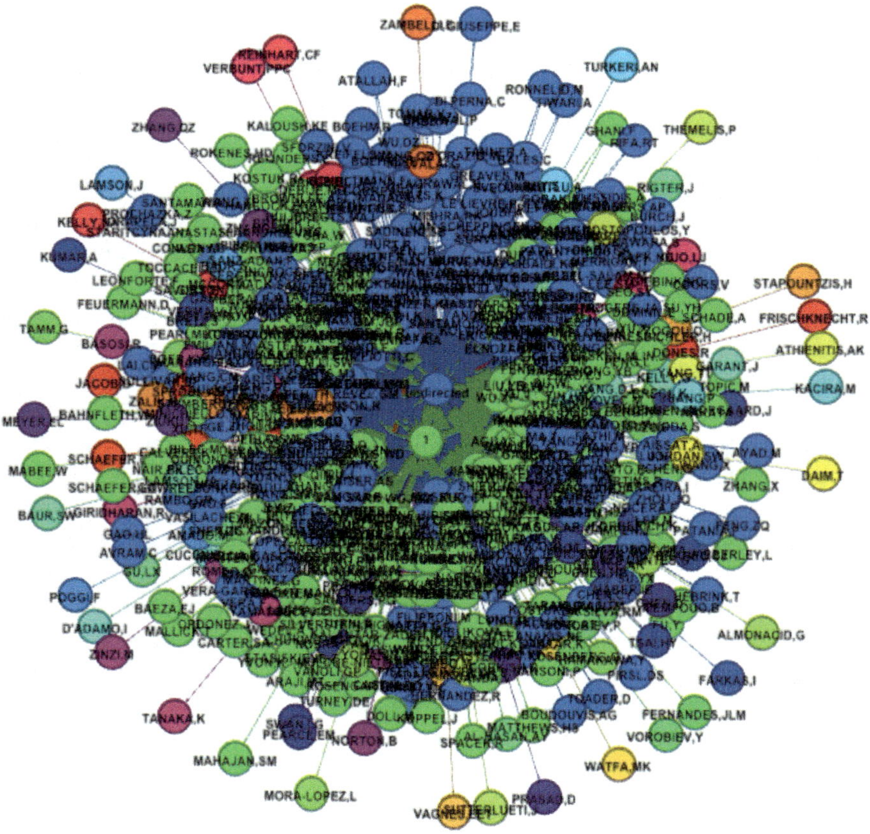


**Fig. 15.13** The overall picture of the whole network with 186 data entries (all data)

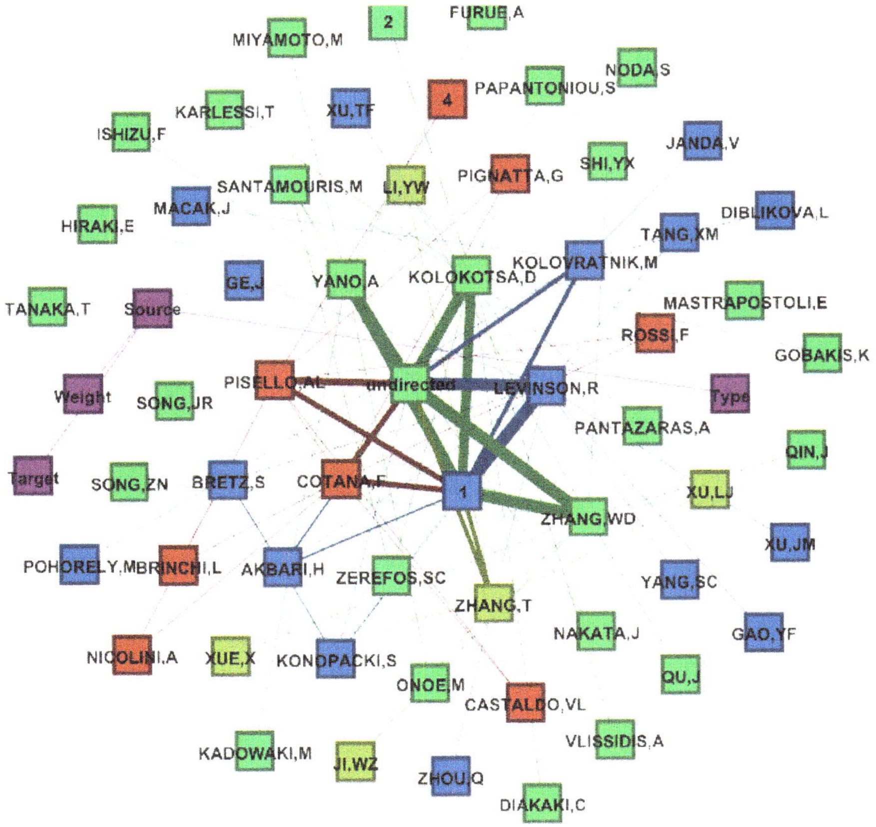


**Fig. 15.14** The overall picture of the key experts

#### 15.4.2.2 Structure of Each Clusters Within Network

As we mentioned above, there are four different clusters of this network: green, red, blue, and yellow (divided by modularity value).

- ***Cluster with green nodes***. The green nodes are the highest in number in this network. If we look deeply into the graph, we could see that there are three small clusters embedded in this green cluster (see in Figs. 15.16, 15.17, and 15.18).

From these above three graphs, we could see that there is a core or central node inside every green cluster. The core node of the 1st green cluster is *Kolokatsa-D*, the core node of the 2nd green cluster is *Zhang-WD,* and the core node of the 3rd green cluster is *Yano-A*. Without these core nodes, experts in green nodes cannot connect to each other and form a cluster. That is why the core or central node is very important because they help to form the network by connecting other nodes.

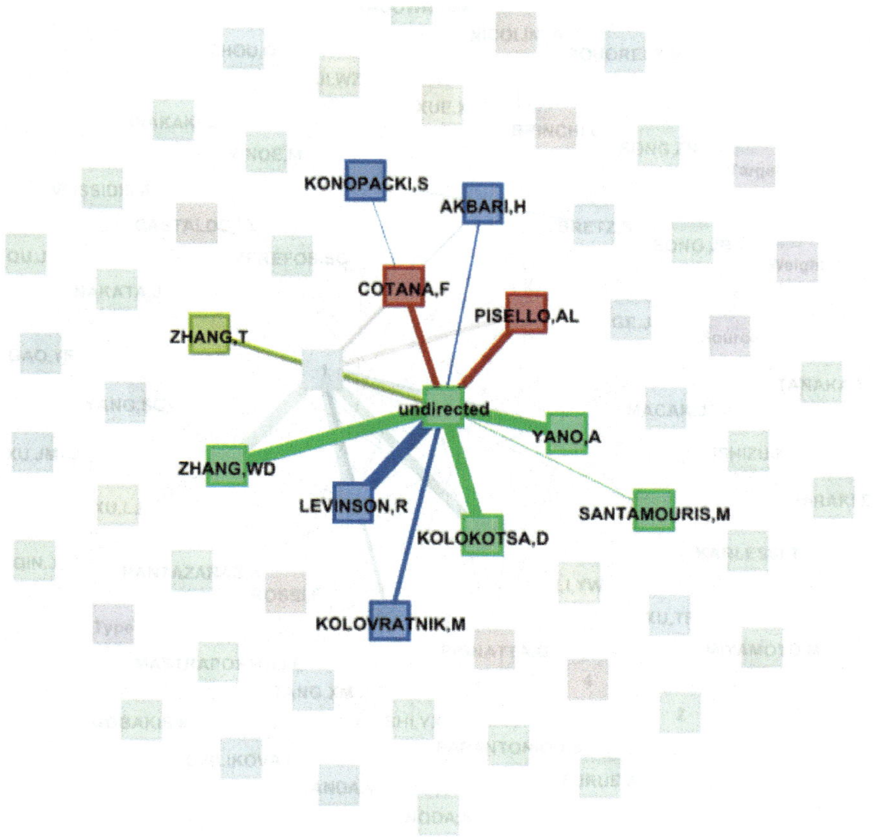


Fig. 15.15 The nodes with strong edges in the whole network

Without them, communication cannot pass to other experts and cluster cannot be formed.

In addition, *Kolokatsa-D and Yano-A* are linked to each other and these two people help to connect the 1st green cluster together with the 3rd green cluster and make the network larger than before. This linkage can benefit all in many aspects. They expand the size of the network and increase the possibility of receiving more information. However, this might slow the communication rate within the network.

- ***Cluster with blue nodes***. This cluster has medium number of nodes. If we look carefully into the graph, we could see that there are two small clusters embedded in this blue cluster (see in Figs. 15.19, and 15.20).

The core node of the 1st blue cluster is *Kolovratnik-M*, and the core node of the 2nd blue cluster is *Levnson-R*. Without these core nodes, experts in blue nodes cannot connect to each other and form a cluster. However, it seems like Levnson-R

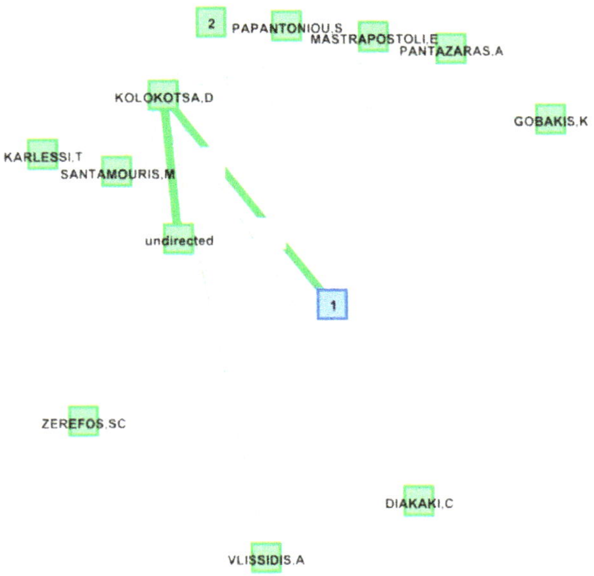


**Fig. 15.16** Shows 1st small *green* cluster with Kolokotsa-D node

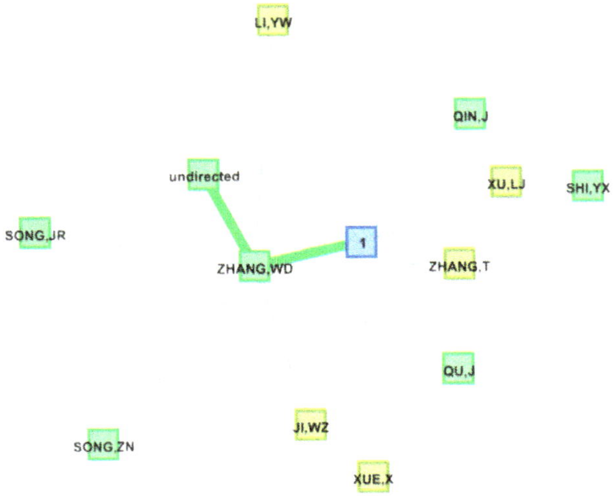


**Fig. 15.17** Shows 2nd small *green* cluster with Zhang-WD node

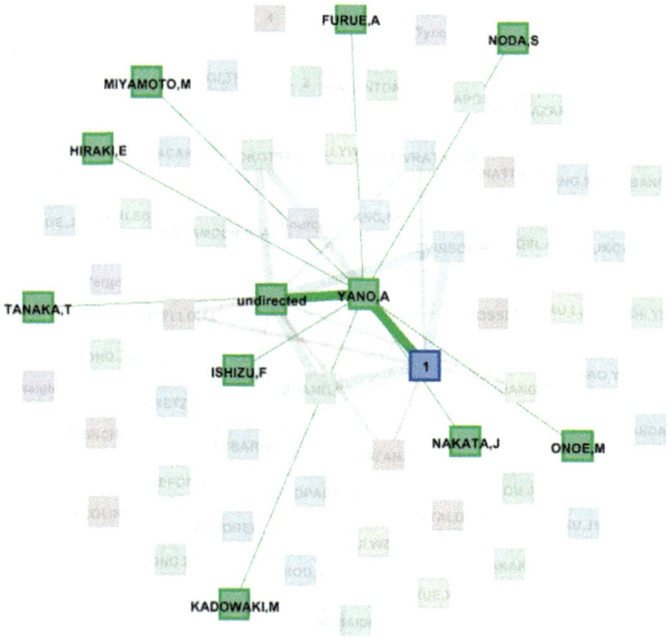


**Fig. 15.18** Shows 3rd small *green* cluster with Yano-A node

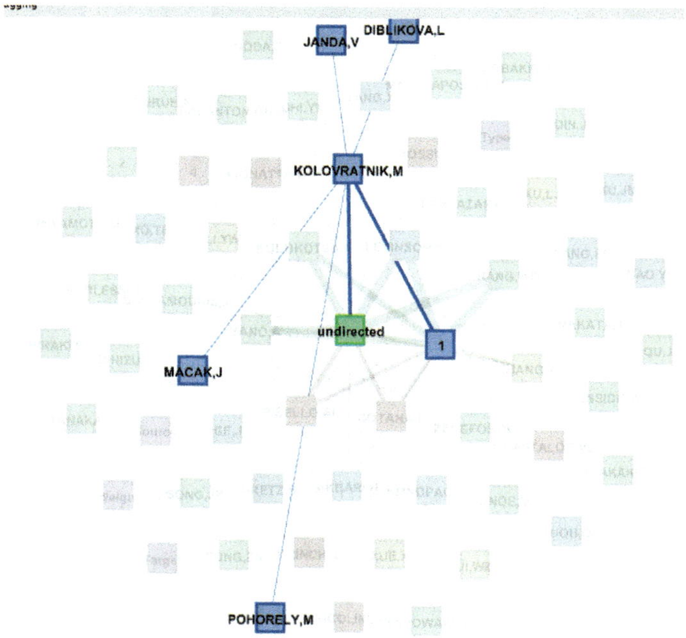


**Fig. 15.19** Shows 1st small *blue* cluster with Kolovratnik-M

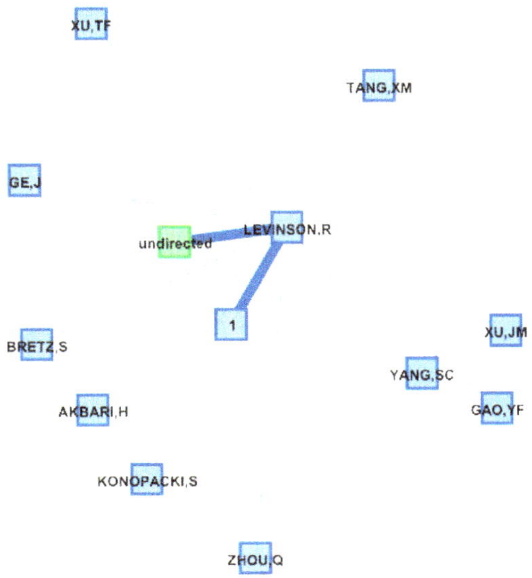


**Fig. 15.20** Shows 2nd small *blue* cluster with Levnson-R

has more power than Kolovrntnik-M because he is connected to many more nodes than Kolovrntnik-M.

- ***Red cluster***

This group of cluster has a small population. There is only one red cluster. The picture of red cluster can be seen in Fig. 15.20.

From above graph, we could see that *Piesello-AL* is the core or central node of this red cluster.

- ***Yellow cluster***

This group of cluster has also a small population and has the similar pattern with red cluster. There is only one yellow cluster, which is shown in Fig. 15.21.

From the above graph, we could see that *Zhang-T* is the core or central node of this yellow cluster.

#### 15.4.2.3 Relations Among Clusters in Network

We tried to analyze the relationships among these four clusters but what we found is that there is none except the relation between green cluster and yellow cluster (see in Fig. 15.22). We found that Zhang-T, yellow node, and Zhang-WD, green node,

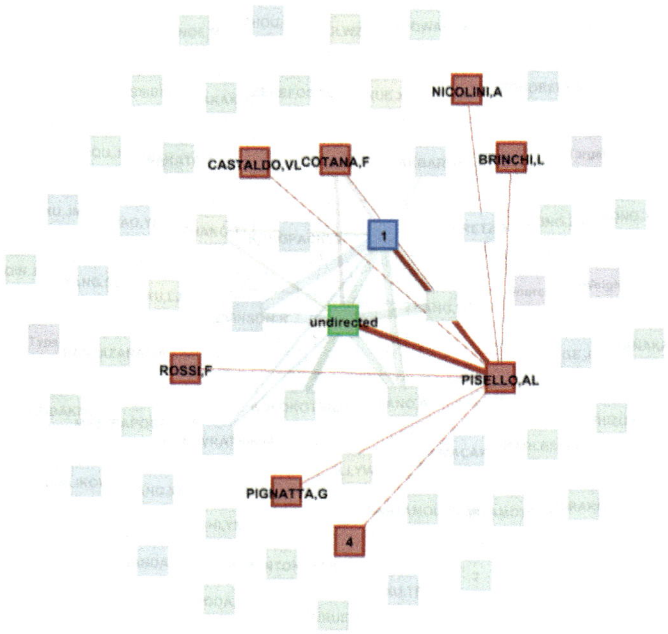


**Fig. 15.21** Shows *red* cluster with Piesello-AL

are connected to each other so they bridge yellow cluster and green cluster together. They connected these two clusters together and made the clusters a bigger cluster. Without them, yellow cluster will stay separate from the network.

#### 15.4.2.4 Group of Important Experts Identify by Using Visualization

Based on all information we got from visualization, we could identify the top ten powerful experts in Table 15.9.

And from the visualization, we could see that Levinson-R, Kolokotsa-D, and Zhang-WD are the most three powerful experts in this field because they are connected to too many experts. Moreover, Zhang-WD also helps to connect the yellow and green cluster together.

When we compared results from using centrality elements (Sect. 4.1) and visualization (Sect. 4.2), we found that they give slightly different insights.

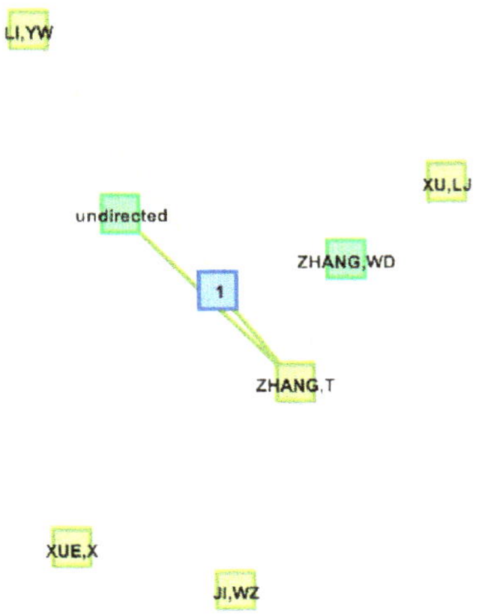


**Fig. 15.22** Shows *yellow* cluster with Zhang-T

**Table 15.9** Shows the most important experts using visualization analyzed by author

| ID | Author | Reason |
|---|---|---|
| 1 | LEVINSON, R | He has strong edges and he is the central node of blue cluster |
| 2 | KOLOKOTSA, D | He has strong edges and he is the central node of green cluster |
| 3 | YANO, A | He has strong edges and he is the central node of green cluster |
| 4 | COTANA, F | He has strong edges |
| 5 | PISELLO, AL | He has strong edges and he is the central node of red cluster |
| 6 | ZHANG, T | He is the central node of yellow cluster and he also connect yellow cluster together with green |
| 7 | ZHANG, WD | He has strong edges. He is the central node of green cluster and he also connect green cluster together with yellow |
| 8 | KOLOVRTINI-M | He is the central node of blue cluster |

## 15.5 Conclusion and Future Research

Social network analysis is a good analysis tool that can be applied in many fields. This chapter shows one of the specific examples of SNA application that can be used in the real world. We can identify who the important experts in the field are. With the visualization, we could explain the picture of the network, the pattern of relations, the structure of the network, and also the linkage between clusters.

## References

Borgatti, S. P. (2005) Centrality and network flow. *Social Networks, 27*(1), 55–71.

Borgatti, S. P., & Everett, M. G. (1997). Network analysis of 2-mode data. *Social networks, 19*(3), 243–269.

Daim, T. U., & Oliver, T. (2008). Implementing technology roadmap process in the energy services sector: A case study of a government agency. *Technological Forecasting and Social Change, 75*(5), 687–720.

Durland, M. M., & Fredericks, K. A. (2005). An introduction to social network analysis. *New Directions for Evaluation, 2005*(107), 5–13.

Freeman, L. C. (1979). Centrality in social networks conceptual clarification. *Social networks, 1*(3), 215–239.

Fritsch, M., & Kauffeld-Mon, M. (2010). The impact of network structure on knowledge transfer: An application of social network analysis in the context of regional innovation networks. *The Annals of Regional Science, 44*(1), 21–38.

Hanneman, R. A., & Riddle, M. (2005) *Introduction to social network methods* (pp. 1–115).

Hansen, D. L. (2009). Do you know the way to SNA? A process model for analyzing and visualizing social media data. University of Maryland Tech Report: HCIL-2009-17, pp. 1–10.

Huisman, M., & Van Duijn, M. A. J. (2005). Software for social network analysis. *Models and Methods in Social Network Analysis*, 270–316.

Knoke, D., & Kuklinski, J. H. (1982). *Network analysis, Sage University paper series on quantitative applications in the social sciences* (pp. 7–28).

Krause, J., Croft, D. P., & James R. (2007) Social network theory in the behavioral sciences: potential applications. *Behavioral Ecology and Sociobiology*, 15–27.

Müller, M. O., Groesser, S. N., & Ulli-Bee, S. (2012). How do we know who to include in collaborative research? Toward a method for the identification of experts. *European Journal of Operational Research, 216*(2), 495–502.

Otte, E., & Rousseau, R. (2002). Social network analysis: A powerful strategy, also for the information sciences. *Journal of information Science, 28*(6), 441–453.

Scott, J. (1988). Social network analysis. *Sociology, 22*(1), 109–127.

Staab, S. (2005). Social networks applied. *Intelligent Systems, IEEE, 20*(1), 80–93.

Wellman, B., & Berkowitz, S. D. (1988). *Structural analysis in the social sciences 2, social structures: A network approach* (pp. 62–82). Cambridge: Cambridge University Press.

Wetherell, C., Plakans, A., & Wellman, B. (1994). Social networks, kinship, and community in Eastern Europe. *Journal of Interdisciplinary History, 24*, 639–663.

White, H. D. (2000). Toward ego-centered citation analysis. In B. Cronin & H. B. Atkins (Eds.), *The web of knowledge* (pp. 475–496). Information Today, Medford: NJ.

Yang, K.-W., & Huh, S.-Y. (2008). Automatic expert identification using a text categorization technique in knowledge management systems. *Expert Systems with Applications, 34*(2), 1445–1455.

# Chapter 16
# Building a View of the Future of Antibiotics Through the Analysis of Primary Patents

**Cristina d'Urso de Souza Mendes and Adelaide Maria de Souza Antunes**

**Abstract** The primary patent application of a new drug is the application that claims the chemical structure of that new compound and is usually the first patent filled regarding that new drug. Therefore, the analysis of the recent primary patents in a therapeutic group reflects the results of the research toward finding new compounds and allows building a view of the future of that therapeutic class. The identification of the primary patents is challenging and requires data treatment, visualization, and analysis tools. In order to address this matter, this paper presents a method for gathering and analyzing the primary patent applications for new antibiotics using the VantagePoint software. This therapeutic class was chosen due to the continuous rise of resistant bacteria, the critical need for new antibiotics that, combined with the lack new drugs in the market, leads to an urgent need for public and private policies to improve research in the field. The method resulted in 1333 primary patent applications of new antibiotics that were analyzed regarding the discovery strategy, the chemical classes, and mechanisms of action and according to the bacteria for which they are active. This analysis was made using different VantagePoint resources and allowed view of the new compounds that might reach the market in the future.

**Keywords** Antibiotics · Patent search · Patent landscape · Pharmaceutical R&D

C.d. de Souza Mendes (✉) · A.M. de Souza Antunes
Technology of Chemical and Biochemical Processes, Technology Center, Federal University of Rio de Janeiro (UFRJ), EQ/UFRJ, Centro de Tecnologia, Bloco E, Ilha Do Fundão, Rio de Janeiro, RJ 21949-900, Brazil
e-mail: cmendes@inpi.gov.br

A.M. de Souza Antunes
e-mail: adelaide@eq.ufrj.br

C.d. de Souza Mendes · A.M. de Souza Antunes
Brazilian National Institute of Industrial Property, INPI/Rua Mayrink Veiga No. 9/19 Andar, CEP 20090-910, Rio de Janeiro, RJ 20090-910, Brazil



T.U. Daim et al. (eds.), *Anticipating Future Innovation Pathways Through Large Data Analysis*, Innovation, Technology, and Knowledge Management,
DOI 10.1007/978-3-319-39056-7_16

## 16.1 Introduction

The discovery and innovation of antibiotics is regarded as one of humankind's foremost achievements, yet the emergence and spread of antibiotic-resistant pathogenic bacteria has put in jeopardy the possibility of treating infections commonly found in communities and hospitals to the point of becoming a major public health problem (Barrett 2005; Fernandes 2006; Hogberg et al. 2010).

There is a critical need to develop new antibacterial agents (Payne and Tomasz 2004). As a way of supporting decision-making processes for policies to foster the development of new antibiotics, this article presents a methodology for selecting and analyzing patent applications for new antibiotics, whose analysis can indicate which compounds could reach the market in the future.

It can take from ten to fifteen years to develop new antibiotics, and only one out of every 5000 to 10,000 compounds that enter the research and development (R&D) pipeline actually receive approval. The cost of developing a new drug is very high—somewhere between US$800 million and U$1.3 billion (IFPMA 2010). Therefore, the return on capital invested in R&D and the profitability of a new antibiotic depends heavily on market exclusivity. Bearing in mind that the cost of imitating these new molecules is far lower than the cost of their development, this market exclusivity is normally sought through the protection conferred by patenting (Simmons 1998).

The pharmaceutical industry is one of the most patent-intensive industries, as industry players attempt to obtain market exclusivity for their inventions for as long as possible. A single drug may be protected by multiple patents making different types of claims, filed at different stages of its development (Table 16.1).

**Table 16.1** Type of patent applications filed during the R&D of new drugs

| Primary patents | Secondary patents |
|---|---|
| 1. Patents for compounds: are patents that claim protection for new drugs (active ingredients) that are not yet considered the state of the art. They do not include claims for new forms of known compounds (such as isomers or polymorphs), which can only be protected by secondary patents | 1. New versions of known polymorphs, salts, hydrates or crystalline forms, isomers, and enantiomers |
| | 2. Formulations, including new compositions or formulations with a view to improving the delivery of a compound |
| | 3. Fixed-dose combinations: combinations of two or more active ingredients taken in a single dose, benefitting patients by reducing the number of pills per day |
| | 4. Synthesis and process methods: New manufacturing processes that improve purity and yields and reduce costs, contributing to economies of scale of the new compounds |
| | 5. Treatment methods/uses: claim for a method for treating disease X with compound Y or using compound A to treat disease B |

*Source* Adapted from data from Instituto Nacional De Propriedade Industrial [INPI] (2010), Kapczynski et al. (2012), and Sternitzke (2013)

According to Sternitzke (2013) and Kapczynski et al. (2012), applications for primary patents (for compounds) are generally filed before drug development begins, which may be ten to 12 years before it is marketable. Meanwhile, secondary patent applications are filed later, often once the drug has been approved by the regulatory agency, thereby adding further years to the period of market exclusivity. Therefore, analyzing primary patent applications in a therapeutic class can give a picture of which drugs may reach the market in the future.

## 16.2 Methodology: Search and Identification of Patents Applications for New Antibiotics

The methodology for identifying primary patent applications was divided into three stages: search for patent applications for antibiotics, select patent applications for new molecules, and remove patent applications unrelated to antibiotics.

### *16.2.1 Search for Patent Applications*

We searched for patent documents filed in the Derwent Innovations Index from 2008 to November 2013 (when the search was done) using classifications designating antibacterial properties of pharmaceutical compounds for human use.

Initially, the idea was to use "antibacterial" or "antibiotic" as the keywords to search for the target patent documents, but we found that this retrieved too many that had nothing to do with the subject, such as the example cited in Table 16.2. We therefore opted for using the International Patent Classification (IPC) codes and the Derwent Manual Codes for antibacterial therapeutic activity.

The main words in all headings (even run-in headings) begin with a capital letter. Articles, conjunctions, and prepositions are the only words which should begin with a lower case letter.

The IPC is organized by the World Intellectual Property Organization and is used by over 70 countries. Subclass A61P covers the therapeutic activity of chemical compounds or medical preparations. We selected the following subgroups from this subclass: "A61P-031/04—antibacterial agents," "A61P-031/06—for tuberculosis," and "A61P-031/08—for leprosy," all under "A61P-031—Antiinfectives, i.e., antibiotics, antiseptics, chemotherapeutics" (WIPO 2010; EPO 2010).

The Derwent Manual Code is a classification system developed by Derwent that divides patents in the Derwent Innovations Index into three broad areas: chemical, engineering, and electronic and electrical engineering. This classification is attributed by the Derwent indexers when the patents are inputted into the database. The codes selected to be searched were all under the umbrella code "B14-A01 Antibacterial," namely B14-A01 Antibacterial general; B14-A01A Gram-negative genera, general,

**Table 16.2** Example of a patent application unrelated to the development of new antibiotics retrieved using "antibacterial" or "antibiotic" as the keywords

| Example: Use of antibiotic in agar plates to select colonies of microorganisms | | | |
|---|---|---|---|
| Application number | Assignee | Title in Derwent | Abstract in Derwent |
| WO2013019647-A1 | LS9 INC | New recombinant microorganism culture useful for producing composition of fatty acid derivatives having target aliphatic chain length and preferred percent saturation with high titer, comprises engineered recombinant microorganisms | EXAMPLE—Thioesterase A (tes A) expression was optimized by modulating activity of 5′ non-coding polynucleotide sequence adjacent to the 5′ end of the open reading frame of the tes A gene via randomization of the regulatory sequences. The regulatory sequences operably-linked to the thioesterase coding sequence were modified by randomization of the non-coding polynucleotide sequences to create a plasmid library. The plasmid library was transformed into cloning strain and colonies selected using Luria-Bertani (LB) agar plates containing suitable quantity antibiotic. The library was transformed into strain DV2 to obtain recombinant microorganisms. Selected colonies were picked and inoculated into glass culture tubes containing 2 ml of LB medium. After overnight growth 50 μl of each tube was transferred to new tube of fresh LB medium. The clones were cultured for 3 h after which each culture was used to inoculate 20 ml of V-9 media in a 125 ml flask. Isopropylthio-beta-galactoside (IPTG) (1 mM) was added to the culture to induce protein expression. After 20 h of fermentation recombinant microorganisms cultures were extracted with butyl acetate |

and other; B14-A01A1 Bordetella; B14-A01A2 Borrelia; B14-A01A3 Escherichia; B14-A01A4 Mycoplasma; B14-A01A5 Neisseria; B14-A01A6 Pseudomonas; B14-A01A7 Rickettsia; B14-A01A8 Salmonella; B14-A01A9 Vibrio; B14-A01B Gram-positive genera, general, and other; B14-A01B1 mycobacteria; B14-A01B1A antibacterial M. tuberculosis; B14-A01B1B antibacterial—M. leprae; B14-A01B2 Streptococcus; B14-A01B3 Streptomyces; B14-A01B4 Staphylococcus; B14-A01B5 Bacillus; B14-A01X combating resistant bacteria.

The search strategy presented in Table 16.3 yielded 32,068 patent documents. Each of these represents a patent family, meaning a “set of patents (or applications) filed in several countries [i.e., patent offices] which are related to each other by one or several common priority filings” (OECD 2009). As each patent record relates to one invention, we shall here refer to each registration in the Derwent database as a patent application.

Figure 16.1 shows that the use of both classifications (IPC and Derwent Manual Codes) was effective, in that some patent filings only used one or other of the codes. Indeed, only 8247 of the 32,068 patent applications retrieved cited both classification codes.

Because so many patent applications were retrieved, we used VantagePoint text mining software to process the data.

**Table 16.3** P search for patent applications for antibiotics in the Derwent innovations index—2008 to November 2013

| Stage | Number of patent applications | Strategy | Description |
|---|---|---|---|
| 1 | 27,409 | [a]MAN = (B14-A01*) | Search using Derwent Manual Codes |
| 2 | 12,906 | [a]IP = (A61P-031/04 or A61P-031/06 or A61P-031/08) | Search using IPC subclasses |
| 2” | 32,068 | 1 or 2 | Sum of results (removing duplicates) |

[a]MAN and IP are the names of fields in the database

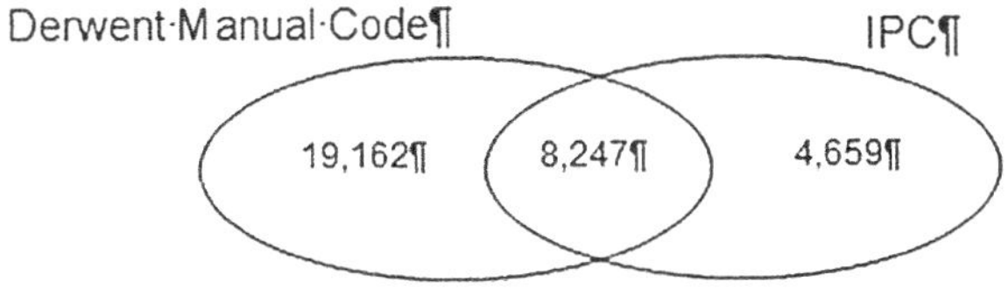


**Fig. 16.1** Number of patent applications retrieved using both classification codes (Derwent manual code and IPC)

### *16.2.2 Selection of Patent Applications for New Molecules*

As the aim of this study is to give a perspective on the future of new antibiotics (new molecules), the analysis of the patents revolves around identifying primary patents, meaning ones for new molecules with antibacterial activity. Therefore, the first challenge in the data processing stage was to identify which patent applications were for new compounds (new molecules), i.e., to separate out from the total of 32,068 the ones of interest to our research.

A specific identifier is designated for each compound described in a patent, called the Derwent Chemistry Resource (DCR) Number. Each DCR Number designates the role of the substance in the patent. Table 16.4 presents the number of patent applications retrieved for each role.

By this process, we identified 8188 patent applications for new molecules. We should note that 2920 of the patents had no DCR Number or Markush Number. In these cases, the titles were read to select the ones that were for new compounds. Through this process, another 166 patents were added to the original group, bringing the total to **8354** to be taken through to the next stage of the analysis.

### *16.2.3 Removal of Patent Applications Unrelated to Antibiotics—For Biological, Phytotherapeutic, and Probiotic Products*

The definition of antibiotics used in this study is the one proposed by Davies and Davies (2010): "**Any class of organic molecule that inhibits or kills microbes by**

**Table 16.4** Number of patents retrieved per role of the substances they describe

| Role | Number of patent applications retrieved |
|---|---|
| K—known compound | 23,165 |
| M—component of a mixture | 19,155 |
| U—use of a single compound | 6655 |
| P—known compound produced | 4895 |
| Q—product defined in terms of starting materials | 1562 |
| S—starting material | 1395 |
| V—reagent | 1143 |
| A—substance analyzed/detected | 614 |
| C—catalyst | 429 |
| D—detecting agent | 203 |
| E—excipient | 58 |
| R—removing/purifying agent | 25 |
| X—substance removed | 17 |
| T—therapeutically active | 1 |
| **N—new compound** | **8188** |

**specific interactions with bacterial targets, without any consideration of the source of the particular compound or class.**"

As such, all the medicines of biological origin that do not act directly on the microorganism, but activate some response in the immune system to combat and/or prevent bacterial infections do not therefore fit into the definition. To remove these patent applications from the sample, we prepared a thesaurus in VantagePoint whose strategy is described in Table 16.5. After removing those patent applications that did not fit the definition of antibiotics, the set of patent documents was reduced to 5999.

**Table 16.5** Strategy for identifying patent applications for biological agents

| Type | Selection strategy | No. of documents |
|---|---|---|
| Antibodies and monoclonal antibodies | Keywords: Antibody or immunoglobulin (552)<br>Title: 301, Use: 401, Novelty: 310, Mechanism of action: 55<br>Manual Codes (922)<br>B04-G01 to B04-G11, B04-G21 to B04-G24, C04-G01 to C04-G11, C04-G21 to C04-G24, D05-H15, D05-H17C1<br>IPC (799)<br>A61K39/00[a], A61K-039/395 to A61K-039/44 | 1311 |
| Vaccines or antigens | Keyword: Vaccine (898)<br>Title: 366, Use: 482, Novelty: 107, Mechanism of action: 829<br>Manual Code: 877<br>B14-S11 and subclasses,—C14-S11 and subclasses, D05-H07<br>IPC: 854<br>A61K39/00[a]: 323, A61K39/02 to A61K-039/39: 804 | 1138 |
| Gene therapy | Keyword: Gene Therapy (253)<br>Title: 5, Use: 30, Novelty: 0, Mechanism of action: 245<br>Manual Code: 266<br>B14-S03, C14-S03<br>IPC: 316<br>A61K-048/00 | 529 |
| Probiotic | Keyword: Probiotic (49)<br>Title: 32, Use: 42, Novelty: 12, Mechanism of action: 0<br>Manual Code: 88<br>D03-H01T2A<br>IPC: no specific class | 122 |
| Medicines containing microorganisms | IPC: classified under A61K-35 but not under A61K31 or A61K38. | 335 |
| Bacteriophages | Keyword<br>bacteriophage<br>IPC<br>C12N-007/00 e C12N-007/01 | 140 |
| Total | | 2355 |

*Source* Own research based on data obtained from the Derwent Innovations Index, 2008—Nov. 2013

[a]The documents from A61K-039/00—medicinal preparations containing antigens or antibodies were included in the antibodies and vaccines groups (both were removed from the sample)

### 16.2.4 Removal of Secondary Patents for New Compounds (Formulations or Intermediates)

Even though all the patent documents were now for new compounds, some were secondary patents for new compounds, because they were mainly for formulations—i.e., patents for the synthesis of an inert compound from a formulation, new versions of known compounds, or new synthesis methods (when the new compound is an intermediate of a known compound). These patents were removed from the selection. Table 16.6 shows the strategy employed for identifying the secondary patents.

Having removed the applications for secondary patents, we were left with 5751[1] patent applications for antibiotics.

### 16.2.5 Selection of Most Recent Patent Documents, Published Between 2008 and 2013

The time period selected for retrieving patents from the Derwent Innovations Index picks up the years when the first applications for patent families are included in the database. This date has nothing to do with the dates the applications are filed, their priority date, and the publication of the application, which are the dates used to determine the validity of the protection conferred by the patents (if granted) and the dates used to search for prior patents.

This search therefore successfully retrieved patent applications published prior to the date used in the search. To make sure the patents analyzed were the most recent in terms of state of the art, we selected those whose first application for the patent family was published in 2008 or later. This reduced the total number of patents retrieved to **5112**.

**Table 16.6** Strategy for identifying secondary patents for new compounds

| Heading | Strategy | No. of patents |
|---|---|---|
| Patents for formulations for new compounds | IPC: A61K-047—Medicinal preparations characterized by the non-active ingredients used, e.g., carriers, inert additives | 173 |
| Hydrates, crystals and polymorphic forms | Derwent Manual Codes: B12-M11H—polymorphic form; B12-M11H2—crystalline form; C12-M11H—polymorphic form; B12-M11H1—special amorphous | 78 |
| Total | | 248 |

[1]Three of the patent documents were classified as formulations and polymorphic forms.

## 16.3 Selection of Patents for Chemical Compounds Likely to Be Antibiotics

Patent filings are not normally very precise when it comes to describing the potential therapeutic uses of a drug. Even if they are classified in one therapeutic class, this does not rule out their being classified in others. It is therefore a challenge to identify the ones that are really for antibiotics.

One example is patent no. US2008153857-A1, whose Derwent abstract is reproduced in Table 16.7 below.

In order to select the patents that were most likely to be for antibiotics from the **5112** selected in the previous stages, we classified them into four groups:

1. patents for new compounds from chemical classes of antibiotics that already exist on the market;
2. patents for specific antibacterial agents—classified specifically as antibacterial agents and not in any other therapeutic category;
3. patents that mention activity tests or give examples showing the antibacterial activity of the new compounds, presenting minimum inhibitory concentrations against pathogenic bacteria;
4. patents that mention pathogenic bacteria in the title or abstract.

This selection process reduced the total number of patent documents to 2029. The methodology used to select patents from each of these groups is described below.

### *16.3.1 Selection of Patent Applications for New Compounds from Chemical Classes that Already Exist on the Market*

The list of all the antibiotic compounds that are still used or have already been used in medical practice was obtained from the ATC/WHO database.[2] These compounds were classified according to their chemical structure, which enabled the identification of the chemical classes on the market.

For each chemical class identified, we consulted two references to obtain the name of the compounds: the World Health Organisation's Anatomical Therapeutic Chemical (ATC) classification system and MEDLINE's Medical Subject Headings

[2]The World Health Organization's Anatomical Therapeutic Chemical (ATC) classification system divides active substances into different groups according to the organ or system in which they act and their therapeutic, pharmacological and chemical properties (WHO 2014).

**Table 16.7** Example of a patent application for many different therapeutic classes

| Title | Abstract (novelty) | Abstract (use) |
|---|---|---|
| New (7-pyridyl-4-phenylamino-quinazolin-2-yl)-methanol analogues, useful for treating condition responsive to capsaicin receptor modulation including pain, itch, cough or hiccup, burns, autoimmune disorder, arthritis, psoriasis | (7-pyridyl-4-phenylamino-quinazolin-2-yl)-methanol analogues (I) or their salts selected from 133 compounds as given in the specification | For reducing calcium conductance of a cellular capsaicin receptor in vivo in an animal; for inhibiting binding of vanilloid ligand to a capsaicin receptor in vitro and in a patient; for treating a condition responsive to capsaicin receptor modulation in a patient; for treating pain in a patient (including neuropathic pain associated with a condition selected from post mastectomy pain syndrome, stump pain, phantom limb pain, oral neuropathic pain, toothache, postherpetic neuralgia, diabetic neuropathy, reflex sympathetic dystrophy, trigeminal neuralgia, osteoarthritis, … loss in an obese patient (all claimed); and for treating conditions (including burns, autoimmune disorder, arthritis including rheumatoid arthritis, psoriasis, Crohn's disease, lupus erythematosus, irritable bowel syndrome, tissue graft rejection, hyperacute rejection of transplanted organs, trauma including injury to the head or spinal cord, cardio- and cerebo-vascular disease, **infectious diseases,** urinary incontinence and overactive bladder) |

(MeSH).[3] We also associated the classes with IPC classifications (whenever there was a specific one). Table 16.8 shows some examples of the thesaurus created.

With this information, we prepared thesauruses in VantagePoint with the names of the compounds and the related terms as keywords, which were put in the "novelty" field of the abstract in the Derwent database.

When the corresponding IPC chemical class was specific (i.e., it did not encompass other chemical classes), the IPC was also used to select the patents in the chemical classes for antibiotics on the market. Table 16.9 shows the number of patent applications from the sample of 5112 that were selected automatically for relating to chemical classes already existing in the market.

We identified 595 patents for new antibiotics from known chemical classes.

### *16.3.2 Selection of Patent Applications Classified Only as Antibacterial Agents and not in Other Therapeutic Classes*

The patent applications called "specific antibacterial agents" were the ones classified as antibacterial agents by the IPC or the Derwent Manual Code that were not also classified under any other therapeutic classes, totaling 1174 patents.

### *16.3.3 Selection of Patent Applications that Mention Activity Tests of the New Compound Described in the Patent Application*

Other patent applications included in the sample were those that gave examples of the antibacterial activity of the new compound being claimed, meaning those with a minimum inhibitory concentration against pathogenic bacteria in the "activity," "mechanism of action," or "use" fields, which yielded 306 patents from the Derwent database.

### *16.3.4 Selection of Patent Applications that Mention Pathogenic Bacteria in the Title or Abstract*

We prepared a thesaurus in VantagePoint with the names of pathogenic bacteria taken from different scientific articles, and used the thesaurus to identify these

[3]MeSH (Medical Subject Headings) is a controlled vocabulary thesaurus. It is used to index articles on MEDLINE, the US National Library of Medicine's bibliographic database, which contains over 22 million references to journal articles in the life sciences.

**Table 16.8** Example of terms used to build the thesaurus used to identify patents applications for new compounds from chemical classes that already exist on the market

| Chemical class | Keywords | ATC/WHO compounds | MeSH compounds | IPC |
|---|---|---|---|---|
| Glycopeptides | Glycopeptide<br>Glicopeptide | Vancomycin<br>Teicoplanin<br>Telavancin<br>Dalbavancin<br>Oritavancin | Teicoplanin<br>Vancomycin | C07K 9/00 Peptides having up to 20 amino acids, containing saccharide radicals and having a fully defined sequence; Derivatives thereof; A61K 38/14—Peptides containing saccharide radicals; Derivatives thereof |
| Tetracyclines | Tetracycline<br>Tetracycline<br>Glycylcycline | Chlortetracycline<br>Clomocycline<br>Demeclocycline<br>Doxycycline<br>Lymecycline<br>Meclocycline<br>Metacycline<br>Minocycline<br>Oxytetracycline<br>Penimepicycline<br>Rolitetracycline<br>Tetracycline<br>Tigecycline | Chlortetracycline<br>Demeclocycline<br>Doxycycline<br>Lymecycline<br>Methacycline<br>Minocycline<br>Oxytetracycline<br>Rolitetracycline<br>Tetracycline | A61K 31/65· Tetracyclines [2]<br>C07C 237/26—of a ring being part of a condensed ring system formed by at least four rings, e.g. tetracycline |
| Beta-lactams | Beta lactam, beta-lactam, betalactam | – | – | C07D 507/00 Heterocyclic compounds containing a condensed beta-lactam ring system, not provided for by groups C07D 463/00, C07D 477/00 or C07D 499/00 to C07D 505/00; Such ring systems being further condensed<br>C07D 507/02· containing 3-oxa-1-azabicyclo [3.2.0] heptane ring systems<br>C07D 507/04· containing 2-oxa-1-azabicyclo [4.2.0] octane ring systems<br>C07D 507/06· containing 3-oxa-1-azabicyclo [4.2.0] octane ring systems<br>C07D 507/08 · containing 4-oxa-1-azabicyclo [4.2.0] octane ring systems |

**Table 16.9** Number of patent applications for classes of antibiotics known on the market—patents published between 2008 and 2013

| Class of antibiotics | Number of applications |
|---|---|
| Beta-lactams (cephalosporins or cephamycins) | 97 |
| Macrolides | 95 |
| Aminoglycosides or aminocyclitols | 80 |
| Beta-lactams (carbapenems) | 79 |
| Oxazolidinones | 52 |
| Quinolones | 49 |
| Beta-lactams (penicillins or penemes) | 36 |
| Tetracyclines | 30 |
| Glycopeptides | 28 |
| Cyclic polypeptides (polymyxins) | 13 |
| Rifamycins | 10 |
| Pleuromutilins | 9 |
| Cyclic polypeptides (lipopeptides) | 9 |
| Beta-lactams | 6 |
| Beta-lactams (carbacephems) | 5 |
| Lincosamides | 5 |
| Trimethoprim and derivatives | 5 |
| Beta-lactams (oxacephem) | 4 |
| Other antibiotics | 3 |
| Polypeptides (tyrothricins) | 2 |
| Cyclic Polypeptides (bacitracins) | 2 |
| Amphenicols | 1 |
| Anti-tuberculostatics—others | 1 |
| Anti-tuberculostatics—aminosalicylic acid and derivatives | 1 |
| Beta-lactams (derivatives of clavulanic acid) | 1 |

*Source* Own research

bacteria in the titles and abstracts of the patent documents in the Derwent database (in the "activity," "mechanism of action," or "use" fields). This yielded 1260 patent applications. Some examples of the bacteria selected are in Table 16.10.

### *16.3.5 Reading and Validation of Patent Applications for New Antibiotic Compounds*

Having selected 2029 patent documents in the previously described automated phases, we found that some of the documents in the final sample did not fit into the definition of antibiotics adopted in this study, because they were for targets other than bacterial targets. The Derwent titles and abstracts were read to validate the

**Table 16.10** Examples thesaurus built to identify pathogenic bacteria in the title or abstract

| Bacteria |
|---|
| Acinetobacter baumannii; A baumannii; A. baumannii |
| Acinetobacter calcoaceticus; A calcoaceticus; A. calcoaceticus |
| Acinetobacter radioresistans; A radioresistans; A. radioresistans |
| Acinetobacter spp; A baumannii; A calcoaceticus; A radioresistans; A. baumannii; A. calcoaceticus; A. radioresistans; Acinetobacter |
| … |
| Ehrlichia canis; E canis; E. canis; Ehrlichia canis |
| Ehrlichia chaffeensis; E chaffeensis; E. chaffeensis |
| … |
| Yersinia enterocolitica; Y enterocolitica; Y. enterocolitica |

documents and remove the ones that were not valid. When necessary, the complete documents were read. Through this process, 696 patent applications were removed because they were for compounds that act on human and not bacterial targets, compounds for veterinary or plant use only, intermediates, inhibitors of virulence factors and quorum sensing, and other reasons.

Completing this process, we retrieved a total of **1333** patent applications for antibiotics. Figure 16.2 summarizes the methodology used to identify the primary patent applications for antibiotics in this study.

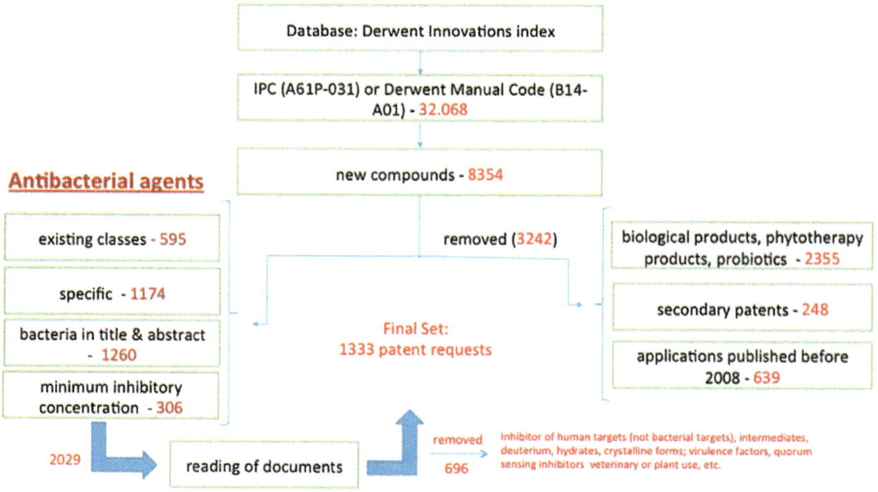


**Fig. 16.2** Stages in the identification of primary patent applications for antibiotics—2008–2013

## 16.4 Use of Primary Patent Applications for Antibiotics to Identify Future Trends

The final set of 1333 patent applications for new antibiotics published in the five-year period from 2008 to 2013 reflects the recent results obtained in the discovery phase of new antibiotics.

In many of the primary patent applications, the mechanism of action or specific activity of the compounds is not described. Often, the patent will say that the compound is active against specific bacteria, but will not present activity data (MIC value) that can be compared with the state of the art. In this discovery phase of new compounds, in vitro and in vivo tests are still being done, so little information on the clinical action of the compounds is available.

These data are provided at later stages of the product development process, normally soon before and during the clinical trials. As such, many of the compounds described in the patents analyzed may not even reach the market, but the analysis of these documents still gives a picture of the discovery strategies being used and potential future compounds.

Examples of analyses of these 1333 patent applications for foresight purposes are presented below.

(a) **Identify the chemical class and mechanism of action of new antibiotics**. This was done by reading the abstracts of the patent applications in the sample. This revealed 195 applications for new compounds in new chemical classes and their mechanism of action (known or not) and 434 applications for new compounds for known chemical classes and mechanisms of action.

Figure 16.3 shows the number of patent applications per mechanism of action and whether the chemical classes are new or known. Many of the patent applications are for protein synthesis inhibitors from existing classes.

(b) **Identify the use of new compounds, i.e., the bacteria the antibiotics act on**. To do this, we prepared a thesaurus of pathogenic bacteria in VantagePoint, which yielded 459 patent applications for broad spectrum antibiotics (for Gram-positive and Gram-negative bacteria), 309 specifically for Gram-positive bacteria, and 157 specifically for Gram-negative bacteria.
(c) **Identify the strategy employed to discover new antibiotics**: To do this, we sorted the 1333 patent applications by the IPC classes they quoted. We found that 563 were obtained from natural sources or using biotechnology, 434 were obtained by altering existing chemical classes, and 336 were obtained by synthesizing new compounds.

The patents identified using the strategies described in a and b served to identify the types of antibiotics and types of bacteria they act on that are currently at the discovery phase and which could be taken to clinical trials and potentially be marketed in the future.

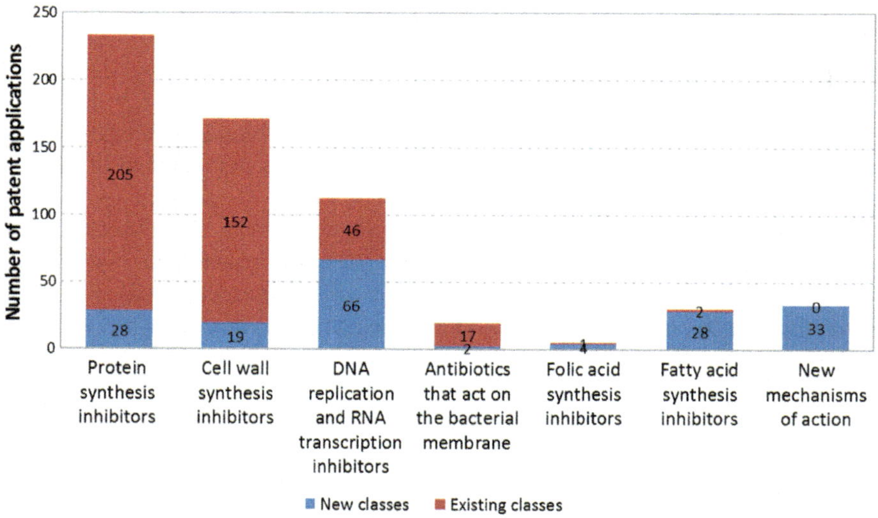


**Fig. 16.3** Number of patent applications for new antibiotics per mechanism of action. *Source* Own research using data from the Derwent innovations index

With these results, it is possible to ascertain whether the new antibiotics from new classes will meet current medical needs and whether the discovery strategies (item c) are effective or even new routes to be explored.

These three approaches could be used together with the results of the clinical trials underway to analyze the R&D process of new antibiotics as a whole. One example is the analysis of the R&D of antibiotics for the Gram-negative microorganisms of the greatest concern shown in Fig. 16.4.

| Bacteria | Patent Applications | Phase 1 Clinical Trials | Phase II Clinical Trials | Phase III Clinical Trials |
|---|---|---|---|---|
| **Enterobacteriaceae (CRE, ESBL)*** | 307 (107 chemical synthesis, 103 natural or biotechnological, 69 existing classes) | 6 (4 beta-lactams, 1 fluoroquinolone) | 1 (beta-lactam/beta-lactamase inhibitor) | 5 (1 aminoglycoside, 2 tetracyclines, 2 beta-lactams/beta-lactamase inhibitors) |
| ***P. aeruginosa*** | 191 (82 chemical synthesis, 66 natural or biotechnological, 43 existing classes) | 2 (1 beta-lactam, 1 new) | 1 (beta-lactam/beta-lactamase inhibitor) | 2 (beta-lactams/beta-lactamase inhibitors) |
| ***N. gonorrhoeae*** | 74 (41 chemical synthesis, 19 natural or biotechnological, 14 existing classes | 0 | 0 | 1 (solityromycin – ketolide) |

**Fig. 16.4** R&D of antibiotics for the Gram-negative microorganisms of the greatest concern

This shows that in the near future, there will be an increasingly limited range of options for fighting Gram-negative bacteria, because the clinical trial pipeline is increasingly scant, and mostly for compounds from existing chemical classes, despite the pressing need for new classes of antibiotics. In our study of patent applications, we identified some research of new compounds with new mechanisms of action, but these are yet to progress to clinical trials, which is necessary before they can be introduced as safe, effective options in the future.

## 16.5 Conclusions

In this article, we presented a methodology for searching and selecting primary patent applications for new antibiotics. The first part of the methodology was developed in order to separate the primary patents from the secondary patents. Taking into account that the therapeutic classes are not well described in primary patents, the second part of the methodology aimed at selecting the patents with the highest chance to be of a new antibiotic.

Using this methodology, we selected a group of 1333 patent applications that can be further analyzed regarding patent applicant that will allow identifying the institutions that are still doing research in the field and the object of its research. These patent applications can also be analyzed regarding its technical matter, the results can show if the new antibiotics are from old or new chemical classes, for which bacteria they are active against, if they have new mechanism of action, it also the strategy used to discover new compounds and many other analysis.

The results of the further study will point out the new antibiotics that could reach the market in the future, and therefore, the study be used in the decision-making processes in order to prioritize the critical bacterial diseases that need new antibiotics, meaning the diseases with higher medical need and less new products being researched to treat it.

## References

Barrett, J. (2005). Can biotech deliver new antibiotics? *Current Opinion in Microbiology, 8*, 498–503.

Davies, J., & Davies, D. (2010). Origins and evolution of antibiotic resistance. *Microbiology and Molecular Biology Reviews, 74*, 417–433.

European Patent Office [EPO]. (2010). IPC (International Patent Classification). Available: http://www.epo.org/patents/patent-information/ipc-reform.html. Cited: May 30, 2010.

Fernandes, P. (2006). Antibacterial discovery and development—The failure of success? *Nature Biotechnology, 12*, 1497–1503.

Hogberg, L. D., Heddini, A., & Cars, O. (2010). The global need for effective antibiotics: Challenges and recent advances. *Trends in Pharmacological Sciences, 31*, 509–515.

IFPMA. (2010). The pharmaceutical innovation platform: Sustaining better health for patients worldwide. Available http://www.ifpma.org/documents/NR1916/PIP_final.pdf. Cited: 20 May 2010.

Instituto Nacional da Propriedade Industrial [INPI]. (2010). O que é patente?. Available: http://www.inpi.gov.br/menu-esquerdo/patente/pasta_oquee. Cited: May 30, 2010.

Kapczynski, A., Park, C., & Sampat, B. (2012). Polymorphs and prodrugs and salts (oh my!): An empirical analysis of "secondary" pharmaceutical patents. *PLoS ONE, 7*, e49470.

OECD. (2009). OECD Patent Statistics Manual. Available: http://browse.oecdbookshop.org/oecd/pdfs/browseit/9209021E.PDF. Cited: Jul 20, 2009.

Payne, D., & Tomasz, A. (2004). Antimicrobials—The challenge of antibiotic resistant bacterial pathogens: the medical need, the market and prospects for new antimicrobial agents. *Current Opinion in Microbiology, 7*, 435–438.

Simmons, E. (1998). Prior art searching in the preparation of pharmaceutical patent applications. *Drug Discovery Today, 3*, 52–60.

Sternitzke, C. (2013). An exploratory analysis of patent fencing in pharmaceuticals: The case of PDE5 inhibitors. *Research Policy, 42*, 542–551.

World Health Organization [WHO]. (2014). International language for drug utilization research - ATC/DDD. Available: http://www.whocc.no/. Cited: Jun 10, 2014.

World Intellectual Property Organization [WIPO]. (2010). International Classifications. Available: http://www.wipo.int/classifications/ipc/en. Cited: May 30, 2010.

# Chapter 17
# Combining Scientometrics with Patent-Metrics for CTI Service in R&D Decision-Making: Practices of National Science Library of CAS

**X. Liu, Y. Sun, H. Xu, P. Jia, S. Wang, L. Dong and X. Chen**

**Abstract** Scientometric analysis and text-mining have been applied to scientific and technological trend-tracking and related scientific performance evaluations for several years in China. Since 2012, NSL-CAS provides CTI (competitive technical intelligence) services based on metrics for supporting R&D decision-making. NSL helps technology-based firms improve their innovation capabilities via CTI, for technology novelty review, selection of innovation paths, product development evaluation, competitor monitoring, identification of potential R&D partners, and support for industrial technology and development strategizing. Scientometric methods have established many indicators for technology analysis that can be applied individually or in combinations. Composite indexes are another useful option. For CTI services, we choose or customize layer or level indexes schemas for different purposes. For supporting industrial technological strategy decision-making and innovation path identification, scientometric indicators can be used for R&D

X. Liu (✉) · Y. Sun · H. Xu · P. Jia · S. Wang · L. Dong · X. Chen
National Science Library, Chinese Academy of Sciences, Beijing 100190, China
e-mail: liuxw@mail.las.ac.cn

Y. Sun
e-mail: sunyl@mail.las.ac.cn

H. Xu
e-mail: xuhf@mail.las.ac.cn

P. Jia
e-mail: jiap@mail.las.ac.cn

S. Wang
e-mail: wangs@mail.las.ac.cn

L. Dong
e-mail: donglu@mail.las.ac.cn

X. Chen
e-mail: chenxl@mail.las.ac.cn

X. Liu
Center for the Study of Information Resources, Wuhan University, Wuhan 430072, China



T.U. Daim et al. (eds.), *Anticipating Future Innovation Pathways Through Large Data Analysis*, Innovation, Technology, and Knowledge Management,
DOI 10.1007/978-3-319-39056-7_17

trend analysis. Specifically, in meso-technology analysis, bibliometrics and patent analysis indicators can be combined in accord with different subjects or stages of an emerging technology, whose characteristics can then be reflected by these mixed indicators. Scientometric indicators can profile the framework for research subjects, and patent metrics can describe the technology development trends. In micro-technology analysis, technology trends analysis is used for new technological product development in planning strategy for technology-based firms, and bibliometric indicators can identify directions of related scientific subjects and research directions. In fact, when a client expresses a CTI need, they request the meso- and micro-, and even macro-technology analysis. So when we execute a CTI service, we run an iteration and loop analysis through bibliometric and patent metrics. We focus theme tracing or subject analysis by tech-mining and co-wording. For macro analysis, such as competition from institutions or countries and regions, we pay close attention to the combination of scientometric and patent indicators and appropriate schemas for CTI services.

**Keywords** CTI for industrial technology · Iteration analysis · Scientometric indicators · Patent technology analysis · National science library of CAS (NSL-CAS)

## 17.1 Introduction

The National Science Library of the Chinese Academy of Sciences (NSL-CAS) was established in 1949, and it is the biggest academic library and information institution among China's numerous scientific and technological information institutions. The integration of traditional library functions and information consulting services has been its fundamental strategy for some time. Generally, NSL-CAS provides three types of information consulting services, which include S&T information monitoring, technology intelligence analysis at different levels—such as technology forecasting, technology evaluations, and policy-making support for S&T (science and technology) administration, and traditional library services for end users (for researchers and scientists). As early as 1978, the NSL-CAS established an Information Analysis Department for handling the increasing intelligence service needs. Its main task was to serve the needs of R&D management, supporting the service needs of relevant policy-making and exploring the intelligence service model for needs drawn from the scientific disciplines.

Since 1985, NSL-CAS has been working on scientific and technological information and intelligence analysis, which closely follows the needs of S&T management, decision-making, technology commercialization, in-house R&D of firms, and R&D research projects. In 2001, NSL-CAS experienced a significant transformation, in which it began to exploit the ICT and digital content technology to meet the universal needs for supporting scientific research literature. NSL-CAS took on trend analysis in scientific fields, technology development, and competitive technical intelligence for industrial sectors. In these services, NSL-CAS applies scientometric indicators and text-mining techniques to technological trend tracing and provides professional analytic reports on scientific fields and R&D trends, supervised by scientists and technology specialists in those fields.

So far, NSL-CAS's scientific and technological trend analysis and dynamic monitoring services have focused on 13 key scientific fields and provide situation analyses and monitoring of those fields, technology monitoring and profiling, support for science and technology planning and strategy, and for industrial technology development. Its customer base has expanded to include policy-makers, R&D institutions, R&D firms, and technological innovation teams. The services and products of NSL-CAS now include technology dynamic monitoring reports, technology subject analysis and scanning, analysis of technology development trends, and strategy and planning consulting report.

In recent years, NSL-CAS introduced the concept of CTI into its services and has combined logistics, analysis method, and data mining in technology, market, and business intelligence, aiming to provide the best possible services for strategic decision-making related to S&T, R&D management, technology commercialization, technology transfer, and competitive technical intelligence for enterprises. In short, NSL-CAS is trying to change its identity and intent, to realize a strategic transformation, from a traditional library that primarily serves as an information or literature provider, to that of an information hub, whose job is to provide knowledge, rather than just raw information.

Methodologically, NSL-CAS has been exploring the application of scientometric indicators to decision-making support and scientific subject monitoring. It also combines different methods—bibliometrics, patent-metrics, text-mining, and expert review—for improving the quality of its competitive technical intelligence services. The purpose of this article was to explain how these methods are being used in the CTI service in NSL-CAS.

## 17.2 Practices of CTI in NSL-CAS

In practice, NSL-CAS has three primary types of CTI services for clients—novelty review for the development of a specific technology or scientific subject; CTI for the development of a particular technology theme; and CTI for an industry. They address three levels of needs—micro-level, meso-level, and macro-level. We will introduce them with examples for each.

**Example 1**: Novelty evaluation service for development of specific technology topics.[1] For specific technology topics—an example of micro-level services—our CTI services are provided mainly in the form of novelty search and review reports

[1]Source: NSL-CAS novelty review report of project for funding, The Application of Hydrodynamic Cavitation Technique in the Wastewater Processing, 2014.

for technology research and development. There are two kinds of novelty review services in NSL-CAS. One is for evaluating the novelty and necessity of an R&D project proposal before it is formally started and funded—a proposal checkup service. Another is for assessing a finished project's performance and achievements, for, say, project acceptance. As its name indicates, novelty review serves the purpose of scientific project evaluation predominantly through the indicators of novelty and is primarily for researchers of universities, research institutes, and firms. The first kind of novelty service, for project proposals, is essential to the projects' management, because it relates to funding allocations. But proposal novelty examination services are not prevalent in the past twenty years, because government's huge scientific budget is being poured into S&T projects, and efficiency has been neglected. The governmental S&T management departments have paid more attention on project acceptance. In the past two decades, NSL-CAS kept improving its work and has obtained the ability to complete about 300 novelty review reports per year, most of which are novelty examinations for project acceptance. In this paper's discussions, we focus on the novelty examination service for project proposals.

For instance, the company A plans to develop a hydrodynamic cavitation technology for wastewater processing, which is for protecting the environment. The company asked NSL-CAS to provide the novelty examination service for the proposal. NSL-CAS created a novelty review report for the technology project proposal with the bibliometrics and patent metrics data from the WOS database or others, and it serves as reference material with the project proposal for funding and technology development strategies. In preparing the report, we found that the research points and topics of hydraulic cavitation technology's application in sewage treatment mainly concentrate on the hydraulic cavitation mechanism, processing methods, and devices, and the key factors of cavitation, by mapping and profiling the technology topics. What's more, we have illustrated, hydrodynamic cavitation technology can also be applied to biological cell wall breaking, sterilization, and biodiesel preparation.

We achieved these findings primarily by means of keyword clustering of articles and patent theme mapping. The first step was to search relevant scientific papers and form a research article dataset.[2] With the help of TDA® (Thomson Data Analyzer), we chose keywords from those research articles about hydrodynamic cavitation technology, and then clustered them. The result we obtained is shown in Fig. 17.1 as a network of domains.

The second step is to create a patent application or document dataset, which we get by searching patent data regarding hydrodynamic cavitation and applications. We identified more than 290 active patent assignees in this area, and individual patentees are 129, accounting for 44.48 % in all applicants. Also, we found that among the top eight assignees, four are American. The countries that ranked as the top five are Russia, US, China, Canada, and Japan. By analyzing the titles and

[2]Search formula = ("water cavitation*" or "hydrodynamic cavitation*") in WOK database.

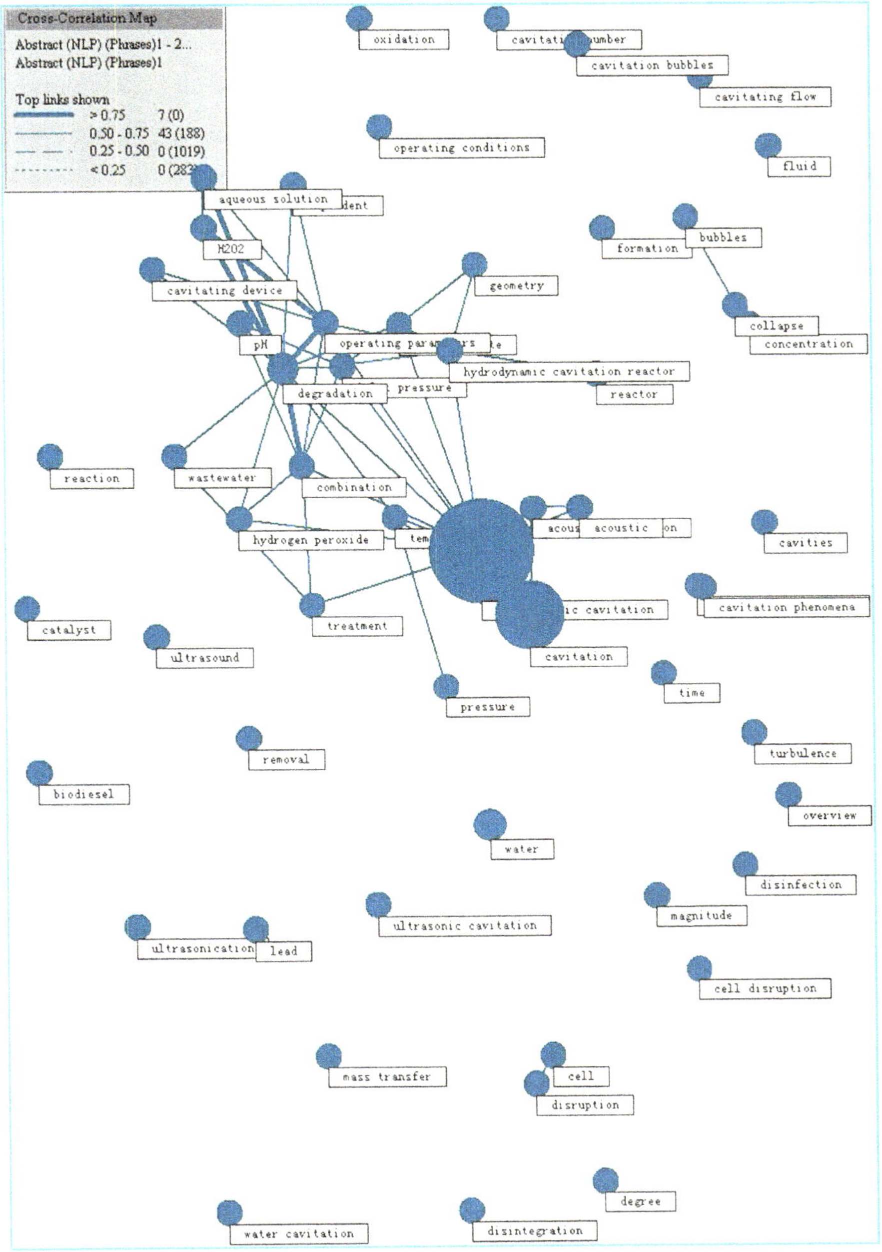


Fig. 17.1 Relationship between research subjects in hydrodynamic cavitation technology

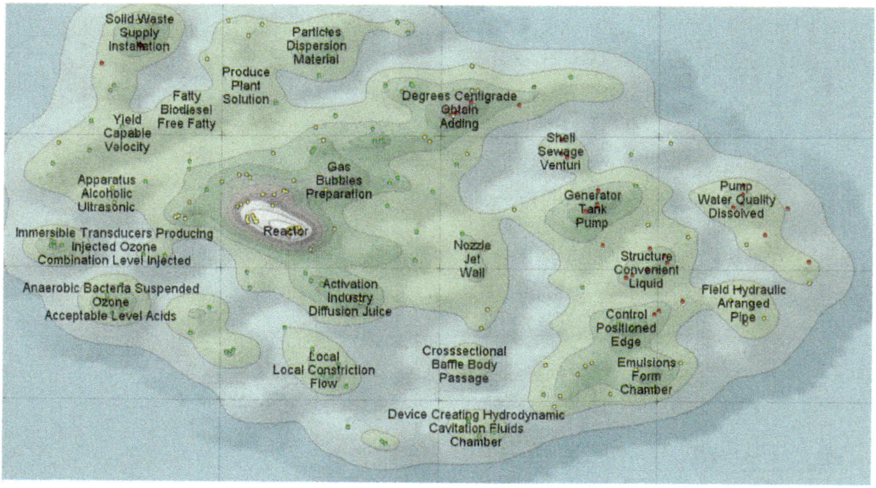


**Fig. 17.2** Patent technology subjects layout. *Source* NSL-CAS project novelty review report, the application of hydrodynamic cavitation technique in the wastewater processing, 2014. In the figure, the *yellow points* represent the Russian patents' subject distribution, which concentrate on reactors; the *green points* reflect the American ones, no apparent concentration; while the *red points* are Chinese patentees, which mainly focus on applying hydrodynamic cavitation techniques to wastewater processing

keywords of these patents, with the help of TDA, we achieved a patent theme map (Fig. 17.2). It is apparent that the majority of patents concentrated on the field of hydrodynamic cavitation reactors.

By scientometric methods, in Fig. 17.1, we count up the frequencies of keywords in the research papers, cluster the research subjects, and then draw out the keyword relations visually. Other than usage of scientometric indicators to analyze the development of hydrodynamic cavitation technology, we also use the IPC (International Patent Classifications) of patents to pattern the technology fields and important points, and draw the layout of patents in Fig. 17.2, showing that the subject of most of the patent applications is hydrodynamic cavitation reactors. By checking the map of patent subjects, we found the major patents of hydraulic cavitation devices center around the use of porous plates and venturi tubes, and that hydraulic cavitation is coupled with oxidation reagents to strengthen the quantity of free radicals through which the effectiveness of removing organic pollutants can be improved. In conclusion, as an emerging technology, hydrodynamic cavitation, with its relevant technology, has been applied in organic wastewater treatment. It is a valuable technology and worthy of further development.

**Example 2**: CTI for technology fields.[3] When facing a broader technology area, rather than a "technological topic," which is at the meso-level that we define, we

---

[3]Source: NSL-CAS project novelty review report, Serial Reports of Swine Vaccine Technology Analysis, 2014.

also execute the competitive intelligence service for clients, particularly for industrial clients. Contrasted with the novelty examination service for the specific technology, CTI reports comprise a full-range analysis of related technology topics, in which the reports integrate a scientometric index, patent technology analysis, and professional reviews of technological development. All these reports are organized systematically.

Company B is a medical technology company whose main products are pig vaccines. In 2014, upon the company's request, we accepted the task to analyze the development of swine vaccine technologies, including pig pseudorabies vaccine, swine fever virus vaccine, swine Japanese encephalitis vaccine, and pig transmissible gastroenteritis vaccine. Note that this subject's range is broader than the company A example, which is relatively narrow, and concerns only one specific technology (hydrodynamic cavitation technology). In contrast, this one is about an area that contains several sub-techniques, so the output of our work will be a series of reports rather than just one.

In the example of company B, one sub-report of the serial analysis report is about research and technology development of a swine fever virus vaccine. We analyze R&D advances in this subfield with the help of bibliometric indicators such as article publications and citation data. We describe past evolution and current situations, and predict the trend of the swine fever virus vaccine technology's development. We perform a cluster analysis to identify the key techniques and their distribution, and indicate the main competition and potential R&D cooperation in this area. After an analysis based on academic articles, we perform a patent analysis, which includes patent application and development trends, patent distribution in different countries (regions), patentee and technology subject distribution analysis, and competition and cooperation among patent assignees.

Specifically, to produce an analysis of pig pseudorabies vaccine, which is one sub-report of the series, we firstly collect relevant research papers. By searching the databases of the ISI Web of Science and ISI Medline,[4] we form our dataset. Then, we roughly identify the topics of these papers and index the subject words, relying mainly on machine reading. Through a simple statistical analysis, we find the most popular research subjects about pig pseudorabies vaccines are vaccine effectiveness, vaccine preparation, effects of maternal antibodies, vaccine immunity pathways, vaccine immunogens, vaccine immunity adjuvants, passive immunity, immune modulators, vaccine security, vaccine carriers, nasal cavity immunity, and so on. Based on analysis of these subject words, most research concentrates on immune effectiveness, vaccine preparation, and maternal antibodies.

We then use the ISI Derwent Innovations Index database to create a patent dataset, and obtain approximate 164 results relevant to our subject of pig pseudorabies vaccine. China and the USA are two predominant sources of patent applications. After scanning and selecting these data using computer-based

[4]Search formula = (porcine or pig or swine) and ((pseudorabies or aujeszky disease) near vaccine*).

techniques, we get a dataset with 89 patents that are closely associated with the technology of pig pseudorabies vaccines. These data indicates that the primary patent topics also include vaccine preparation, vaccine's immune effectiveness, vaccine carriers, vaccine immunity adjuvants, testing methods for the vaccine's immune response effectiveness, vaccine immunogens, and passive immunity. Similar to what was revealed in the article dataset, the most frequently appearing study themes are vaccine preparation, immune effectiveness, and vaccine carriers.

**Example 3**: CTI for an industry sector's strategic decision research[5]: Apart from CTI services at the micro- (particular technology topics) and meso- (technology subject) levels, NSL-CAS also provides analysis services for an entire industry, which we call macro-level service. Currently, we provide consulting services to local government agencies and support their decision-making process, for which we produce several reports, such as *Strategic Intelligence of Ionic Rare Earth Industry, Technical Intelligence Analysis of Effective Development and Utilization of Tungsten Resources Industry, Industry Technology and Economic Analysis of Coal Glycol*. In these works, we combine different analysis methods and tools, such as literature reviews (tertiary information or documents), scientometric indicators, patent analysis, and text-mining, to make our reports as comprehensive and referable as possible. To perform such tasks, we organize an operation team, construct the list of key intelligence topics (KIT) according to our discussions with the researchers, and perform bibliometric analyses and patent technology theme (or core-tech) analyses.

Here, we cite the Ionic Rare Earth Industry report as an example. We first identify and review the key technology fields or topics of the ionic rare earth industry, and find it is mainly associated with extractions and separations, Nd–Fe–B magnetic material, white LED rare earth phosphor powder, rare earth hydrogen storage materials, rare earth ceramics, etc. We perform an analysis for each of these fields. For example, in rare earth extraction and separation, based on the academic article and patent publication data from WOS, via scientometric indicators that were mentioned above, we identify the most important countries and top five organizations that perform well in relevant research and development. And we identify the hot topics and find core-tech patentees. More specific analyses follow:

1. Research organization analysis: For organizations that are involved in research of rare earth extraction and separation techniques, we do a statistical analysis of the top five's publications. The results show that they are most interested in: solvent extraction, liquid–liquid extraction, synergistic extraction, ion exchange, Cyanex 272 and 923, crown ethers, fractionated extraction, rare earth element extraction (cerium, scandium, yttrium, ytterbium, lanthanum, samarium, erbium, phosphate). When focusing on four organizations, the Chinese Academy of Sciences, the Russian Academy of Sciences, India's Bhanha Atomic Research Center, and Japan's Atomic Energy Research Institute, keyword frequency statistics reveal their research emphases. The Chinese Academy of Sciences

[5]Source: NSL-CAS Industry Analysis Report, 2014.

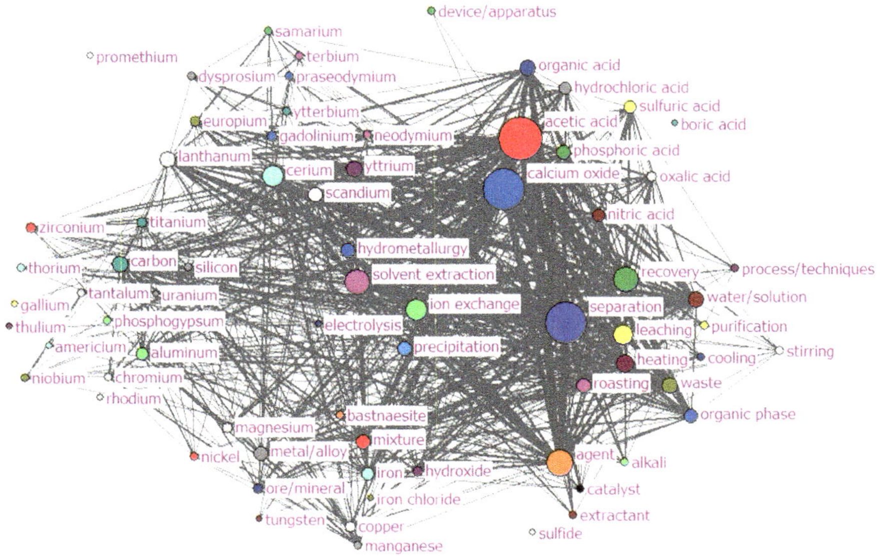


**Fig. 17.3** Technology theme clusters for rare earth extraction and separation

focuses on trace analysis, europium, neodymium, lanthanum, film, carboxylic acid, nitric acid, and nitric acid. The Russian Academy of Sciences mainly pays attention to nitrate, yttrium, europium, cerium, lanthanum, nitrate molten salt, Cyanex 272, and crown ethers. The Bhanha Atomic Research Center prefers to do more research on liquid–liquid extraction, selective extraction, thenoyl trifluoroacetone (TTA), TODGA, Cyanex 923, nitrate, yttrium, terbium, ytterbium, solvent extraction, luminescence, and circulation. The Japan Atomic Energy Research Institute emphasizes researching circulation and rare earth elements such as europium, neodymium, yttrium, lanthanum, erbium, and dysprosium.

2. Cluster analysis of subfields of the current research areas: Based on the ontology and knowledge of rare earth science, rare earth extraction and separation techniques have the following subfields: ore decomposition (with different acids, such as nitric acid, hydrochloric acid, and sulfuric acid), extraction of rare earth oxides and elements (extraction from liquid or roasting, applying various reagents, P204, and TBP), extraction and solvent filter devices (including reaction devices, extraction equipment, and ion exchange devices). In the analysis process, we perform natural language processing (NLP) for keywords of rare earth extraction and filter technology, and clean and merge the thesauruses and classifications. We then apply the relevance analysis UCINET software package. Finally, we obtain a keyword clustering figure like that shown in Fig. 17.3.

The clustering analysis shows that research topics regarding rare earths extraction and separation primarily concentrate on extraction processes and methods, aids and solvents used, extraction and separation of different elements, relevant equipment and devices, sources of rare earth elements, and wastewater treatment in the processing. The hottest topics:

(a) Rare earth extraction and separation technology: Major methods: ①hydrometallurgy, including the ion exchange method, a solvent extraction method, and a precipitation method; ②pyrometallurgy, purifying the calcined rare earth's oxides or alloys. The major processes are leaching, ion exchange, or precipitation separation, calcining and purifying, wastewater treatment. Sometimes heating or cooling is required.
(b) Acids and solvents used in extraction and separation: The clustering analysis indicates that the extraction and separation now mainly rely on hydrometallurgy, in which the most frequently used acid solvents are hydrochloric acid, sulfuric acid, nitric acid, oxalic acid, and phosphoric acid. Commonly used extracting agents are amine compounds.
(c) Rare earth element separation: The most important part of the extraction and separation techniques is to separate cerium, yttrium, and scandium from the rare earth oxides or ores and purify them.
(d) Devices for extraction and separation: The primary devices and equipment include those for reacting, precipitating and washing, filtration, extracting, ion-exchanging, evaporating, crystallizing, and drying.
(e) Rare earth ore and sources: Specific extraction and separation processes always depend on the rare earth sources. From these technique themes, it is apparent that the most frequently appearing ores are kaolin, monazite, and baiyuneboite. Other rare earth sources include wastes of phosphor powder, and magnetic materials and alloys. We find that most of the patents about rare earth extraction belong to Japanese organizations.
(f) Wastewater treatment in separation processes: Research in this area focuses on acidic, alkaline, and radiological wastewater processing. This includes the treatment of exhaust gas, to purify the sulfuric acid mist, chlorine gas, and hydrogen chloride gas.

## 17.3 Experiences and Discussion

In keeping with its CTI service goals and client needs, NSL-CAS defines its service objectives and clients as R&D management, research teams, research projects, technological innovation of firms, and industrial technology development (focusing on firms' technology innovations). In this framework, NSL-CAS has developed three kinds of CTI services, including R&D novelty review for technology developers, technological innovation pathway selection, product technological advancement assessment, technology competitor monitoring, collaborator identification, and

S&T strategy decision-making. To provide the best consulting and decision-making support, we guarantee the quality of the intelligence analysis reports. In the following section of the article, we discuss the experiences of NSL-CAS making our intelligence analysis more reliable and accurate, even valuable. We conclude by enunciating principles for high-quality technology intelligence reports.

### *17.3.1 To Classify CTI Services into Three Levels and to Create Different Tactics for Applying Scientometric Indicators Accordingly*

Competitive technical intelligence is undertaken to monitor and interpret key events that impact the technology strategy of client institutions and to provide continuous awareness of science and technology trends.[6] Key events include scientific breakthroughs, R&D design, strategic initiatives, and advancement of key technology commercialization. CTI is widely used for monitoring and interpreting key scientific and technological developments, and it is a useful method to track science and technology development trends. Before W. Bradford Ashton (2009) first introduced competitive intelligence into laboratories, research institutes, and government agencies, CTIs were employed only as part of competitive intelligence for business technology strategy and firms' R&D management. But now, with the help of bibliometric and scientometric tools, a comprehensive mechanism has been developed for applying the CTI to serve different customers and help them maintain their competitive advantages.[7]

As presented in this article, upon receiving customer needs, NSL-CAS classifies the required services in three levels.

The first is micro-level services. Such services focus on particular research or technology topics (for example, the hydrodynamic cavitation technique mentioned above). The CTI service at this level mainly serves the purposes of assessing advancement and feasibility of the technology, and identifying key competitors. We usually do this through novelty review for R&D project proposals and evaluation of product technology advances.

The second is at the meso-level services. Services focus on a technology subject or field, which might consist of many subfields or topics (for example, the swine vaccine technology mentioned above, composed of many different kind of vaccines—but all for pigs), rather than just a narrow technical point. CTI services at this level are mainly for R&D management, such as the management of R&D projects, R&D teams, and for the firm's new product development. Specifically, NSL-CAS provides trends analysis of the scientific research field, core technology analysis, product technology trend analysis, and submits individual reports to the clients.

[6]Murphy (2001).

[7]Ashton and Hohhof (2009).

The last one is at the macro-level services. The services are related to business and strategic intelligence, in which we analyze technology development of a particular industry (this covers an entire industry, such as the rare earth instance shown above). This CTI service is mainly for strategic planning of research institutes and firms, technological trends of emerging industries, science and technology dynamics or trends (of different countries and regions), and other uses on this level. The service subjects or contents include discipline strategies in research institutes, technology trends analysis of the particular industry, even competitiveness comparisons of the science and technology in different regions or countries.

For these different levels of CTI services, we employ different analysis methods of articles and patent applications.

At the micro-level, the CTI of particular technology topics, we rely on precise retrieval techniques to collect the information and create a database, in which the standard datasets of academic articles and patents are organized. Then, we read the research papers and patent applications one by one and classify them. With the help of clustering analysis, we can detect the important research themes, directions, research communities (teams and individuals), collaboration, and competitive relationship or situations. In this intelligence mining process, we employ general scientometric and statistical indicators of technology related to the patents to explore the state of technology development, such as the publications, authorship, classification and category, patentee, and citation. We use these indicators to reveal the state of the technology. Then, we summarize and draw conclusions from the comprehensive review of scientific advancements written by specialists or scientists in related professional organizations.

At the meso-level of CTI services, we seek to establish a framework for technology topic analysis, firstly by discussing with the clients. We execute retrieval actions for collecting the scientific research and patent applications by iteration, and create a database. According to the analysis framework, we choose the bibliometrics and scientometrics and patent-metrics indicators to analyze the stages of technology development, to find the important research institutions (important research teams and researchers), to highlight the hottest technology subjects or topics, and to reveal the relationship of keywords. And by means of topic and keyword clustering, IPC, patent technology function analysis, and co-word analysis, we can identify the development stages, important specialists, core technology, and relationships between research themes in a given technology subject.

For the macro level of CTI services, different indicators and methods of scientometrics, text-analysis, and text-mining, based on big data, are used comprehensively. In the analysis process, we pay attention to the features of the technology evolution, hot technology topics, relevance of technology topics, the relationship of technology to industries, and the technology competitive landscape. In short, depending on clients' specific requirements and service needs, we figure out different relationships via analyzing the metadata of academic articles and patents, and mining the information and word relationships of the full text.

### 17.3.2 Analyzing the Technology Trends of an Industry via the Combination of Bibliometric and Patent-Metric Indicators

The achievements of scientific researches are usually present in the form of academic articles and patents. The bibliometrics or scientometrics, which are based on academic articles, rely on the metadata of articles to reflect the scientific research activities in outline. The quantity of publications could reflect the vitality of the research field, the number of citations could reflect the importance of an article, and the quantity of publications or patents shows the research abilities of the countries, regions, and organizations that produce them. Metadata analysis based on patents could imply the relationship between scientific research and industry. It also shows the technology innovation capability of firms (research institutes, countries, and regions), partnership and competition among them, and the evolution of relevant technologies. Moreover, by using patent citation, we can identify the core technologies in an industry, form a patent pool, and develop cooperation in R&D.

In the CTI practice of NSL-CAS, to exploit the different features of scientometrics (bibliometrics) and patent-metrics, we have built an integrated and complex analysis framework, to show the R&D situation of an entire industry. How do we operate the service? Firstly, we employ bibliometric methods to describe the research topic distribution in the scientific field and use article citations to profile the evolution paths of research topics. Exploring publications, we are able grasp the details of the field in terms of the most advanced research and the leading researchers.

Secondly, we conduct a patent analysis in which the focus is to reveal the most significant technological topics, their current situations, research capabilities of key R&D organizations, and from these, to derive the major technology directions of the field in question.

Thirdly, we choose technology subjects closely related to the patent technology, working from the subject map we create based on a bibliometric analysis of research papers. Then, we turn back to patent technology analysis for the technology subject and work out the key directions of the industrial technology.

Alternatively, we can also reverse part of the process, by doing further bibliometric analysis on the patent technology topics. We choose technology subjects from the patent technology topic map, then perform bibliometric analyses of the research papers around the patent topics. From these analyses, we identify "hot" research subjects and subject details, and show the research power distribution and competitiveness. No matter which way we take, the main purpose of this analysis was to find the "hot" research topics, their distribution, and research organizations' relationships and competitiveness.

Similarly, when using the CTI service in response to the industrial sector's requests for assistance with technology and innovation path selection, scientometric indicators are very effective for displaying the technology development directions from a macro-perspective. With patent-metric analysis, we can find the most important technology details. For micro-level technology analysis, integrating bibliometric and patent-metric indicators allows us to establish different analysis frameworks over business R&D orientations for different issues in different technology development stages (discussed below). Such a composite index is good for describing basic features of the technology's development. For meso-level analysis, patent technology trend analysis, in which scientometric indicators are used for technology monitoring, can be widely applied to new product technology development.

### 17.3.3 Establishing a CTI Analysis "Iteration" Mode for Science and Technology Monitoring

Based on the technology S-curve, Brenner (1996) discussed patents, seeing them as a boundary point between research and technology development. He explained the relationship between technology intelligence and competitive intelligence, and he illustrated how CTI services work in product lifecycles and in technology lifecycles. Murphy (2001) classified a product's lifecycle as conceptualization of new product, maturation of new techniques, commercialization of the technology, and commercialization of the product. He identified the information flow of the entire process, which includes gray literature (information that falls outside the mainstream of published journal and monograph literature, not controlled by commercial publishers), research articles, patent applications, development of technical processes in enterprises, products releases, and product sales. Industrial technology and basic research are naturally related, and patent literature bridges between them. Thus, CTI services have two parts, the monitoring of research subjects and of technology development.

In its CTI services, for supporting its scientific subject analysis and technological topic analysis, NSL-CAS established an iteration mode—"bibliometric analysis + patent technology analysis." A scientific subject (domain or subject or topic) framework is constructed by bibliometric analysis; then key industrial technologies are chosen as topics from the bibliometric subject framework, for subsequent patent technology analysis. Core patent technologies are selected afterward, and bibliometric analysis is used again, to analyze the core patent technologies. Content analysis and mining are also introduced.

In many cases, consulting agencies (such as the NSL-CAS) suffer from the lack of specific professionals who have detailed scientific and technological knowledge or subject background in the relevant areas, which makes their CTI services not as professional as we would like. However, science and technology advance so rapidly that it is very hard for intelligence analysts to find suitable professional for every subject. But scientometric analysis is able to provide trend information even in the absence of analysts expert in the particular subject, and it is bolstered by the patent-metrics analyses.

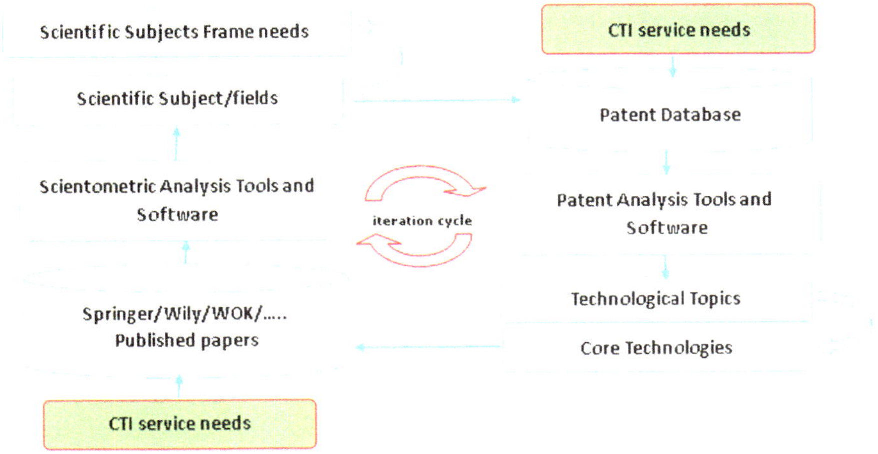


**Fig. 17.4** The iteration cycle of scientometrics and patent-metrics in the NSL-CAS CTI service

To overcome these limitations, at the beginning of our CTI analysis work, other than merely using bibliometric methods, we also work with our clients (researchers) to create a framework of retrieval words or keywords for technology topics, to identify the most important themes in the research subject area. Then, we choose technology topics from the well-defined frameworks that have been developed enough to enable us to conduct patent-metric analysis, which helps to determine the core technology in this subject. Ultimately, we conduct research-article-based bibliometric analysis for the core technology and to find its key scientific and technological points, and uncover the details of the R&D situation for the technology subject. The entire process mentioned above is iterated, with both bibliometric and patent-metric methods and indicators involved (as Fig. 17.4 shows).

### *17.3.4 Creating the CTI Service Procedures, Ensuring the Involvement of Our Professionals*

The core point and importance of the CTI service is the high-quality technology trends analysis reports and novelty examination. To guarantee the reliability of our analysis reports, NSL-CAS has quality control measures and principles in the analysis processes. In the novelty examination service, there are some principles that must be obeyed by the staff. First, training programs for novelty review are organized. The training emphasis is on improving skills for using science and engineering databases, and staff are required to be familiar with major databases of different subjects and exercise precise searching. Second, standards for writing novelty review reports have been established. NSL-CAS has formulated clear rules regarding the novelty examination report form, writing style, content, and way of expression, and all staff are required to obey them strictly. Third, to maintain the independence of the novelty review service, working procedures, especially about how to communicate with the clients, are also in place, to guarantee our work results are free from customer intervention. Fourth, to make sure the report conclusions are sufficiently precise, we also have regular communication mechanism, to allow our workers to contact technicians and professionals. There are also regulations to handle the situation whenever differences between us and customers over the report conclusions arise. Our workers must follow the regulations to reply and improve the work, to ensure the examination results are scientific and independent.

In the process of providing CTI services for technology subjects, we try to ensure our work is scientific and guarantee its efficiency from four aspects. First, we set up a procedure for requirement collection, in which our staffs maintains close communication with the client researchers and developers. Researchers provide keywords or subject words for searching for articles or patents. After obtaining preliminary results, we exchange ideas with researchers and clients, improve and optimize the preliminary searching results, and create a database of full-text articles, abstracts, and patent applications. Then, we perform a scientific-keyword-based clustering analysis, which helps us construct a research subject or technology topic framework. We maintain constant contact with outside researchers and professionals until we obtain satisfactory results.

Second, when we are working on the important technology subjects, we organize consulting meetings to create the analysis framework. If necessary, we invite outside experts to give suggestions and to assist us in creating a technology topic framework (see Fig. 17.5, the subject frame for technology analysis), conduct clustering analysis, and revise our final report.

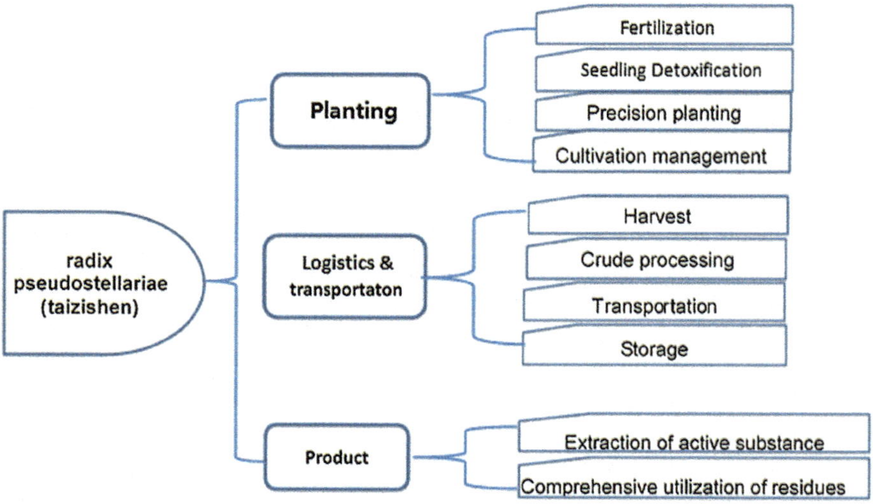


**Fig. 17.5** Subject or domain framework for technology analysis (e.g., Taizishen industry chains)

Third, we have set up a new service model, performing the CTI based on key intelligence topics (KIT). We let the KIT be the core of the technology subject analysis. We divide the client's requirements into several groups based on the technology areas of industries, such as the medicine and health sectors, agriculture and food sectors, IT and equipment manufacturing sectors, and the new material and energy industry, to organize specialized technology intelligence teams in the form of communities of practice (COP). The members of each team learn knowledge of the technology areas of which their team is in charge. These teams also study "hot" technology topics, major industry technology subjects, and core technology and products of their industry. All these help our staff accumulate information and professional knowledge for future use.

Fourth, we establish stable collaboration with experts of many technology and industry areas to help us in CTI services.

### 17.3.5 *To Integrate Sources of Business Information and Technology Intelligence, and to Provide Customized Services According to Needs*

NSL-CAS usually faces diverse needs, including from either S&T decision-makers or firms, and can help with R&D management, specific research projects, or business R&D. With the recent reform of China's R&D system, China's research institutes are increasingly involved with the development of industrial sectors and business R&D, while the firms are also engaged in R&D activities, and want to establish links between their products, marketing activities, and academic research. All these changes and developments bring more diverse demands for technology intelligence services. Thus, it is necessary for us to associate trends of technology development trends, technology competitiveness intelligence, industrial technology development, and business intelligence, and to provide differentiated CTI services to meet different decision-making needs.

When we provide technology analysis for enterprise R&D, we should pay more attention to the CTI—especially for marketing intelligence, collecting and analyzing market capacities, production scales and potentials, and product upgrading. When supporting decision-making at the meso- or macro-level—as when supporting important government S&T projects—there is a clear need to analyze R&D activities of key R&D institutes. We should also have access to project funding information for our intelligence analysis, and both domestic and international R&D achievements and strategies need to be considered. Also, the information of potential partners and rivals is important.

For a long time, NSL-CAS has served the needs of technology innovations by providing services such as novelty review for technology development, technology innovation path selection, support for business R&D projects, product technique assessment, competitor monitoring, potential partner identification, and R&D strategy suggestions. For CTI services, we can choose or construct (customize) different indicator schemas for different analysis purposes. For industrial technological strategy support and technology innovation path identification (or selection), scientometric indicators are particularly useful. Specifically, in meso-technology analysis, bibliometrics, and patent analysis, indicators should be mixed in accordance with different subjects or stages of the emerging technology whose characteristics could then be indicated thereby. Whereas in micro-technology analysis, as patent technology and core technology analysis are mainly for new product development, some bibliometric indicators that reflect technology trends should be added.

## 17.4 Conclusions

In the experience of NSL-CAS CTI services that support R&D decision-making of enterprises or scientific institutions, we should form a complex scientometric indicator schema to profile technology topics, and then select the commercial technology by patent-metrics from the topics, or set up technology mapping based on the patent-metrics to analyze the core technology and trace the technology advancement by scientometrics. In practice, a good CTI report includes the technology topics, selection of a technology innovation pathway, future technology directions, market and business intelligence, competitor intelligence, and production intelligence. In CTI reports, bibliometric indicators, patent metrics indicators (including text-mining for themes or subjects), and local investigations of competitors, should be included.

Although we have not been able to fulfill 100 % of the needs of Chinese firms and R&D administrators so far, our exploration of CTI services still yielded much valuable experience and many lessons. In the process of learning our clients' needs, we find they have a very strong preference for quantitative analysis of technology trends. Yet, different clients also have their own preferences. Decision-makers of R&D projects are more likely to appreciate visualization and explicit analysis (with charts and graphs), while firms tend to associate technology trends with their products and market closely, and are more interested in potential competitors and future development of technologies. Moreover, with the ongoing Chinese economic reform and structural adjustment, more and more research and technology transformation organizations will demand CTI services for technology maturity, feasibility of commercialization and industrialization, rival monitoring, and so on. To meet these growing demands, with the help of bibliometric and patent-metric methods, we will keep exploring ways to improve our CTI services.

## References

Ashton, W. B., & Hohhof, B. (Eds.). (2009). *Competitive technical intelligence*. Alexandria, VA: Competitive Intelligence Foundation, Society of Competitive Intelligence Professionals.

Brenner, M.S. (1996). Technology intelligence and technology scouting. *Competitive Intelligence Review*, *7* (3), 1–14.

Herring, J. P.(2006). KITs revisited: Their use and problems. SCIP online. http://www.imakenews.com/scip2/e_article000069099.cfm

Murphy, J. (2001). Using competitive technical intelligence techniques to complement research-and-development processes. In C. S. Fleisher & D. L. Blenkhorn (Eds.), *Managing frontiers in competitive intelligence* (pp. 136–148). ISBN: 1-56720-384-1

# Chapter 18
# Tech Mining for Emerging STI Trends Through Dynamic Term Clustering and Semantic Analysis: The Case of Photonics

**Pavel Bakhtin and Ozcan Saritas**

**Abstract** Tech mining (TM) helps to acquire intelligence about the evolution of research and development (R&D), technologies, products, and markets for various STI areas and what is likely to emerge in the future by identifying trends. The present chapter introduces a methodology for the identification of trends through a combination of "thematic clustering" based on the co-occurrence of terms, and "dynamic term clustering" based on the correlation of their dynamics across time. In this way, it is possible to identify and distinguish four patterns in the evolution of terms, which eventually lead to (i) weak signals of future trends, as well as (ii) emerging, (iii) maturing, and (iv) declining trends. Key trends identified are then further analyzed by looking at the semantic connections between terms identified through TM. This helps to understand the context and further features of the trend. The proposed approach is demonstrated in the field photonics as an emerging technology with a number of potential application areas.

**Keywords** Tech mining · Trend analysis · Foresight · Horizon scanning · Clustering · Co-occurrence analysis · Photonics

## 18.1 Introduction

In the world of rapidly developing science, technology and innovation (STI), with increasing volumes of STI-related data, greater interdisciplinary and collaborative research, tech mining (TM) helps to acquire intelligence about emerging trends and future STI developments. The task is becoming crucial not only for high-tech

P. Bakhtin (✉) · O. Saritas (✉)
Institute for Statistical Studies and Economics of Knowledge, National Research University Higher School of Economics, Moscow, Russian Federation
e-mail: pbakhtin@hse.ru

O. Saritas
e-mail: osaritas@hse.ru



T.U. Daim et al. (eds.), *Anticipating Future Innovation Pathways Through Large Data Analysis*, Innovation, Technology, and Knowledge Management,
DOI 10.1007/978-3-319-39056-7_18

start-ups and large organizations, but also for venture capitalists and other companies, which make decisions about STI investments. Governments and public research institutions are also among the main stakeholders and potential users of TM to set up R&D priorities, plans, and programs according to the current and future state of STI development.

Information collected through TM-analysis demonstrates the state of the art of research and development (R&D), technologies, products, and markets for various areas of interest. Spotting emerging, maturing, or declining trends related to STI within that data allows tracking STI advancements, identifying new research topics and collaboration networks, and prioritizing further activities.

From a quantitative perspective of tech mining STI trend, spotting is generally performed based on term distribution dynamics over the time in scientific publications, patent applications, grants, conference papers, and other sources (Li et al. 2011; Yoon and Kim 2012; Saritas and Burmaoglu 2015). Documents are extracted from databases (e.g., Web of Science, Scopus, WIPO) using existing classifications (research areas, IPC, CPC, and others) or expert-defined queries (Mikova and Sokolova 2014). Trends represented in the form of term statistics show high level of bias due to complex nature of any R&D topic. For example, the dynamics of term "synthetic biology" in publications demonstrates to some extent the popularity of the field, but does not take into account all other terminology classifying and describing it. Moreover, novel concepts (e.g., "regenerative medicine" or "big data analytics") tend to emerge in scientific and business taxonomies that may seem revolutionary. In reality, however, the majority of such concepts only agglomerate existing topics and developments that have existed for many years. Analysis of such terms alone does not justify the novelty and the speed of change if others related concepts are not taken into account.

In order to spot trends based on research topics, this chapter suggests a joint analysis of scientific terms in publications from the perspective of their relation to some data-determined topic (thematic clustering) and popularity over the time period (dynamic clustering).

The approach proposed utilizes existing TM, bibliometrics, natural language processing (NLP) methods, including the analysis of term linguistic dependencies (Marnefe and Manning 2008), and dynamic pattern analysis in order to study changes of objects over the time along with their semantic meanings.

A case analysis is performed with the example of photonics, which as a relatively novel field agglomerates various studies about light. The chapter demonstrates application of proposed methodology to identify main thematic areas of photonics and determine main weak signals (early developments), emerging trends, maturing (stable) developments, and declining trends for each thematic area.

Presenting the results of the study, the chapter consists of three sections. The first section describes the state-of-the-art developments in tech mining, bibliometrics, and natural language processing with applications in foresight studies, especially at the horizon scanning stage. Modern methods and approaches described help to grasp the main ideas proposed by the chapter. The second section describes methodology proposed for the study. The third section demonstrates the applications of the approach for the photonics field. It contains results of the bibliometric analysis of the field, thematic and dynamic clustering, joint analysis of results with some semantic interpretations along with further research in the field through the Web for a better understanding of the photonics as a scientific domain. The chapter is rounded up with a conclusion section with a discussion on the future work to be undertaken.

## 18.2 Literature Review

It has become more crucial for policy and strategy makers to understand what is likely to emerge in the future. Horizon scanning has been used for the purpose of analyzing the current stage of STI development (with the relation to the future) and alert policy makers and has become one of the first steps in Foresight activities (Amanatidou et al. 2012). Further integration of social, technological, economic, environmental, political, and value systems, their interdependence and constant change into the model allowed for mapping of STI state-of-the-art, existing issues and future scenarios in the form of evolutionary approach (Saritas and Nugroho 2012).

The ever-growing amount of STI-related publications needed for horizon scanning made many researchers consider contents of the text (abstracts or full text) as main object of their study. Tech mining has been introduced as one of the main tools for the extraction of intelligence needed to spot various entities, such as technologies, products, and markets. Porter (2005) identified tech mining or "tech mining" as "text mining[1] of science and technology information resources."

Tech mining uses bibliometrics and different natural language processing (NLP) tools to collect, process, and represent competitive technological intelligence (CTI) for various stakeholders: from researchers and developers to policy makers (Porter and Cunningham 2004; Yoon 2008). Processing, analyzing, and interpreting STI-related information helps to identify early warning about threats (Sun et al. 2015), opportunities to solve problems and challenges, such as various diseases (Carvalho et al. 2015), and key players to cooperate or compete with private companies (Kerr et al. 2006).

In order to identify the search strategies to find, download, and process relevant documents that potentially contain information about emerging technologies, iterative processes from general search categories and terms to more concrete words and phrases are generally used (Haung et al. 2015). Key terms and phrases, identified with the help of tech mining, demonstrate the most relevant objects of the research area. However, in order to identify and visualize development (or, in other words, trends) of technologies, products, methods, and other solutions researchers use bibliometric data to link key terms and phrases with their occurrence on the

[1]Text mining—process of extraction of information from text.

timeline (Saritas and Burmaoglu 2015). Finally, pattern analysis in time series (Assfalg et al. 2009) allows further trend identification and clustering using results of tech mining and bibliometrics.

The latest developments in the semantic analysis related to STI documents brought even more opportunities to tech mining and trend analysis. NLP is exploited to process publication, patent abstracts and other textual data to extract text structure, dependencies between tokens and identification of part of speech, dependency types (Manning et al. 2014), technology properties, and functions (Yoon and Kim 2012) and even predict possible future trends based on TRIZ[2] classification (Park et al. 2013). Wang et al. (2015) developed methodology to identify trends based on subject–action–object analysis (SAO). Such advancement made it possible to move forward from pure co-occurrence analysis of terms to identification of more sophisticated relationships between objects (technologies, products, and other entities) enabling ontology development for the research field.

Recent studies dedicated to comparing existing STI trends and road maps also employed tech mining making it possible to validate and refine technology road maps (Lahoti et al. 2015).

Overall, the mentioned advancements created the basis for further development of methodology for STI trend, spotting by combining bibliometric, TM, NLP, and pattern analysis in time series, in other words, dynamic clustering.

Photonics has become one of the most important research fields for the development of such areas as telecommunications, electronics, and solar energy. It drives advancement in the speed of data processing and transmission, efficiency of energy conversion, and many more (Yeh 2012). The case of photonics is used as the main application area to spot emerging trends and identify existing thematic areas.

## 18.3 Methodology

The chapter proposes a systemic approach to identify STI trends that can be described in several steps.

The scope of STI trends and developments research is generally identified in terms of keywords, patent technology domains, international patent classifications (IPCs), publication research areas, and categories. In case of photonics, the main idea was to understand the evolution of the field by looking at the collective intelligence emerging from the publications in the field. Hence, search strategy included general terms related to photon, photonic, and different variations. The study involved the analysis of the Web of Science scientific publications in the last 25 years (1990–2014, both years inclusive) and covered the top 10 % mostly cited

[2]TRIZ ("theory of inventive problem solving")—formal forecasting approach to identification of possible development paths of inventions developed by Genrich Altschuller.

articles in each year. In total of 10,571 were analyzed within the scope of the bibliometric analysis and TM study.

The data sample was processed by the VantagePoint, high performance TM and bibliometric tool. Keywords and phrases were extracted from the abstracts, titles, and authors' keywords. Relevancy of terms was estimated using Stanford dependencies with the help of Stanford CoreNLP toolkit. As a result, thematic clusters were identified based on network analysis and graph clustering using term co-occurrences as the measure of similarity between the nodes with the help of VOSviewer.

Afterward, all extracted terms were analyzed from the perspective of their occurrences in the abstracts of publications over the time period (1990–2014). Such representation of data was perceived as time series where the main task was to divide all terms by their trend patterns as part of the dynamic clustering. To achieve the task, Pearson's correlation coefficient was considered as similarity criteria of terms' dynamics:

$$\mathrm{corr}\left(\mathrm{term}_i, \mathrm{term}_j\right) = \frac{\sum\left(\mathrm{term}_{i,n} - \overline{\mathrm{term}_i}\right) * \sum\left(\mathrm{term}_{j,n} - \overline{\mathrm{term}_j}\right)}{\sqrt{\sum\left(\mathrm{term}_{i,n} - \overline{\mathrm{term}_i}\right)^2 * \left(\mathrm{term}_{j,n} - \overline{\mathrm{term}_j}\right)^2}}$$

where $\mathrm{term}_i$, $\mathrm{term}_j$—are two different terms; $\mathrm{term}_{i,n}$—the amount of publication abstracts in the year $n$ where $\mathrm{term}_i$ occurs; $\overline{\mathrm{term}_i}$—average amount of publication abstracts over all time period where $\mathrm{term}_i$ occurs.

All terms were then clustered based on their similarity criteria of dynamics using VOSviewer. That resulted in the two cluster lists: (i) thematic clusters generated during the first stage and (ii) dynamic clusters of terms. Joint cross-cluster analysis was performed based on estimation of intersection of terms in two types of clusters. That generated distribution of various trend patterns for each thematic area. Same distribution was also calculated for each term based on its co-occurrence with words in various types of trend patterns.

Finally, some examples of terms associated with STI trends were mapped as subjects and objects of various relationships as part of semantic analysis using Stanford CoreNLP toolkit. Such semantic relationships were visualized as the network that could be assessed by an expert to validate STI trends and developments in the area.

The results of applied methodology in photonics generated the following findings:

1. State of the art in the evolution of photonics,
2. Thematic clusters,
3. Dynamic term clusters,
4. Cross-cluster linkages,
5. Mapping of terms and STI trend patterns, and
6. Semantic analysis of results and interpretations.

These are described in the next section.

## 18.4 Findings

### *18.4.1 State of the Art in the Evolution of Photonics*

As rapidly growing area, photonics' number of publications in the WoS database has increased from a few thousand in the 1990s to over 100,000 as of 2014. Figure 18.1 illustrates the evolution of the number of publications across years, considering the top 10 % of the most highly cited articles.

### *18.4.2 Thematic Clusters*

Tech mining phase of the study generated 181 terms. Following the identification of the terms and their networks, they were clustered to understand the main themes studied in photonics. A thematic cluster analysis was conducted based on the co-occurrence of the terms. These are shown in Fig. 18.2 with color-coded clusters with numbers that are explained in the next paragraph.

The analysis of the terms identified resulted with the generation of eight clusters in photonics. These include (numbers are shown in Fig. 18.2):

1. Fiber optics (fiber laser, holey find, semiconductor optical amplifier, optical network)
2. Bio- and nanophotonics (golden nanoparticle, thin film, metal nanoparticle, drug delivery, material science)
3. Optical metamaterials (photonic crystal, chalcogenide glass, glass substrate, organic light-emitting diode)
4. Optoelectronics (integrated photonic circuit, optoelectronic device, silicon chip, microdisk laser, optical memory, optical antenna)

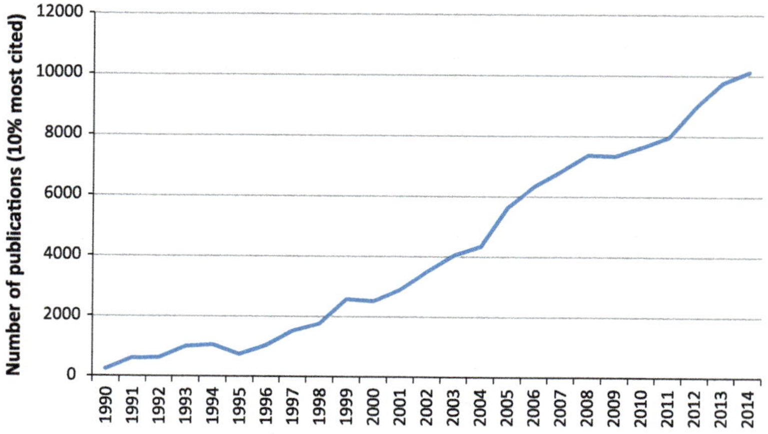


**Fig. 18.1** Number of publications on "photonics" in WoS (top 10 % cited for each year)

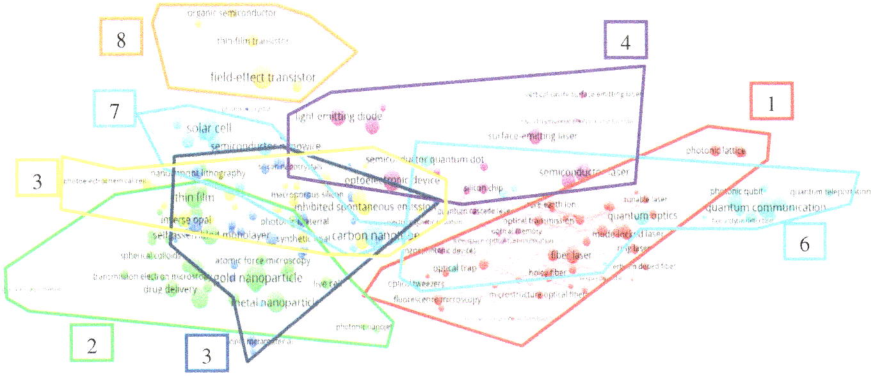


Fig. 18.2 Thematic areas identified in photonics

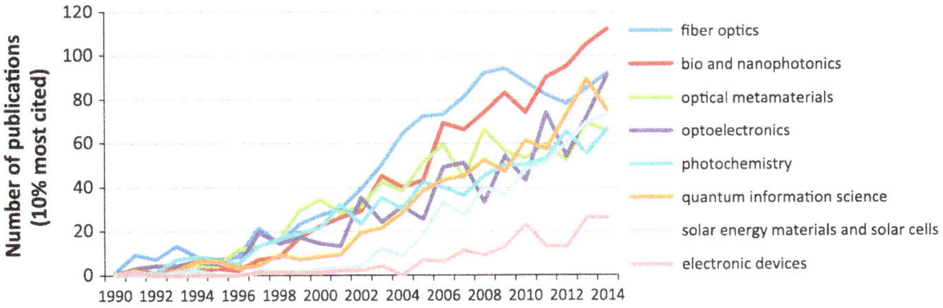


Fig. 18.3 Cluster dynamics

5. Photochemistry (inverse opal, liquid crystal, energy transfer, dye molecule, synthetic opal, fluorescent dye)
6. Quantum information science (quantum communication, quantum optics, quantum computer, quantum memory, quantum cryptography)
7. Solar energy materials and solar cells (carbon nanotube, solar cell, photovoltaic device)
8. Electronic devices (field-effect transistor, light-emitting device, thin-film transistor).

Figure. 18.3 illustrates the dynamics of the clusters over time.

As an example, topics under "quantum information science" cluster are presented (Fig. 18.4).

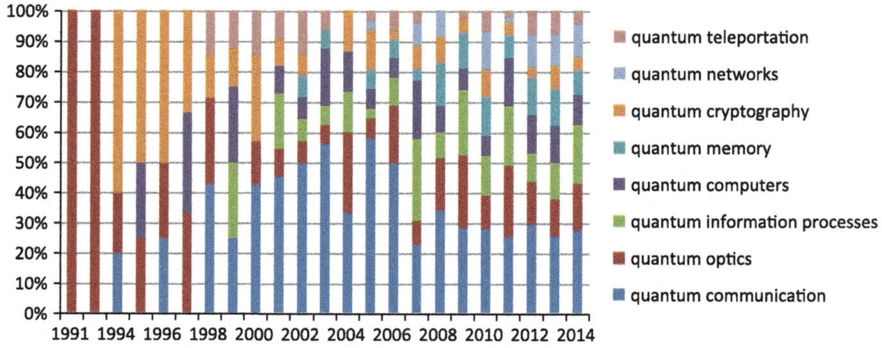


**Fig. 18.4** Percentage distribution of topics (sub-clusters) across years in "quantum information science"

## *18.4.3 Dynamic Term Clusters*

Results of the dynamic term clustering introduced and described in the methodology section are presented in the form of four major clusters. Terms with increasing frequencies of occurrence were clustered as "emerging trends." The ones correlated based on their stable occurrence at the peak level across time were clustered as "mature trends." A group of terms, which indicate a fluctuating characteristic, were named as "weak signals." These may either become a trend in the longer term or may decline in the future. Due to their uncertain characteristic, they are worth analyzing among the other trends. Finally, some terms were correlated due to their declining occurrence across time. These were labelled as "declining" developments. Figures 18.5, 18.6, 18.7, and 18.8 illustrate the evolution of terms under these clusters collectively.

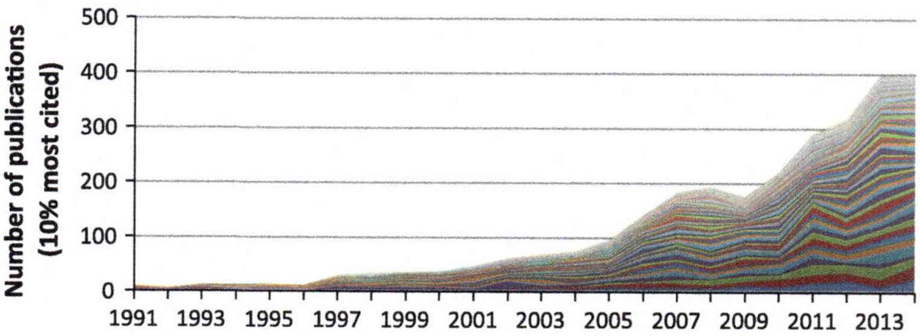


**Fig. 18.5** Emerging trends

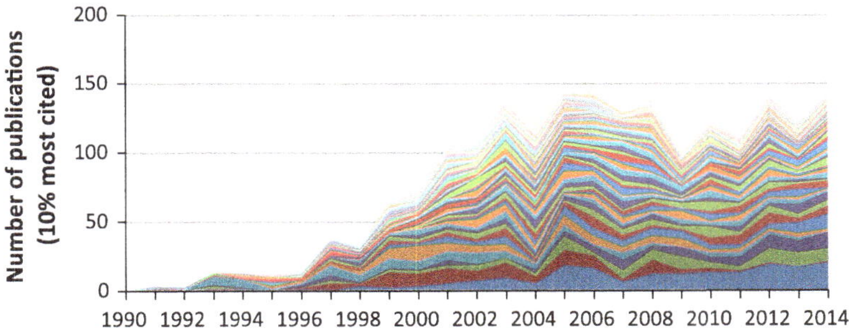


**Fig. 18.6** Mature (stable) trends

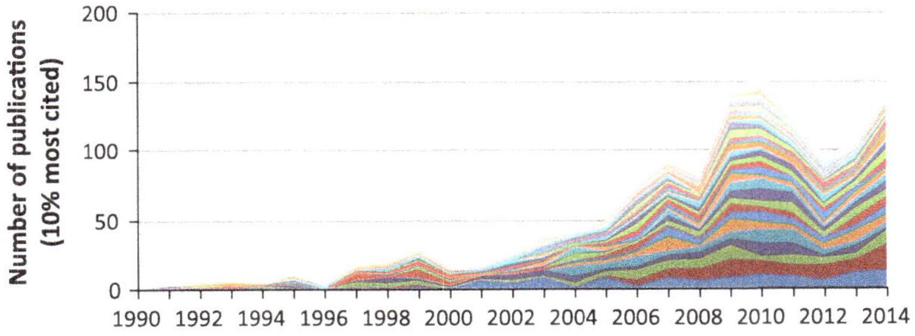


**Fig. 18.7** Weak signals (early developments)

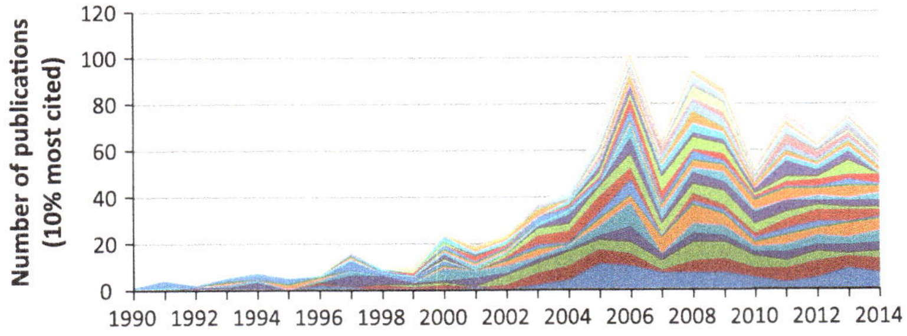


**Fig. 18.8** Declining trends

### *18.4.4 Cross-Cluster Linkages*

This section combines the results of the join analysis of thematic and dynamic clusters. Each of the eight thematic clusters presented earlier is distributed over emerging, stable, and declining trends as well as weak signals. These are represented in Fig. 18.9.

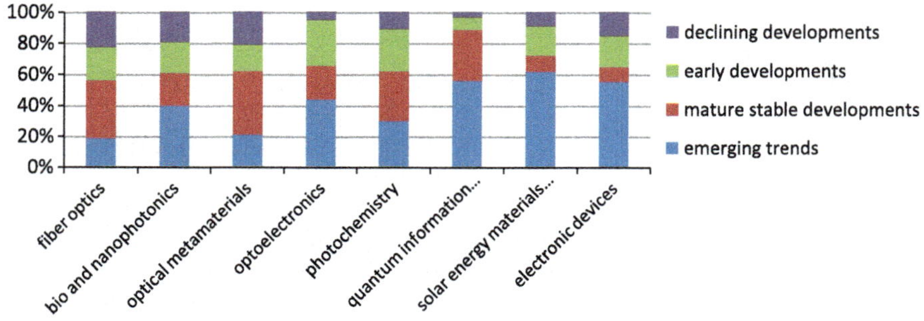


**Fig. 18.9** Joint analysis of term and dynamic clusters

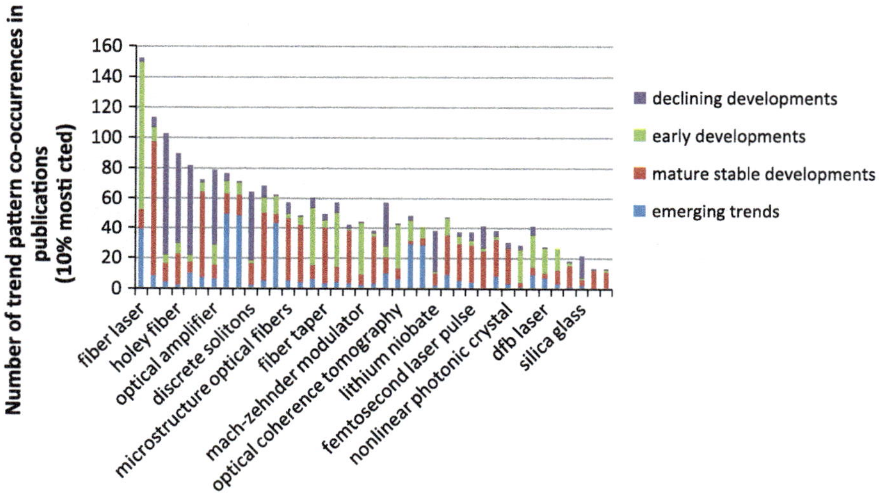


**Fig. 18.10** Fiber optics—co-occurrence of terms and dynamic clusters' elements

Figure 18.9 illustrates that the area of fiber optics consists of largely mature trends. This is also the case for the optical metamaterials. The clusters such as quantum information science, solar energy and solar cells, and electronic devices indicate a more vibrant composition by largely consisting of emerging trends.

The following Figs. 18.10, 18.11, 18.12, 18.13, 18.14, 18.15, 18.16, and 18.17 demonstrate the relation of the terms (involving R&D areas, technologies, products, and markets) in each of the eight clusters to four types of trend patterns.

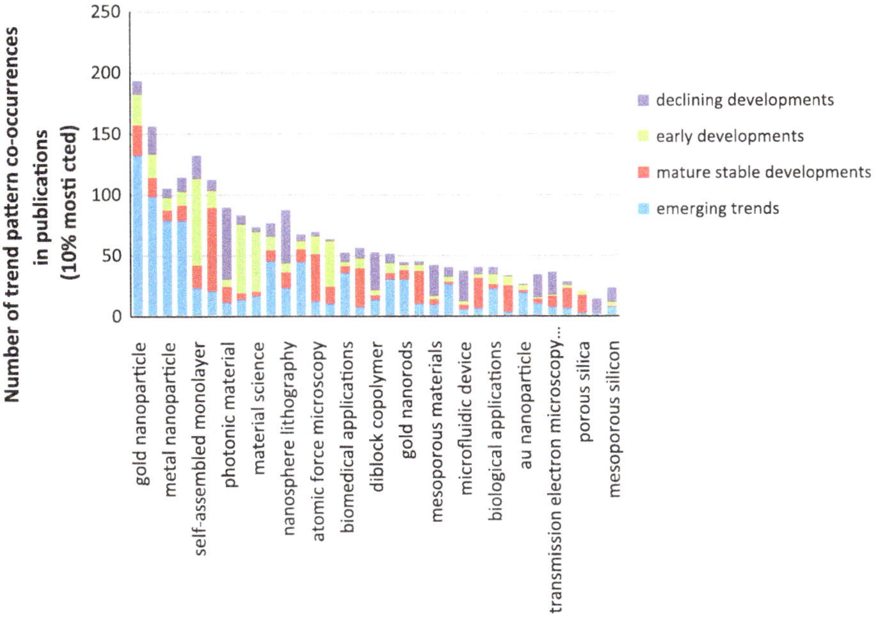


**Fig. 18.11** Bio- and nanophotonics—co-occurrence of terms and dynamic clusters' elements

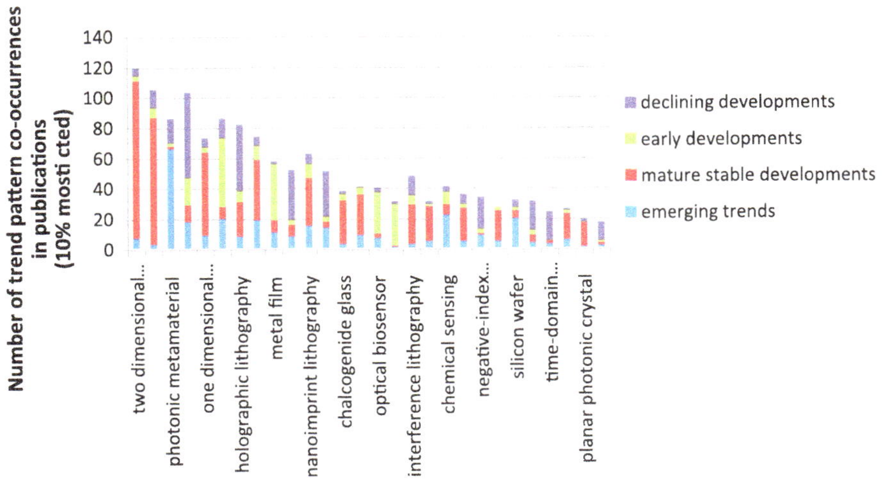


**Fig. 18.12** Optical metamaterials—co-occurrence of terms and dynamic clusters' elements

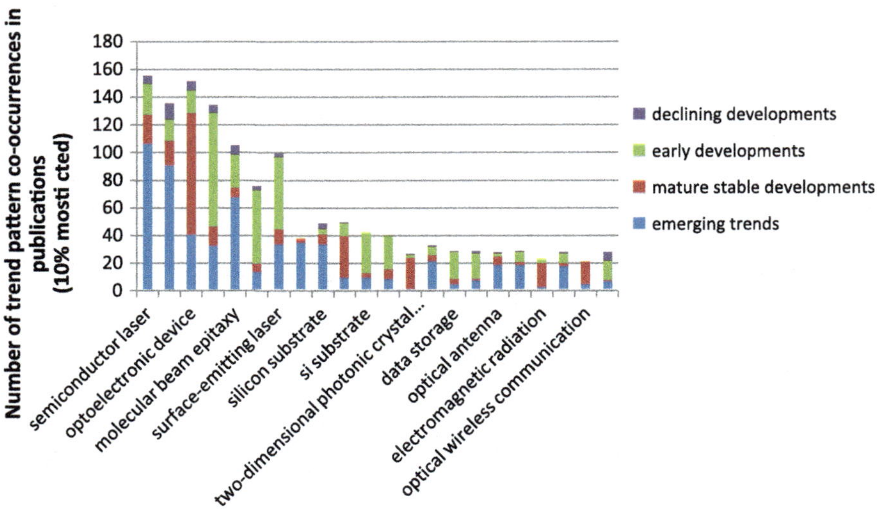


Fig. 18.13 Optoelectronics—co-occurrence of terms and dynamic clusters' elements

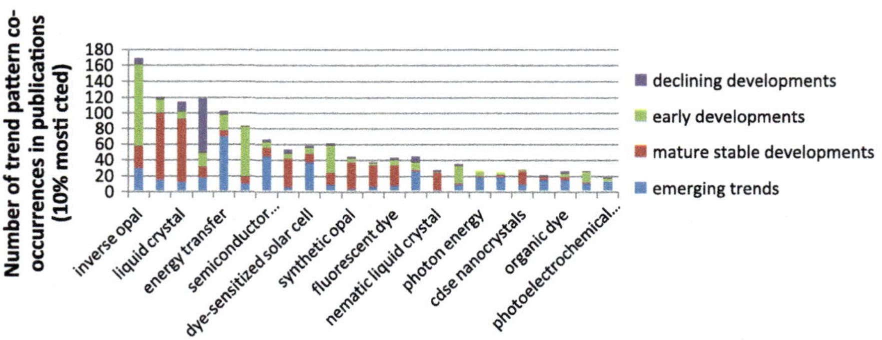


Fig. 18.14 Photochemistry—co-occurrence of terms and dynamic clusters' elements

## 18.4.5 Mapping of Terms and STI Trend Patterns

Table 18.1 provide lists of terms related to emerging and mature trends, weak signals of likely future trends and declining trends. These trends cover R&D areas, technologies, products, and markets.

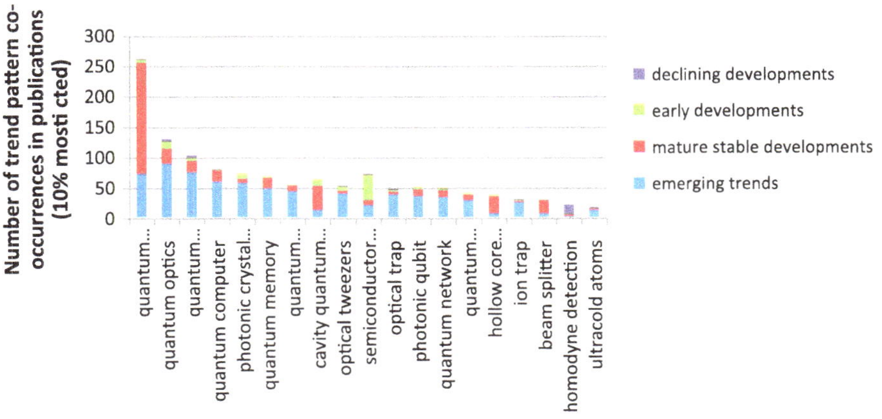


**Fig. 18.15** Quantum information science—co-occurrence of terms and dynamic clusters' elements

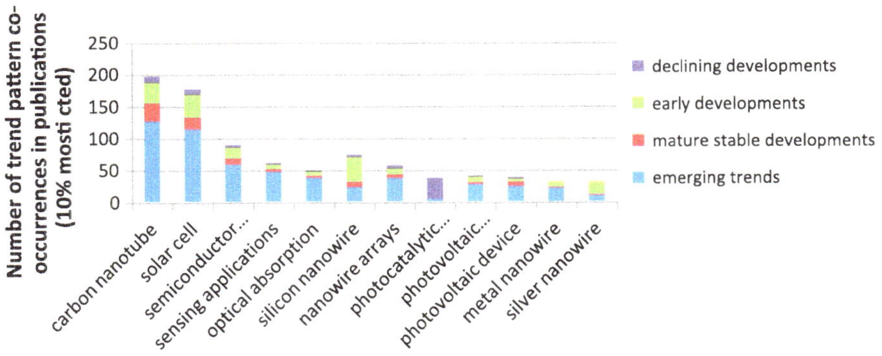


**Fig. 18.16** Solar energy materials and solar cells—co-occurrence of terms and dynamic clusters' elements

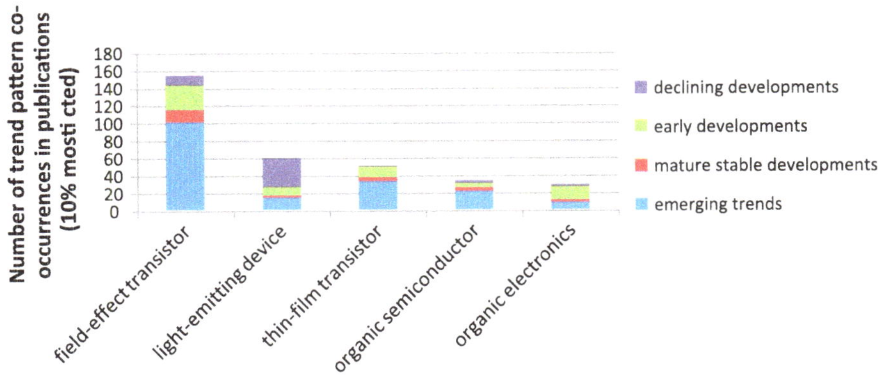


**Fig. 18.17** Electronic devices—co-occurrence of terms and dynamic clusters' elements

**Table 18.1** Distribution of terms over trend patterns

| Emerging trends | Mature trends | Weak signals | Trends in decline |
|---|---|---|---|
| Gold nanoparticle | Quantum communication | Inverse opal | Photonic lattice |
| Carbon nanotube | Two-dimensional photonic crystal | Fiber laser | Optic devices |
| Solar cell | Single mode fiber | Light-emitting diode | Holey fiber |
| Semiconductor laser | Optoelectronic device | Self-assembled monolayer | Semiconductor optical amplifier |
| Field-effect transistor | Inhibited spontaneous emission | Photonic wire | Photonic material |
| Thin film | Three dimensional photonic crystal | Live cell | Porous silicon |
| Integrated photonic circuit | Liquid crystal | Silicon chip | Optical amplifier |
| Quantum optics | Electron microscopy | Surface-emitting laser | Discrete solutions |
| Metal nanoparticle | Optical network | Material science | Holographic lithography |
| Silver nanoparticle | One dimensional photonic crystal | Chemical sensor | Nanosphere lithography |
| Quantum information processes | Optical communication system | Semiconductor quantum dot | 3D photonic crystal |
| Energy transfer | Microstructure optical fibers | Optical soliton | Light-emitting device |
| Molecular beam epitaxy | Cavity quantum electrodynamics | Silicon nanowire | Photocatalytic degradation |
| Photonic metamaterial | Electron beam lithography | Metal film | Diblock copolymer |
| Quantum computer | Atomic force microscopy | Transmission electron microscopy | Fluorescence microscopy |
| Semiconductor nanowire | Femtosecond pulse | Nonlinear fiber | Split-ring resonator |
| Photonic crystal nanocavity | Fiber taper | Dye molecule | Lithium niobate |
| Femtosecond laser | Colloidal photonic crystal | Mach-zehnder modulator | Mesoporous materials |
| Quantum memory | Photonic crystal fiber (PCF) | Optical coherency tomography | Microfluidic device |
| Optical transmission | Synthetic opal | Si substrate | Negative-index metamaterial |
| Sensing applications | Nanoimprint lithography | Optical biosensor | Bragg reflector |

(continued)

**Table 18.1** (continued)

| Emerging trends | Mature trends | Weak signals | Trends in decline |
|---|---|---|---|
| Drug delivery | Soft lithography | Photonic crystal | Dielectric microsphere |
| Optical microscopy | Photonic crystal fiber | Photonic crystal laser | Time-domain spectroscopy |
| Quantum cryptography | Microdisk laser | Dispersive fiber | Transmission electron microscopy tem |
| Semiconductor nanocrystal | Chalcogenide glass | Rare earth ion | Homodyne detection |
| Mode-locked laser | Hollow-core photonic crystal fiber | Supramolecular chemistry | Silica glass |
| Optical tweezers | Core-shell particles | Data storage | Glucose concentration |
| Optical trap | Glass substrate | Silver nanowire | Mesoporous silicon |
| Optical absorption | Interference lithography | Nanophotonic devices | Smart dust |
| Nanowire arrays | Photonic crystal pc | DFB laser | |
| Dye-sensitized solar cell | Ring laser | Organic electronics | |
| Photonic qubit | Fluorescent dye | Erbium doped fiber | |
| Biomedical applications | Spherical colloids | Silicon nanocrystals | |
| Quantum network | Confocal microscopy | Efficient energy transfer | |
| Free-space optical communication | Femtosecond laser pulse | | |
| Silicon substrate | Gap soliton | | |
| Thin-film transistor | Tunable laser | | |
| Functional materials | Nematic liquid crystal | | |
| Gold nanorods | Nonlinear photonic crystal | | |
| Pulse generation | Photonic bandgap material | | |
| Quantum teleportation | Macroporous silicon | | |
| Photovoltaic applications | Opal photonic crystal | | |
| Pulse energy | Two-dimensional photonic crystal slab | | |
| Ion trap | Beam splitter | | |
| Photonic nanojet | Organic light-emitting diode | | |
| Mesoporous silica | Electromagnetic radiation | | |
| Photovoltaic device | Metallic photonic crystal | | |

(continued)

**Table 18.1** (continued)

| Emerging trends | Mature trends | Weak signals | Trends in decline |
|---|---|---|---|
| Biological applications | Thick films | | |
| Chemical sensing | CdSe nanocrystals | | |
| Metal nanowire | Optical wireless communication | | |
| Organic semiconductor | Planar photonic crystal | | |
| Quantum cascade laser | Porous silica | | |
| Silicon wafer | Fiber-optic communication systems | | |
| Au nanoparticle | Hollow-core photonic bandgap fiber | | |
| Optical antenna | Nonlinear photonic crystal fiber | | |
| Optical memory | | | |
| Photon energy | | | |
| Solar energy conversion | | | |
| Vertical cavity surface-emitting laser | | | |
| Efficient energy conversion | | | |
| Organic dye | | | |
| Ultracold atoms | | | |
| Photoelectrochemical cell | | | |

### *18.4.6 Semantic Analysis of Results and Interpretations*

Finally, the subject–action–object (SAO) analysis allows deep studying of the content of abstracts of various publications, patents, and other STI-related documents in order to determine existing relationships between words and phrases. The subject of each sentence is identified along with affected objects, as well as the action which subject commits to objects within the sentence. As a result, an oriented graph is built with all relevant subjects and objects of the network extracted from all documents of the data sample. Edges between subjects and objects represent various types of action and semantic meanings (e.g., verbs, prepositions). Below is an SAO analysis for one of the key emerging technologies "gold nanoparticles," which was identified in the thematic cluster of bio- and nanophotonics. A closer look is taken on the nanoparticle group with a focus on "gold nanoparticles" to understand the nature of the relationships between the components. As an example, the

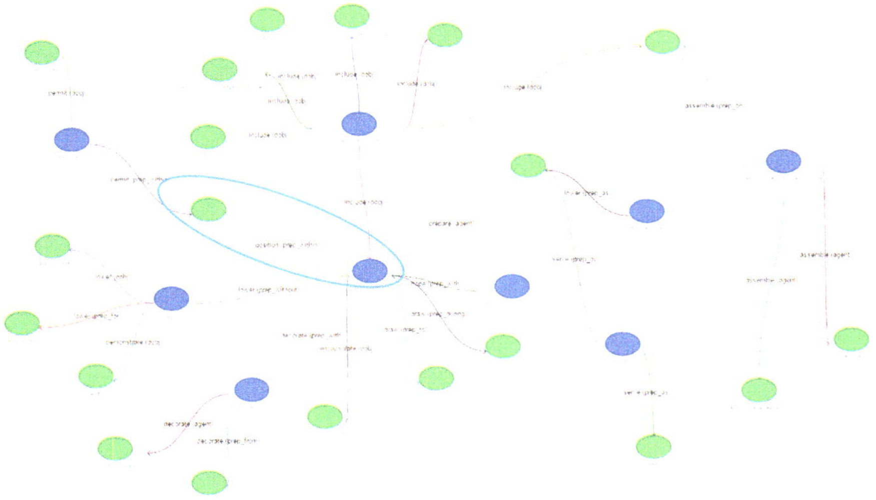

**Fig. 18.18** Subject action object analysis for gold nanoparticles

relationship between "gold nanoparticle" and "photonic nanojet" (marked with a circle in Fig. 18.18 above) is investigated through the SAO analysis.

The SAO analysis indicates "the positioning of gold nanoparticle within nanojet." The original article is then referred to understand the context and purpose of this relationship (see highlighted text below).

| | |
|---|---|
| **Title** | Experimental confirmation at visible light wavelengths of the backscattering enhancement phenomenon of the photonic nanojet |
| **Authors** | Backman, Vadim |
| | Taflove, Allen |
| | Yang, Seungmoo |
| **Abstract** | We report what we believe is the first experimental confirmation at visible light wavelengths of the backscattering enhancement phenomenon of the photonic nanojet. A specially designed sample stage consisting of a multilayered sandwich of glass, solid polydimethylsiloxane (PDMS), and liquid PDMS, permitted the precise positioning of a gold nanoparticle of diameter between 50 and 100 nm within the nanojet emitted by a 4.4 μm diameter $BaTiO_3$ microsphere embedded within the PDMS. We determined that, ***when the gold nanoparticle is optimally positioned within the nanojet, the backscattering of the microsphere can greatly increase***: for example, by 3: 1 (200 %) for the 50 nm gold nanoparticle. The increased backscattering is strongly dependent upon the illumination wavelength and the numerical aperture of the imaging system, and occurs for nonresonant illuminations of the isolated microsphere. Low objective numerical apertures of approximately 0.075 yield the maximum observed increases in backscattering. The measured data agree well with numerical calculations incorporating Mie-based theory and Fourier optics. (C) 2011 Optical Society of America |
| **Source** | Optics express |

The SAO analysis also helps to identify weak signals in particular domains, such as in the case of gold nanoparticles presented above. The following examples can be mentioned with evidence of their emergence in various other publications.

## 18.5 Gold-Nanoparticle-Based Biosensors for Detection of Enzyme Activity

- Gold nanoparticles (GNPs) are used for the construction of sensitive biosensors.
- GNP-based biosensors are used for the detection and measurement of enzyme activity.
- Most of the enzyme activity measurements are achieved by colorimetry and FRET.

http://www.sciencedirect.com/science/article/pii/S0165614713001193.

## 18.6 3D Graphene Oxide-Encapsulated Gold Nanoparticles to Detect Neural Stem Cell Differentiation

- Monitoring of stem cell differentiation and pluripotency is an important step for the practical use of stem cells in the field of regenerative medicine.
- A new nondestructive detection tool capable of in situ monitoring of stem cell differentiation is highly needed.
- A 3D graphene oxide-encapsulated gold nanoparticle that is very effective for the detection of the differentiation potential of neural stem cells (NSCs) based on surface-enhanced Raman spectroscopy (SERS).
- A new material, 3D GO-encapsulated gold nanoparticle, is developed to induce the double enhancement effect of graphene oxide and gold nanoparticle on SERS signals which is only effective for undifferentiated NSCs.

http://www.ncbi.nlm.nih.gov/pubmed/23937915.

## 18.7 Conclusions

Fast dynamics of technological change, growing amount of data related to STI sphere, technology diffusion, and knowledge exchange between different research areas make it a very challenging task for researchers, technology analysts, managers, and companies in general to keep up with the latest progress and achievements in STI. Traditional methods of gathering technology intelligence by communicating with various organizations and experts in the same field are very expensive, time-consuming, and generally do not represent the whole state of developments at the very moment.

Tech mining is the solution to listed problems and provides methods and tools to process, analyze, visualize, and present STI information. However, spotting the trends within the data is still crucial and very sensitive stage of the analysis. In order to spot trends based on the research topics rather than separate terms that cannot fully represent the field or bring high level of expert bias, this chapter introduced joint analysis of scientific terms in publications from the perspective of their relation to some data-determined topic (thematic clustering) and popularity over the time period (dynamic clustering).

The proposed approach was demonstrated on the prospective field of study photonics. The chapter presented the main photonics thematic areas based on co-occurrence data, distribution of relevant terms (R&D directions, technologies, products, materials, etc.) over dynamic term clusters (or trend patterns), and combined the results of the join cross-cluster analysis. Finally, some of the interesting relationships between terms were analyzed using semantic tools and validation based on external sources.

The future work will focus on the full implementation of semantic analysis during the stage of thematic clustering along with application of profound term dependencies and ontologies, as well as studying and comparing different methods and metrics for analyzing similarity based on dynamic behaviors of terms.

**Acknowledgments** The chapter was prepared within the framework of the Basic Research Program at the National Research University Higher School of Economics (HSE) and supported within the framework of the subsidy granted to the HSE by the Government of the Russian Federation for the implementation of the Global Competitiveness Program.

## References

Amanatidou, E., et al. (2012). On concepts and methods in horizon scanning: Lessons from initiating policy dialogues on emerging issues. *Science and Public Policy, 39*(2012), 208–221.

Assfalg, J., et al. (2009). Periodic pattern analysis in time series databases. In *Proceedings of the 14th International Conference, DASFAA*, Brisbane, Australia, pp. 354–364.

Carvalho, K., et al. (2015). Analysis of technological developments in the treatment of Alzheimer's disease through patent documents. *Intelligent Information Management, 2015*(7), 268–281.

Haung, Y., et al. (2015). A systemic method to create search strategies for emerging technologies based on the web of science for 'big data'. Scientometrics.

Kerr, C., et al. (2006). A conceptual model for technology intelligence. *International Journal of Technology Intelligence and Planning, 2*(1), 73–93.

Lahoti, G., et al. (2015). Tech mining to validate and refine a technology roadmap. In *Proceedings of the 5th Global Tech Mining Conference*. Atlanta, USA.

Li, H., et al. (2011). TechWatchTool: Innovation and trend monitoring. In *Proceedings of the International Conference on Recent Advances in Natural Language Processing 2011 (RANLP 2011)*. Hissar, Bulgaria.

Manning, C., et al. (2014). The Stanford CoreNLP natural language processing toolkit. In *Proceedings of 52nd Annual Meeting of the Association for Computational Linguistics: System Demonstrations* (pp. 55–60).

Marnefe, M., & Manning, C. (2008). *Stanford typed dependencies manual*. Stanford University.

Mikova, N., & Sokolova, A. (2014). Global technology trends monitoring: Theoretical frameworks and best practices. *Foresight-Russia, 8*(4), 64–83.

Park, H., et al. (2013). Identification of promising patents for technology transfers using TRIZ. *Expert Systems with Applications, 40*, 736–743.

Porter, A. (2005). Tech mining. Competitive Intelligence Magazine 8, n. 1.

Porter, A., & Cunningham, S. (2004). *Tech mining: Exploiting new technologies for competitive advantage*. Wiley.

Saritas, O., & Burmaoglu, S. (2015). Future of sustainable military operations under emerging energy and security considerations. *Technological Forecasting and Social Change, 102*(2015), 331–343.

Saritas, O., & Nugroho, Y. (2012). Mapping issues and envisaging futures: An evolutionary scenario approach. *Technological Forecasting and Social Change, 79*(2012), 509–529.

Sun, G., et al. (2015). Technology early warning model: A new approach based on patent data. In *Proceedings of the Second International Workshop on Patent Mining and its Applications (IPAMIN)*. Beijing, China.

Wang, X., et al. (2015). Identification of technology development trends based on subject-action-object analysis: The case of dye-sensitized solar cells. *Technological Forecasting and Social Change, 98*(2015), 24–46.

Yeh, C. (2012). *Applied photonics*. ISBN 978-0-08-049926-0.

Yoon, B. (2008). On the development of a technology intelligence tool for identifying technology opportunity. *Expert Systems with Applications, 35*, 124–135.

Yoon, J., & Kim, K. (2012). TrendPerceptor: A property–function based technology intelligence system for identifying technology trends from patents. *Expert Systems with Applications, 39*, 2927–2938.

CPSIA information can be obtained
at www.ICGtesting.com
Printed in the USA
LVOW05*2020230916
505935LV00002B/28/P